Introduction
to plant
physiology

Introduction to plant physiology

Jacob Levitt, Ph.D.

Professor of Botany,
University of Missouri,
Columbia, Missouri

With 153 illustrations

Saint Louis
The C. V. Mosby Company
1969

Copyright © 1969 by
The C. V. Mosby Company

Printed in the United States of America

Standard Book Number 8016-2992-6

Library of Congress Catalog Card Number
69-15219

Distributed in Great Britain by Henry Kimpton,
London

Preface

This text is intended for use in a first course in plant physiology. But what excuse is there for adding another to the already long list of plant physiology texts? In recent years I have tried several of these texts without satisfying either myself or my students. A basic reason is the unprecedented information explosion, which has left all of them woefully out of date. True, there is a rapidly expanding list of monographs, symposium volumes, and short or long treatises, each of which covers selected topics in depth. They are invaluable to the instructor. But what of the student? Can he be expected to go to the twenty-odd volume *Handbook of Plant Physiology* and to the *Annual Review of Plant Physiology?* Or should we be satisfied to have him study in depth selected topics and ignore others? I reject both these alternatives. Symposia, encyclopedias, annual reviews, and treatises never develop a consistent viewpoint and always fail to achieve a balance in covering a whole field. Furthermore, knowledge is expanding at an explosive rate in all fields. It is, therefore, unrealistic to expect the student to read large numbers of volumes for each course. On the other hand, most students of plant physiology are exposed to only a single course. Therefore, they will be forever ignorant of those topics not covered in this introduction to the science.

There is no question, then, of the student's need. To fill this need, I have attempted to build on a course taught for twenty-five years. Though originally founded on standard works (Maximov, Meyer and Anderson, Thomas, Scarth and Lloyd, Gortner), the course has evolved steadily by incorporating the results of modern research. An earlier text that I authored (*Plant Physiology*, Prentice-Hall, 1954), now out of print and out of date, has served as the core; but it has been completely rewritten, new chapters have been added, and new terminology and concepts introduced in place of the outdated ones. The attempt has been made, on the one hand, to give the known facts and, on the other, to leave the student with an appreciation of the many uncertainties that still remain. Since so many texts must be read by the student, the objective has been conciseness without loss of thoroughness. To achieve conciseness, lengthy discussions of irrelevant questions have been omitted. To achieve thoroughness without loss of conciseness, the attempt has been made to combine selectively and critically both the old and the new; for the true is not always new and the new is not always true.

Nevertheless, because of the many new advances, certain sacrifices had to be made: (1) the subject matter has been confined, with few exceptions, to the higher plants, which is logical in view of the tremendous growth of microbiology as a separate science and the paucity of information on the physiology of the "macro" lower plants;

(2) in opposition to some plant physiology texts, which include whole chapters on plant anatomy, neither this nor any other branch of botany is included; and (3) a detailed consideration of each topic is left for more advanced courses (for example, the many products and enzymes of intermediate metabolism).

The prime goal of this text is the presentation and elucidation of the principles and laws of plant physiology, without including much of the scientific literature, since it is not possible in a first course to include all the history of a science. In some cases this goal has not been achieved, since the principles have not yet been adequately established, and it is necessary to attack the problem by a "recent advances" approach—that is, a lengthier consideration of the recent literature. This approach has the advantage of emphasizing to the student that each small advance of science is the result of the painstaking work of many dedicated scientists. It is therefore a pleasure to give credit to many of these scientists in the following pages. But unfortunately, all the important contributions cannot be mentioned, and some of the most important have been omitted. Apologies are due these investigators as well as the reader for not acquainting him with their work. Such slights are not intentional, but are largely the result of circumstances. Some less important contributions are mentioned simply to indicate the present state of an unsolved problem. Some more important contributions are omitted for the sake of brevity and because the problem is sufficiently well understood to explain without reference to these contributions. The length of the bibliography is therefore not to be taken as evidence of the state of knowledge of a subject, since, unfortunately, the understanding of a subject is sometimes inversely related to the volume of its literature.

Another apology is due an unknown number of investigators. Sometimes a concept may be read or heard and then lie dormant in one's brain. At a later date, this concept may germinate into what is apparently one's own idea, but really originated with someone else. Consequently, although I have attempted to give credit in all cases where it is due, I can only apologize in advance for any borrowings that may not be acknowledged. Many of these ideas have their source in the teachings of my former professor, the late G. W. Scarth.

No man can complete a work unaided by others. So many at one time or another have influenced and aided me that I can only say this book is inevitably the creation of the many people in my life, and I am only their mouthpiece. For direct aid in the preparation of the manuscript, I should like to thank my partner in all things, my wife; my artist, Mrs. Judy Ban; and my office helper, Miss Cheryl Zdazinsky.

Jacob Levitt

Contents

part I

INTRODUCTION

chapter 1

Basic principles of plant physiology

A. Physiology as a science
B. Basic laws of plant physiology

PHYSIOLOGY AS A SCIENCE

The first question the student is likely to ask is, "What need does the study of plant physiology fulfill?" The answer, however, will vary with the needs of the student himself.

To some it is of interest primarily as a source of basic information that can be applied practically in such allied fields as horticulture, agronomy, forestry, soil science, etc. They would prefer to study physiology as a physicist studies mathematics—purely as a tool to aid in attacking and understanding the science in which he is really interested. But physiology has nowhere approached the exactness of mathematics and therefore cannot be taught in this way. Plant physiology therefore is not a satisfying study for one who must have pat answers to all questions. Any attempt to write a text that proposes to give categorical answers to most physiological problems will surely, although unintentionally, lead to a falsification of the facts and to a complete denial of the very spirit of the science.

For those students with exploratory and analytical minds, a course in plant physiology has much more than bare facts to offer. It can help them to discover the problems of life that physiology attempts to solve and the methods used to solve them, the facts that have been discovered by these methods, and most important, the hypotheses that have been suggested to explain why the plant behaves in the way it does. By describing experiments currently being performed and suggesting experiments that should be attempted, the course can give these students a feeling of privileged observers watching explorations into unknown fields and learning the techniques and reasoning that are being used to attack the unsolved problems. It should encourage them to examine the known facts critically, to compare them carefully with any conclusions reached, and if reason so dictates, to reject these conclusions, and if sufficiently inspired, to launch attacks of their own in an attempt to add another grain of truth to the, as yet, small pile.

Where does plant physiology fit into the field of human knowledge? This question can be simply and graphically answered as follows:

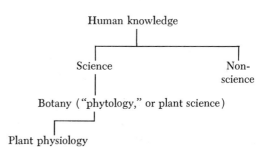

3

By definition, plant physiology is the science that attempts to explain all plant processes by means of physical and chemical principles. This means that the basic assumption of the plant physiologist is that all plant processes *can* be explained by physical and chemical principles. So many processes have been successfully explained in this way that the universality of the assumption is nowadays seldom questioned. From time to time, however, scientists as well as nonscientists have rejected the assumption that physics and chemistry can explain all aspects of plant life, although no one has produced an acceptable reason for discarding it (Schaffner, 1967). Nevertheless, the true scientist will always remember that it is an assumption, although he adopts it in practice. On this basis all explanations not founded on physical and chemical principles are beyond the realm of plant physiology.

Since plant physiology is a science, it cannot be understood without a clear concept of the meaning of science. The aim of science is to discover all the laws of nature. In practice, however, a science is most simply defined as a body of knowledge obtained by the scientific method. The classical, although incomplete, concept of the scientific method is as follows (Simpson, 1963):

1. A problem is stated.
2. Observations relative to the problem are collected. (This includes a search of the literature for recorded observations, as well as additional observations by the investigator.)
3. A hypothetical solution of the problem consistent with the observations is formulated. (Frequently, this may also be found in the literature.)
4. Predictions of other observable phenomena are deduced from the hypothesis.
5. Occurrence or nonoccurrence of the predicted phenomena is observed by means of experiment.

6. The hypothesis is accepted, modified, or rejected in accordance with the degree of fulfillment of the predictions. If accepted, it becomes established as theory and finally as law.

This description of the scientific method is, of course, greatly oversimplified. It does not, for instance, explain the methods that must be used to obtain scientifically valid observations or to test the hypothesis—the controls that must always be included in an experiment, the statistical analysis of data to determine whether the differences are significant, the critical analysis of the data, etc. In practice, furthermore, it is very useful to suggest alternative hypotheses and to devise crucial experiments that will, as nearly as possible, exclude one or more of the hypotheses (Platt, 1964). It may not, in fact, be possible to prove any one hypothesis, since a later, alternative explanation may be just as good or better. Furthermore, it must also be remembered that every hypothesis involves some assumptions, and no matter how reasonable these may be, their existence must never be forgotten.

Although this method is not always strictly followed, it does emphasize that at least part of the "body of scientific knowledge" is not necessarily the truth, since it includes calculated guesses (hypotheses). In other words we must clearly distinguish between the aims and the actual state of a science. This "deficiency" of science must be faced squarely; but, on the other hand, it does not detract from the usefulness and even the necessity of such hypotheses, without which science would be at a standstill. Unfortunately, not only students but even sophisticated scientists themselves frequently speak contemptuously of hypotheses and plead, "Let us be practical. Let us talk sense and give only the experimentally determined facts." But these "facts" by themselves are at best half truths and may even be completely false if the experiments were not performed carefully. Furthermore, the most carefully performed experiments may not reveal the truth, since as Heisenberg's uncertainty principle points out, the very performance of an experiment changes the conditions; and the observed phenomena may be quite different from what they were before the test.

Thus the deficiencies of both hypothesis and experiment must be recognized. It is only when the two are analyzed critically and fused into a logical explanation, that the hypothesis can begin to approach scientific truth. New facts may appear to contradict the hypothesis, but it must not be discarded hurriedly. The wavelength theory of light has continued to be used successfully long after the discovery of the photon and quantum, which seemed to oppose this theory. The hypothesis can therefore be considered as a bridge that must be crossed in order to arrive at the truth, and experiments can be considered as pathways that lead to either new or old bridges.

Science, therefore, consists of many guesses that are constantly being tested by experimental research. For this reason we cannot expect cut-and-dried answers to all questions. Nevertheless, a good scientist should be expected to make only educated and critically sound guesses. He should not suggest a hypothesis if evidence is already available that proves the hypothesis untenable. This requires familiarity with, and understanding of, the already existing body of knowledge—that is, the "literature" on the subject. Unfortunately, scientists are human, and sometimes their hypotheses can be proved untenable on the basis of already available information without performing a single new experiment.

The purpose of research is therefore twofold: (1) to uncover the untruths and (2) to discover new truths. But every new research is in danger of adding not only truths but untruths. Thus of all the pursuits of scientists, the most important (and usually, alas, least practiced) is the critical analysis of the methods, the data, and the conclusions or hypotheses proposed to explain the data. If a scientist devotes 1 year to accumulating the data, he may have to devote 2 years to a critical analysis of them in relation to the accumulated body of knowledge; but he seldom does so. It is, therefore, possible for some scientists to make their contributions to science by critically analyzing the work of others without performing experiments themselves, although, of course, there is a limit to this kind of contribution, and one who has not himself performed experiments is usually unable to understand the experiments of others.

The older the science the more closely it may be expected to approach its aim. Yet even in the case of the one most generally accepted to have progressed the farthest, physics, some confusion still exists (Hubbert, 1963). For, as Hubbert points out, in many cases the truths are available but the scientists themselves are unaware of them. This is because of the great concentration of efforts on the discovery and mastery of the newer truths, leading to a neglect of the older truths.

The point that must be driven home to every student is that the body of knowledge in this text, as well as in any other, must not be accepted as absolute truth. Eternal vigilance is needed by all scientists, and each must receive every hypothesis (including his own) with skepticism. It behooves the student to show this same skepticism. Every statement should be analyzed critically by the student and accepted or rejected on the basis of such careful analysis. Since the author and other scientists are doing this constantly, a new edition of any science text differs from the old not only by including new truths (and perhaps untruths) but also by having weeded out old untruths. It does not necessarily follow, however, that the newer is true and the older false. Truths must be identified on their own merits without regard to whether they are new or old. But in every case a principle, law, or hypothesis must be thoroughly understood before exposing it to the test of experiment or criticism. Only too often rejection of a principle has resulted from its misinterpretation—by student and mature scientist alike. It is for this reason that a text must concentrate on presenting principles, laws, and even hypotheses clearly, thoroughly, and quantitatively.

BASIC LAWS OF PLANT PHYSIOLOGY

In the physical sciences, rigorous application of the scientific method has established many natural laws, for example, Newton's laws of motion, the gas laws, the laws of thermodynamics, etc. In most of the biological sciences, theory is still in its early stages, and fewer natural laws have been established. In this respect, physiology has some advantages over other biological sciences, since its main purpose is to explain all biological phenomena in terms of physics and chemistry. Thus, on the basis of our definition of plant physiology, the physiologist in search of the laws of physiology has two main sources of information: (1) physical and chemical laws and (2) direct experimental evidence. In many cases the physical and chemical laws by themselves clearly indicate which of several conclusions are plausible and which are untenable, without the necessity of the scientist having to perform a single experiment. But this does not, by any means, free the physiologist of the necessity of turning to experimental evidence. On the contrary, in order to determine which laws control a physiological process, it is usually necessary to perform many carefully planned experiments. For example, anyone with the merest rudiments of physical knowledge is certain to explain freezing injury in plants as a result of cell rupture because everyone knows that water expands on freezing. Yet experiment has shown that not only does no cell rupture occur on freezing but that the cell actually contracts because of ice formation in the intercellular (air-filled) spaces at the expense of water drawn out of the cells. This does not mean that the plant changes the physical properties of water. It means that other properties of water, besides its expansion on freezing, must be considered in order to explain the behavior of plants on freezing. The properties that must be applied can be discovered only by experiment. In another series of experiments physiologists succeeded in discovering that the plant actually "pumps" substances into itself instead of waiting for them to diffuse in. Again, this must not be taken to mean that the plant evades the laws of diffusion. On the contrary, both the laws of diffusion and the laws that control the operation of pumps must be made use of to explain the process of absorption by plants.

Thus by applying physical and chemical laws and by performing experiments, the physiologist is gradually succeeding in his aim to develop physiological laws. As a by-product of his experiments, he has also discovered purely physical and chemical phenomena not previously known to the physicist or chemist, for example, osmosis, brownian movement, chromatography, etc.

Unfortunately, despite the yeoman service of physiologists, much experimental work remains to be done before we can hope for a complete understanding of most physiological processes. Consequently, most of them can be discussed only somewhat superficially in the light of the incomplete evidence that is available, and it must always be remembered that the explanations are usually only working hypotheses, the main purposes of which are (1) to describe the known facts and (2) to predict the plant's behavior. These predictions point to the direction in which further experimental work must be pushed.

It is not, then, scientifically permissible to teach even an elementary physiology course in the same cut-and-dried fashion as is often used in a course in the physical sciences. This deficiency of physiology, however, has the advantage of exposing the student not only to the precepts of science but to its methods as well. Consequently, a well-taught course in physiology may quite conceivably give a better insight into scientific methods than those that are usually considered to be well-taught courses in physics and chemistry. For example, the scientific method, as just described, involves both inductive and deductive reasoning—inductive in proceeding from the individual observations to the more general hypothesis and deductive in proceeding from the hypothesis to the individual predictions. One of the major deficiencies of science courses is the lack of this twofold approach. Time does not permit it. Frequently, both the inductive and deductive reasoning are eliminated, and the student memorizes the theory or law as dogma. Furthermore, the more highly developed the field the more attention is devoted by a text to theory and the less to experiment since a sound theory is able to predict experimental results correctly. But even in a plant physiology course, time does not permit use of the scientific method in all phases of the teaching. The basic physical and chemical principles must be taught as dogma on the assumption that they have been firmly established by a rigorous application of the scientific method. For the student who is not satisfied by this dogmatic approach, the only solution is to take those basic physical and chemical courses in which the derivation of these principles is taught.

Must we, then, be physicists and chemists first before studying plant physiology? Unfortunately, the answer to this question is "yes." But, fortunately, a relatively small number of physical and chemical laws and principles may be sufficient for a relatively large fraction of present-day plant physiology. As pointed out by Hubbert (1963), in the whole field of science the "master generalizations number at most but a few tens." He includes among these the "three Newtonian laws of motion, the law of universal gravitation, the three laws of thermodynamics and the associated thermodynamics of irreversible processes,

the two Maxwellian laws of electromagnetism, the law of conservation of matter, and the concept of the atomic and molecular nature of the chemical elements and their compounds." And these can serve as a basis for both student and teacher of plant physiology.

In order to determine which of these physical and chemical principles are most important to the physiologist, physiology itself must be reduced to its essentials. Plant physiology is really the study of the plant at work. One basic aim is to determine the precise chemical nature of all the materials found in the plant and the way in which the plant succeeds in obtaining or making them. This is the *biochemistry* of the plant. The other aim is to understand all the kinds of work performed by the plant (mechanical, chemical, osmotic, electrical) and the nature of the energy involved in performing this work. This is the *biophysics* of the plant. Even the biochemistry of the plant always involves energy changes. Therefore, the one common denominator of all aspects of plant physiology is work and energy. This fact has been recognized by many physiologists. According to Calvin and Calvin (1964, cited by Amen, 1966), two fundamental attributes distinguish living systems from inanimate ones: the ability to transfer and transform energy in a directed way and the ability to remember and communicate information. According to Hawkins (1964, cited by Amen, 1966), a living system is one that contains a complete differential description of itself and one in which the free energy of reproduction is comparable to the total free energy of the system.

It is obvious from this fundamental importance of work and energy in plant physiology, that the first laws that must be mastered by the physiologist are those pertaining to work and energy, that is, the laws of thermodynamics. The first and second laws are particularly simple and useful. According to the first law, energy cannot be destroyed or created but is simply converted from one form to another. It is, of course, now known that energy can also be converted into mass and vice versa, but this fact has no bearing on physiological phenomena since the amounts of energy involved would yield immeasurably small increases if converted into mass; and, of course, no conversion of mass to energy can occur at physiological temperatures (in the case of stable isotopes normally found in plants).

The first law of thermodynamics (or the Principle of the Conservation of Energy) is represented by the equation (Getman and Daniels, 1937):

$$\Delta E = q - w$$

where ΔE = change in energy of a body
        q = heat or other energy absorbed by the body
        w = work done by the body on its environment

In the case of the living plant, the first law can be represented schematically (Fig. 1-1). An increase in the plant's energy results if there is an excess of energy absorbed over work done. The plant does work on its environment by growing (e.g., by pushing downward through the soil or by lifting its new cells upward against gravity), by evaporation of its water, etc. But this is more than compensated for by the radiant energy absorbed by the plant. Part of this absorbed radiant energy is converted into heat energy, which is used to evaporate water from the plant; a second part is converted into mechanical energy, which lifts the water up the plant; part of the remainder is converted into chemical energy in the process of photosynthesis. The first law also tells us about work done within the plant. Energy-consuming processes such as growth must be fed by energy-releasing

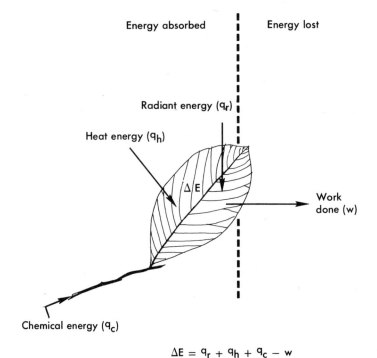

$$\Delta E = q_r + q_h + q_c - w$$

Fig. 1-1. Application of the first law of thermodynamics to the plant. The increase in the plant's energy is equal to the sum of all the forms of energy absorbed minus the work done by the plant on its environment. Each quantity represents the net value; for example, q_h is the heat energy absorbed minus the heat energy given off.

processes such as respiration; and the work done within the plant that results in growth can be no greater than the preceding net increase in energy (ΔE) absorbed from the environment.

The second law of thermodynamics goes further than this. It tells us that not all of this increase in energy is "free" or available for useful work because it consists of two components:

$$\Delta E = \Delta A + \Delta TS$$
where ΔA = isothermally available energy
ΔTS = isothermally unavailable energy
(T = absolute temperature, S = entropy)

The quantity A is frequently called the "Gibbs' free energy." Only this portion of the increase in energy is available to do the work within the plant—the movement of substances and bodies within the plant, the maintenance of concentration, electrical, and other gradients, and most important of all, the synthesis of all the thousands of substances produced in the cell's chemical factories.

According to the second law, heat can pass of itself only from a warmer to a colder body, and therefore, only this fraction of the heat of one body is "free" or available to do work on the other. This law can be applied to all other kinds of energy. The second law is made use of in many cases in the following chapters. It will be seen that no substances can move into a cell of themselves unless they are present in higher activities (or concentrations) in the medium than in the cell. Thus the laws of diffusion follow naturally from the second law of thermodynamics. It is only when the cell expends free energy that it can absorb a nutrient from a medium of lower activity or concentration (with respect to that nutrient) than its own; that is, the cell must then "pump" the nutrient into itself. According to a corollary of the second law of thermodynamics, free energy tends toward a minimum (in the process of doing work). This explains the fact that many substances tend to concentrate at surfaces (a phenom-

enon known as adsorption) because in so doing they reduce the free surface energy. Thus, adsorption follows from the second law.

The dependence of all branches of plant physiology on thermodynamics is shown by the following subtitles for the remaining chapters in this text in terms of the kinds of energy and energy changes involved:

Chapter	Title	Energy subtitles
2	The living cell	The machinery that performs the plant's work
3,4,5	Acidity, specific surface and adsorption, colloids	The work performed in the plant by the electrostatic energy of molecules and particles
6,7,8,11	Diffusion and osmosis, permeability, absorption,* exchange of gases	The work performed in the plant by molecular kinetic energy (or the energy of thermal agitation)
9	Ascent of sap	The conversion of radiant (solar) energy into mechanical work (bulk flow)
8,10	Absorption, translocation of solutes	The conversion of chemical energy into mechanical transfer of solutes
12	Nutrition	The elements needed to control energy transformations in the plant
13,14 15,16	Metabolism, respiration, photosynthesis, other metabolic paths	Energy changes caused by electron transport between molecules
15	Photosynthesis	The conversion of radiant energy into chemical energy

*Electrostatic and electron transport energy are also involved.

17,18	Growth, irritability and movement	The conversion of chemical energy into external work
19,20	Growth regulators, development	The regulation by the plant of the release of chemical energy
21	Plant rhythms	The rhythmic control by the plant of the utilization of its energy
22	Stress resistance	The effect of the environment on the energy-producing machinery of the plant

As new information continues to accumulate in plant physiology, instead of enlarging our texts, it should be possible to compress them by describing the main principles concisely, without all the data and controversial material that was so necessary in the developmental stages. This goal, as yet, has not been achieved. It is possible, however, to compress the subject matter into a small number of topics because the major problems covered by plant physiology fall more or less naturally into three categories: (1) the transfer of materials, (2) nutrition and metabolism, and (3) growth and development.

The first category is mainly biophysical in nature, the second is biochemical, and the third depends on the intimate interrelations between these two. But before these three categories of plant physiology can be investigated, it is first necessary to understand (1) certain physiological characteristics of the living cell and (2) some of the main physicochemical principles that must be applied in attempting to explain the physiology of the plant. These two areas are described in Part I, the previous three in Parts II to IV.

SPECIFIC REFERENCES

Amen, R. D. 1966. A biological systems concept. BioScience **16**:396-401.

Getman, F. H., and F. Daniels. 1937. Outlines of theoretical chemistry. John Wiley & Sons, Inc., New York.

Hubbert, M. K. 1963. Are we retrogressing in science? Science **139**:884-890.

Platt, J. R. 1964. Strong inference. Science **146**: 347-352.

Schaffner, K. F. 1967. Antireductionism and molecular biology. Science **157**:644-647.

Simpson, G. G. 1963. Biology and nature of science. Science **139**:81-88.

GENERAL REFERENCES

Annual review of plant physiology. 1950-.

Biological abstracts.

Chemical abstracts.

Lehninger, A. L. 1965. Bioenergetics. W. A. Benjamin, Inc., New York.

Linford, J. H. 1966. An introduction to energetics. Butterworth & Co. (Publishers), Ltd., London.

Ruhland, W. (Ed.). 1955-. Handbook of plant physiology. [Many volumes.] Springer-Verlag, Berlin.

Spector, W. S. (Ed.). 1956-. Handbook of biological data. W. B. Saunders Co., Philadelphia.

Steward, F. C. (Ed.). 1959-. Plant physiology. [Several volumes.] Academic Press, Inc., New York.

chapter 2

The living cell

THE CELL AS A WHOLE— THE PROTOPLAST

Just as the molecule is the basic unit of a chemical substance, the cell is both the functional and the structural unit of the plant. Thus the molecule is the smallest unit of a substance which still retains the properties of that substance, and the living cell is the smallest unit of the plant which so far has been shown to have all the properties necessary for regenerating the whole plant (Fig. 2-1). This is what is meant when it is said that the living cell is *totipotent*.

Even the nonliving cells usually have definite physiological functions. For instance, the living cells obtain a good share of their raw materials (minerals, water, etc.) by transport through dead cells (vessels and tracheids). But since the life processes are the center of interest for the physiologist, it is primarily these living cells that must be studied to understand the physiology of the plant. How, then, can we distinguish between living and nonliving cells? First we must know the characteristics of living cells and, second, we must develop methods of identifying these. Two characteristics are particularly important to identify.

A living cell does work (Chapter 1)

It is obvious that a living cell works when a plant is growing since it is expanding against its environment. But how can we know whether a mature, nongrowing cell is still alive and doing work or is dead and not doing work? There are many sophisticated methods of measuring chemical or even electrical work, but these require special equipment and complicated techniques. There is one kind of work, however, that can be readily detected by simple observation under the microscope. The protoplasm of a living cell performs mechanical work by continuously moving in streams that carry particles with them. This movement is known as protoplasmic, or *cytoplasmic, streaming,* and it is dependent on a continuous supply of free energy by the living cell. As soon as the cell dies, it ceases to produce this free energy and is no longer able to perform work. Cytoplasmic streaming, therefore, stops. Unfortunately, the absence of streaming is not proof that the cell

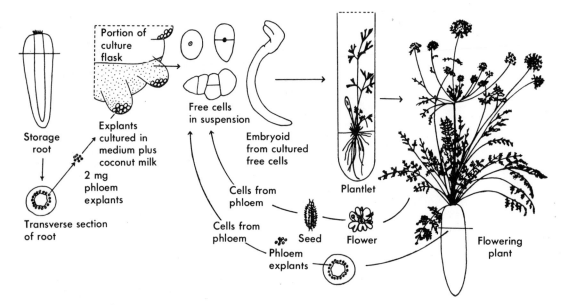

Fig. 2-1. Diagrammatic representation of the cycle of growth of the carrot plant; successive cycles of growth are linked through free, cultured cells derived from phloem or from the embryo. (From Steward, F. C., et al. 1964. Science **143**:20-24. Copyright 1964 by the American Association for the Advancement of Science.)

is doing no work, so negative evidence in this case is no evidence. In both living and dead cells, particles show another kind of movement—*brownian movement*. This does not depend on a supply of free energy released by the cell. It is simply caused by the kinetic energy (or thermal agitation) of the molecules, and since this energy is on the average identical on all sides of a particle, there will be no net movement of the particle in any one direction (as occurs in cytoplasmic streaming) but only a haphazard movement in all directions about a constant average position. Brownian movement is, of course, no evidence of life.

A living cell has semipermeable (differentially permeable) membranes

Approximately 90% of the living cell consists of water in which many substances are dissolved (see later). Some of these substances have been absorbed from the outside; most have been synthesized by the

cell. When the cell is placed in water (or aqueous solutions of other substances), we would expect (from the second law of thermodynamics) that the cell solutes would become equally distributed in the water inside and outside the cell. Since the solutes are essential for supplying the cell with the free energy necessary for the performance of its work, this loss would result in death of the cell. Actually, if the cell is living, there is little or no loss of solutes from it. This is possible only if a barrier surrounds the cell preventing the loss. Since water moves freely into and out of the cell (see later), the barrier must be essentially semipermeable—permeable to solvent but not to solutes. When the cell loses its property of semipermeability, all soluble substances quickly leak out, and the cell immediately loses its ability to do work and is, therefore, dead. There are two simple methods of determining whether or not the cell is semipermeable.

Vital staining. Basic dyes (salts of a dye base and an inorganic acid, see Chapter 5), such as neutral red or methylene blue, penetrate the living cell in the molecular form and pass through the protoplasm into the vacuole, where they are converted to the ionic form (see Chapter 8). Since the semipermeable membranes are impermeable to the ionic form, it accumulates in the vacuole, which therefore becomes colored or "stained" (Fig. 2-2). Such cells may remain alive in the vitally stained state for hours or days. If the cell is subsequently killed, its membranes become freely permeable and the dye quickly escapes. The vacuole of the dead cell is therefore no longer colored. The cell wall may be stained whether the cell is living or dead. Vital staining of the protoplasm by means of other dyes is also possible, but it must be carried out in the dark because the dye sensitizes the protoplasm to light, leading to injury and death within a very few hours. Since light must be used to observe the cells, the least injury results when the staining is produced by a dye that can be observed by means of fluorescent light (e.g., K fluorescein), since this enables the detection of much weaker (and therefore less injurious) staining (Strugger, 1952). A vital stain, therefore, may be defined as a stain (or dye) that colors living but not dead protoplasts (i.e., protoplasm or its inclusions, see later).

Another method of identifying living cells by staining is of practical importance in estimating the "germination potential" of seed lots without waiting for them to germinate (Lindenbein, 1965). Tetrazolium dyes (e.g., TTC, or triphenyltetrazolium chloride) are used for this purpose since they are converted from the colorless to the colored (e.g., red) form by freshly cut surfaces of living cells. This is not a true vital staining since the cells are killed in the process, nor is it evidence of the existence of semipermeable membranes but rather of a chemical reduction of the dye by freshly cut living cells. Cells that are dead before

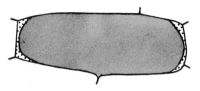

Fig. 2-2. Onion epidermis cell "vitally stained" with neutral red. The dark region occupying most of the cell is the stained vacuole (also called the cell sap). The unstained, granular, terminal caps are the cytoplasm surrounded by a thin cell wall. The lateral cytoplasm is so thin as to be all but invisible. Only the cytoplasm and nucleus (not seen in the diagram) are alive.

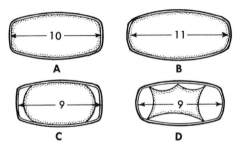

Fig. 2-3. Behavior of a living cell originally 10 units long in solutions of different concentrations. **A,** 9% sugar (slightly hypotonic); **B,** 2% sugar (hypotonic); **C,** 10% sugar (hypertonic); **D,** 20% sugar (hypertonic). Isotonic sugar solution is between 9% and 10%. Plasmolysis is apparent in **C** and **D.**

cutting are unable to produce this chemical reaction.

Plasmolysis. If a living cell (Fig. 2-3, A) is placed in a weak solution, it may expand because of uptake of water (Fig. 2-3, B). If transferred to a moderately strong solution, for example, about 10% sugar (Fig. 2-3, C), it shrinks and *plasmolysis* (a separation of the protoplasm from the cell wall) may occur. Transfer to a still stronger solution (Fig. 2-3, D) fails to cause any further shrinkage of the wall, although the plasmolyzed protoplast contracts still further. If transferred back to a weak solution, the protoplast expands (it deplasmolyzes) to the size it originally possessed in that solution. If

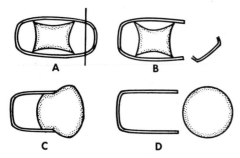

Fig. 2-4. Method of obtaining free protoplast (see Levitt et al., 1936). **A,** Onion cell is plasmolyzed in 20% sugar, then cut with a razor blade. **B,** Open cell. **C,** Protoplast swelling in 10% sugar. **D,** Empty cell wall and spherical free protoplast.

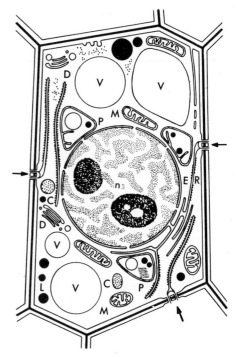

Fig. 2-5. Schematic representation of meristematic (young) plant cell. **n,** Nucleus containing two nucleoli; **V,** vacuole; **M,** mitochondrion; **P,** proplastid (incompletely developed plastid); **C,** lysosome; **ER,** endoplasmic reticulum; **D,** Golgi apparatus containing dictyosomes; **L,** lipid drops. Ribosomes are shown attached to **ER** or free in the groundplasm (at top). Three plasmodesmal connections through the cell walls are shown by the arrows. (From Sitte, P. 1965. Bau und Feinbau der Pflanzenzelle. Gustav Fischer Verlag, Stuttgart.)

the cell shrinks to its minimum size with little or no plasmolysis when placed in a solution, this solution is *isotonic* (of the same strength as the cell sap). Solutions that cause plasmolysis are *hypertonic;* those that fail to produce any trace of plasmolysis (although they may cause a slight shrinkage of the whole cell) are *hypotonic*. If the cell wall is cut during strong plasmolysis and the cell is then transferred to a weak solution, the *protoplast* (the protoplasm plus its vacuole) may actually swell enough to come completely out of the wall (Fig. 2-4). Protoplasts can also be isolated by dissolving the cell wall with the enzyme cellulase in the presence of 0.59 M sucrose (Cocking, 1960). The same process occurs naturally in some fleshy fruit during ripening. Free protoplasts, therefore, can be found in the juice of ripe tomatoes and grapes.

In spite of this ready separation of protoplasts from their cell walls and therefore from each other, the living cells of plant tissues are always connected with each other by protoplasmic strands about 0.2μ in diameter (Livingston, 1964) known as *plasmodesmata* (Fig. 2-5), which pass through openings in the cell walls known as pits. The electron microscope has revealed that there are many more of these than

can be seen with the optical microscope. The foregoing methods must break these strands in order to separate the protoplasts. Therefore, even though the protoplast may deplasmolyze without visible signs of injury, quantitative measurements reveal that its metabolism may be greatly affected (Falk et al., 1966). Similarly, free protoplasts are frequently obviously injured, and in some cases consist only of the vacuole surrounded by the inner protoplasmic membrane (the tonoplast), the rest of the protoplasm having coagulated.

By use of one or more of the foregoing

methods it is possible in all or nearly all cases to determine whether or not a plant cell is alive. Air-dry cells (e.g., of seeds) must be allowed to swell in water before applying any of these methods. Even then, it may be difficult to observe the response of the cells. As mentioned earlier, the most successful method for seeds is staining with tetrazolium dyes.

CELL COMPONENTS

As shown previously, it is possible to release the inside of a living cell from its surrounding wall without loss of life. The empty wall shows no signs of life, but the freed protoplast will expand in weak solutions and contract in strong solutions (Levitt et al., 1936) just as in the plasmolysis and deplasmolysis of the intact cell. Obviously, even the living plant cell consists of a nonliving part (the cell wall) and a living part (the protoplast). The protoplast itself consists of a protoplasmic layer surrounding the vacuole, and since some living cells (e.g., the youngest in meristems) exist without any vacuoles, the protoplasm is the only essential component of a living cell. Direct proof of the nonliving nature of the vacuole has recently been obtained by washing out the vacuole sap and replacing it by an artificial sap without interfering with the normal activities of the cell (Tazawa, 1964). Yet the nonliving cell wall and vacuole have pronounced physiological significance since they greatly affect the protoplasm, for example, its water content, growth, etc.

Cell wall (Siegel, 1962)

The cell wall consists of several substances (Table 2-1), most of which are carbohydrates (cellulose, hemicellulose) or their derivatives (pectins). With few exceptions cell walls also contain a very small amount of protein, more in young cells than in old cells. The name "extensin" has been proposed for the protein found in primary cell walls (Lamport, 1963). It is richer in

Table 2-1. Average analysis for primary cell walls[*]

Substance	Percent of fresh weight
Water	60
Hemicelluloses	5–15
Pectic substances	2–8
Cellulose	10–15
Protein	1–2
Lipids	0.5–3.0

[*]From Setterfield, G., and S. T. Bayley. 1961. Ann. Rev. Plant Physiol. **12:**35-62.

proline and hydroxyproline than other plant proteins (Olson, 1964). Mature cell walls may contain many substances not found in appreciable quantities in young cells, for example, lignin in wood cells, cutin in epidermal cells of leaves, suberin in cork cells, tannin in bark cells, etc. Cellulose-protein complexes may occur during cellulose deposition, after which most of the protein is resorbed (Preston, see Edds, 1961). But the composition varies greatly from one kind of cell to another in the same plant and even in the same cells under different conditions and at different stages of development. In the normal, turgid cell the wall is essentially saturated with water, which may account for 90% or more of the wall (Frey-Wyssling, 1952).

In spite of this high water content, the cell wall has the physical properties of a solid. It is both elastically (reversibly) and plastically (irreversibly) extensible. Cell contraction on water removal before plasmolysis (Fig. 2-3) is a result of the elastic extensibility of the cell wall. In the turgid cell the wall is stretched; in the flaccid cell it returns to the unstretched state. Plastic extensibility of the cell wall results in permanent cell enlargement.

Protoplasm

The most important part of the cell to the physiologist is the living protoplasm because here are synthesized all the multitude of organic substances found in the plant. Many of these substances may subsequently be broken down again in the protoplasm.

The gross analysis of protoplasm is possible in the case of animals and of some lower plants (e.g., slime molds), which yield masses of protoplasm without a cell wall or vacuole (Table 2-2). This old analysis includes the nucleic acids with the proteins. The ribonucleic acid (RNA) content of root tips is 4.5 to 13.8 μg/mg fresh weight, and the deoxyribonucleic acid (DNA) is 10% to 20% as high (Martin, 1966). On this basis the total nucleic acid content would be 0.5% to 1.5% of the root tips. Since these root tips are probably about one half protoplasm, this would mean a nucleic acid content of 1% to 3% in the protoplasm.

Of course, this does not mean that it would be possible to synthesize protoplasm by mixing these substances in the preceding proportions. In other words, the essential physical and chemical properties of protoplasm are not revealed by such gross analyses. If, for instance, the cell sap is mixed with the protoplasm by grinding up cells of leaves, fruit, etc., at least some of the proteins will usually coagulate. This is

Table 2-2. Generalized protoplasmic analysis*

Substance	Percent of fresh weight	Approximate relative number of molecules
Water	85–90	18,000
Proteins†	7–10	1
Fatty substances	1–2	10
Other organic substances	1–1.5	20
Inorganic ions	1–1.5	100

*Adapted from Sponsler, D. L., and J. D. Bath. 1942. *In* W. Seifriz. The structure of protoplasm. Iowa State University Press, Ames, Iowa.
†This undoubtedly includes the nucleic acids, which account for 1% to 3% of the protoplasm.

partly because of the high acidity of the cell sap. Yet this same, acidic cell sap is surrounded by the nonacidic protoplasm in the living cell. Therefore the protoplasm must be able in some way to maintain its own low acidity although in contact with the acid vacuole. It does this by forming a barrier, the semipermeable membrane, which keeps the vacuole contents from penetrating into it.

Vacuole

It is not only in acidity that the vacuole differs from the protoplasm but in many other chemical and physical properties as well. Although the vacuole resembles the protoplasm in its high water content and in consisting of a solution of organic and inorganic substances, unlike the protoplasm it usually does not contain proteins and fatty substances. Salts of organic and inorganic acids and carbohydrates such as sugars, mucilages, etc. are probably always present in the vacuole. It also usually contains some so-called secondary metabolites (see Chapter 16), such as tannins, flavonols, anthocyanins, alkaloids, etc. But the vacuole contents are much more variable than those of the protoplasm and therefore no general gross analysis can be given. This is because the vacuole is the repository of (1) any substances that accumulate in excess from the outside and (2) any substances produced in excess by the protoplasm. Heavily fertilized plants may accumulate nitrates, phosphates, or potassium salts in their vacuoles; halophytes (plants that grow in high-salt soils) store large quantities of sodium salts in their vacuoles; and, as just seen, vital staining with neutral red involves accumulation in the vacuole. Similarly, many plants deposit excess calcium and organic acid as insoluble crystals in their vacuoles, or excess of the breakdown products of protein in the form of asparagine, or in the case of plants with specialized metabolism, alkaloids. Some of the stored substances may be reutilized later (e.g., nitrates, organic acids, asparagine, sugars); others may remain permanently in the vacuoles (e.g., alkaloids, sodium salts). Even cells in the same tissue may vary greatly in their vacuolar contents of such substances. This can be clearly shown by staining a section vitally with neutral red. Cells that stain heavily are called "full," and those that stain lightly are "empty." The former have high tannin contents in their vacuoles; the latter have low contents or none. Even the color of the staining may differ from vacuole to vacuole in the same tissue.

Since the cell wall in a living cell may contain up to 90% water, the protoplasm 85% to 90%, and the vacuole 90% to 95% water, the question arises as to the source of juice that may be squeezed from plant cells for analysis. It should come mainly from that part of the cell which accounts for the largest percent of its volume. This proportion will vary, in the case of the vacuole, from essentially zero in meristematic cells, to over 90% of the cell in mature parenchyma cells (Fig. 2-2). But since meristematic cells account for only a very small fraction of the mature plant and parenchyma cells make up the largest part of the mature (herbaceous) plant, most of the juice obtained from herbaceous plants must come from the cell vacuoles. This is also true of resting organs with high water contents (e.g., bulbs, tubers, corms, etc.).

Organelles

Physically, the protoplasm possesses visible structure as opposed to the optical homogeneity of the vacuole. Discrete bodies can always be seen in it under the optical microscope. The largest is the nucleus and its included nucleoli, the next largest are the plastids (commonly 2μ by 5μ), and the smallest are the mitochondria (commonly 0.5μ by 2μ). The structure of the plastids (Granick, 1963) and of the mitochondria (Green, 1964) is particularly complicated. Because of their small size, it is not surprising that there may be 1000 mitochondria per meristematic cell of 2500 μ^3 volume, about 170 proplastids (incompletely developed plastids), and 225 Golgi bodies (Avers, 1962). Other bodies may be visible in some cells, but for further observation of protoplasmic structure the electron microscope must be used. The biochemist calls these bodies *protoplasmic particulates* and separates them and others (e.g., "microsomes") from the rest of the cell (de Dvue, 1965). The electron microscopist calls them *organelles*. Several kinds occur in all living plant cells (Fig. 2-5). The importance of both their observable structure and their biochemical nature is now recognized, and according to Bell (1965) an organelle should be defined as part of a cell with certain definite functions, distinctive chemical constituents, and char-

Table 2-3. Protoplasmic structures (or organelles) and their physiological functions

Protoplasmic structures	Physiological functions
Plasma membranes	Control of semipermeability, absorption
Nucleus	DNA synthesis or control of code
Nucleolus	RNA synthesis for protein synthesis
Chloroplast*	Photosynthesis
Mitochondrion	Respiration
Ribosome† (clusters)	Protein synthesis
Endoplasmic reticulum (ER)	Membranes, plasmodesmal connections, translocation of solutes
Golgi bodies or apparatus (GA) or dictyosomes	Cell plate, pectins, cellulose of cell wall, origin of vacuoles
Lysosomes (spherosomes)	Digestion
Groundplasm	Glycolysis

*In nongreen cells other kinds of plastids may occur, for example, leucoplasts that synthesize starch from sugar.
†The earlier microsomes of the biochemist were probably pieces of endoplasmic reticulum containing ribosomes.

acteristic morphological features. Thus, each of the bodies performs a specific function in the living cell, although in some cases it is not yet fully understood (Table 2-3). They may even continue to perform these functions when separated from the cell, although it may then be necessary to supply them with substances normally provided by other parts of the protoplasm or lost from them during extraction. The organelles show a considerable degree of independence and may even multiply in the absence of nuclei (Brachet, 1965), although they are presumably originally produced by the nucleus (Frey-Wyssling and Mühlethaler, 1965). It should be emphasized that each organelle may do more than indicated by its main function.

The endoplasmic reticulum (ER), in spite of its name is not a reticulum (a net) but is a system of interconnecting canals within the protoplasm (Frey-Wyssling and Mühlethaler, 1965). The Golgi apparatus (GA) consists of a series of double membranes, sometimes called the dictyosomes. The ribosomes, on the other hand, are simple particles about 15 mμ in diameter, occurring either free in the cytoplasm or adhering to the ER. They frequently occur as aggregates called polysomes. Although the ribosome is the main seat of protein synthesis, the nucleus (James and Richens, 1963), the nucleolus (Birnstiel and Hyde, 1963), the chloroplast (Spencer, 1965), and the mitochondria (Das and Mukherjee, 1964) also synthesize proteins. In the case of the chloroplast (Lyttleton, 1962) and the mitochondrion (Küntzel and Noll, 1967), although apparently not in the case of the nucleus (Flamm and Birnstiel, 1964), this is a result of the presence of ribosomes within it. Similarly, although the nucleus is the seat of DNA synthesis (Ruppel, 1964), DNA has been found in the chloroplast (Goffeau and Brachet, 1965), although only to the extent of one five thousandth the protein content (Gibor and Izawa, 1963). There is evidence that DNA is synthesized in young (but not

old) chloroplasts (Wollgiehn and Mothes, 1964), and although RNA is synthesized in the nucleus, as much as 25% to 35% of the total RNA of a leaf cell may be in its chloroplasts (Heber, 1963). The membranes are also sometimes considered as separate organelles, although, like the ribosomes, they are components of other organelles (the nucleus, plastid, mitochondrion) that they enclose. The whole protoplasm layer is also bounded by its membranes.

It must be admitted that the functions of each organelle are not fully understood. The Golgi apparatus (or dictyosomes) for instance, contain "cisternae" (cavities surrounded by membranes), which may enlarge to form the cell vacuole (Marinos, 1963) or perhaps may form the cell plate (Whaley and Mollenhauer, 1963). Yet they are also believed to be involved in the synthesis of pectins for cell wall formation (Dashek and Rosen, 1966), as well as the cellulose fibrils (Fig. 2-6). They even secrete the mucilage by means of which an insectivorous plant captures insects (Schnepf, 1963). Each kind of protoplasmic body has its own structure (Fig. 2-7), which is related to its function.

Some organelles found in animal cells are just beginning to be identified in plant cells. For instance, although little information is at present available on the role of lysosomes in higher plants, they do seem to be important in lower plants (Brandes et al., 1964). Spherosomes have long been known to occur in plant cells. Recent results indicate that they are equivalent to the lysosomes of animal cells because they have been found to contain hydrolytic enzymes (Semadeni, 1967; Matile et al., 1965), and their function is therefore thought to be digestion of the cell contents when these enzymes are released from the lysosomes (Gahan and Maple, 1966). In the case of *Euglena*, during starvation, portions of the cytoplasm become encapsulated in membrane-bound cavities, similar to cytolysomes of mammalian cells. Progres-

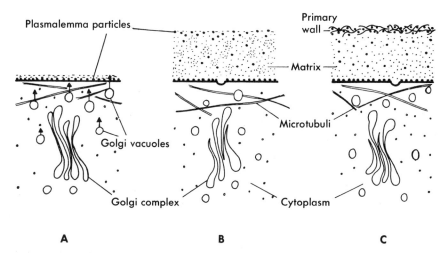

Fig. 2-6. Scheme of the subsequent steps in cell wall formation of the green alga *Chlorella* sp. **A,** Cortical region of the cell with the plasmalemma covered with particles and a subjacent Golgi complex. **B,** Accumulation of matrix material, carried to the cell surface by Golgi vacuoles. Plasmalemma particles become detached and move to the outer periphery of the matrix. **C,** Primary wall formation in the region where the particles are concentrated. (From Mühlethaler, K. 1967. Ann. Rev. Plant Physiol. **18**:1-24.)

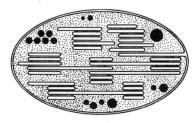

Fig. 2-7. Fine structure of chloroplasts showing schematically the outer membrane, internal lamellae, granular matrix, and several globules. (From Sitte, P. 1965. Bau und Feinbau der Pflanzenzelle. Gustav Fischer Verlag, Stuttgart.)

sive degradation of the encapsulated material may represent a mechanism for providing the cell with breakdown products for utilization in continued maintenance of basic metabolic processes. Strong evidence indicates participation of the Golgi apparatus in the formation of the cytolysomes. Yet the spherosomes of higher plants appear to originate from the ER. Microtubules have been repeatedly described (Fig. 2-6), but their function is unknown.

It is not always certain whether the organelles are self-producing or originate de novo. The Golgi apparatus (dictyosomes), for instance, differentiates de novo from perinuclear bodies, which are synthesized under the action of ribonucleoprotein and are therefore nucleus dependent (at least in *Acetabularia*, Werz, 1964).

Recent high resolution electron microscopy has identified an "elementary particle," or oxysome, in the mitochondrion (Lehninger, 1964). Similarly, the "quantasome" is an oblate sphere with a molecular weight of 2×10^6 (Park and Biggins, 1964) and is the structural and functional unit of the chloroplast. Ribosomes, as already mentioned, occur within the chloroplast (Jacobson et al., 1963), so one particulate or organelle may be a component of another. These subunits have been called "multimolecular assemblies" (MMA) (Schmitt, 1963). The relationship to the cell is shown in Fig. 2-8. The large numbers of such particles is illustrated by the ribosomes. There are about $6000/\mu^3$, and they

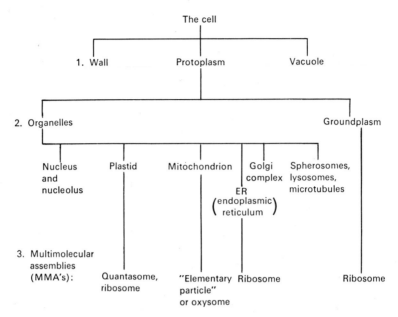

Fig. 2-8. Three levels of organization of cell constituents.

occupy about 0.5% of the protoplasm (Ts'o, 1962).

The fundamental importance of the organelles is emphasized by Palade's (1963) statement that the basic requirements of a living cell are (1) the codes (in the nucleus), (2) the ribosomes, (3) the cell membranes, and (4) an energy-supplying system (e.g., the mitochondria or the chloroplasts). The dynamic state of these organelles cannot be revealed by the electron microscope, since it observes the dead, or at least nonfunctioning, cell. Not only does streaming occur in the cytoplasmic strands of living cells but even the structural components may undergo visible changes. The mitochondria change their shape, fragment, and migrate to the outer rim of the cytoplasm. Even the plasma membrane may show an undulating motion (Gropp, 1963).

It is often forgotten that all the organelles are embedded in a granular matrix, the "groundplasm," which must therefore also have some function. It has a very fine, granular structure, just detectible under the

electron microscope, yet it is also fibrillar (Sitte, 1965). It contains soluble enzymes, which are not bound to the organelles and which function at least in the process of glycolysis (see Chapter 14). The chemical complexity of the structural components of protoplasm is illustrated by analyses of chloroplasts. Each chloroplast (Fig. 2-7) consists of a lamellar phase (mainly in disc-shaped grana) embedded in a matrix (stroma). The whole chloroplast is surrounded by a membrane. The lamellae may be separated experimentally from the matrix (or stroma) and are then found to contain all the chlorophyll and 50% of the protein N. The complete analysis of the lamellae is given in Table 2-4.

In spite of this chemical complexity of the protoplasmic structures, they all consist primarily of three main groups of substances: (1) proteins, (2) nucleic acids (RNA and DNA), and (3) lipids (Table 2-5). It must be emphasized, however, that even the organelles may be largely aqueous; for instance, 80% of the enclosed volume of

Table 2-4. Substances and their proportions in spinach chloroplast lamellae*

	Moles	Parts per million (app.)
I. Lipids		
Chlorophylls	115	103,200
Carotenoids	24	13,700
Quinones	23	15,900
Phospholipids	58	45,400
Glycerides	245	201,000
Sulfolipids	24	20,500
Others		95,300
II. Proteins		464,000
Manganese in protein	1	55
Iron in protein	6	336
Copper in protein	3	159

*From Lichtenthaler, H. K., and R. B. Park. 1963. Nature (London) **198:**1070-1072.

Table 2-5. Distribution (%) of three main groups of organic substances in some protoplasmic structures*

	Protein	Nucleic acid	Lipid
Nucleus	74	26	Insignificant
Mitochondrion	67	1–3	30–33
Ribosome	50	50	

*Adapted from Frey-Wyssling, A., and K. Mühlethaler. 1965. Ultrastructural plant cytology. American Elsevier Publishing Co., Inc., New York.

wheat embryo ribosomes was found to consist of water (Wolfe and Kay, 1967). Although DNA is primarily found in the nucleus, about 9% of the total DNA of the tobacco leaf is in the chloroplast (Tewari and Wildman, 1966). Its molecular weight is approximately 4×10^7. Some of the nucleic acids also occur free in the groundplasm. That the nucleoproteins are the most fundamental substances of life is indicated by the fact that the smallest living entity, the virus, consists solely of nucleoproteins. Similarly, the units of heredity in all organisms, the genes, are also nucleoproteins. Ribosomes are actually ribonucleoprotein particles (30% RNA and 70% protein in the ribosomes of spinach chloroplasts [Biswas and Biswas, 1966]). The published proportions of these two substances vary (see Table 2-5), perhaps depending on their source.

The phospholipids are of particular importance because they are largely confined to the two surface layers (the membranes) of the structural components (the plastids, mitochondria, etc.) and the protoplasm, and they therefore control the semipermeability of the cell. However, the actual proportions of these substances are not the same in the different structures. Furthermore, even the groundplasm in which the organelles are embedded also contains some phospholipids. Because of this reserve, the groundplasm is capable of instantly forming new membranes on wounding the protoplasm (Sitte, 1965).

The matrix of the cytoplasm (the groundplasm) is essentially made up of a protein solution, or gel. This matrix is usually considerably more viscous than the vacuole. Since viscosity and gel formation are characteristic of the colloidal state, protoplasm is obviously a colloid. The nonliving parts of the cell also show colloidal properties. The vacuole, although optically homogeneous and mobile, usually possesses enough colloidal material (e.g., tannin) to accumulate dyes such as neutral red, staining more or less strongly if the cell is alive. The wall is a colloidal gel of cellulose and other large-moleculed organic substances. Although these substances are insoluble in water, they imbibe it readily in the same way that a gelatin gel does.

All three components of the living cell—wall, protoplasm, and vacuole—contain ample water that can pass readily from one to the other. The physical chemistry of aqueous systems (e.g., diffusion, osmosis, etc.) is therefore of paramount importance in plant physiology. The lines of separation between these three components of the cell are known as interfaces. Special physical forces known as surface tension, or interfacial tension, operate at such interfaces. Substances tend to become concentrated there, producing the phenomenon of adsorption.

It is now apparent that in order to understand more about the properties of protoplasm and of the cell as a whole, we must consider such purely physicochemical concepts as acidity, colloids, permeability, diffusion, osmosis, adsorption, etc.

QUESTIONS

1. What is the smallest portion of the plant that is capable of regenerating the whole plant?
2. Does the living cell do work when growing or when not growing?
3. What is the theoretical basis for this conclusion?
4. What experimental evidence is there for this conclusion?
5. Does brownian movement occur in living or dead cells?
6. How can living cells be distinguished from dead cells?
7. If a cell shows cytoplasmic streaming at room temperature but not when its temperature drops to $+1°$ C, does this indicate it is now dead?
8. When a living cell is placed in water, do the solutes leak out? Explain.
9. How can the semipermeability of living cells be demonstrated?
10. What is vital staining?
11. Do vital stains color cell walls of living or dead cells?
12. Are tetrazolium dyes vital stains? Explain.
13. What is meant by plasmolysis?
14. Does the cell wall remain unchanged when a cell plasmolyzes?
15. If a cell plasmolyzes, is the solution hypertonic, hypotonic, or isotonic?
16. What is the protoplast?
17. What is a free protoplast?
18. How can free protoplasts be obtained?
19. What chemical substances occur in the cell wall?
20. Of the cell's wall, protoplasm, and vacuole, which part is alive?
21. How does a cell wall react to a stretching force?
22. What chemical substances does protoplasm consist of?
23. What chemical substances does the vacuole consist of?
24. If excess fertilizer is absorbed, in what part of the cell is it accumulated?
25. If an excess of some substances is produced by the cell, in what part of the cell is it most likely to accumulate?
26. From what part of the cell does plant juice mainly originate?
27. What are the protoplasmic particulates?
28. What are the cell organelles?
29. What are their functions?
30. What are the basic structural requirements for a living cell?
31. What chemical substances do organelles consist of?
32. What is the chemical nature of the protoplasmic surface?
33. Which organelles are components of other organelles?
34. In what protoplasmic material do the organelles occur?
35. What are MMA's?
36. With what cell function are phospholipids associated?
37. What is required for regeneration of wounded membranes?

SPECIFIC REFERENCES

Avers, C. J. 1962. Fine structure studies of *Phleum* root meristem cells. I. Mitochondria. Amer. J. Bot. 49:996-1003.

Bell, P. R. 1965. The structure and origin of mitochondria. Sci. Progr. (London) 53(209):33-44.

Birnstiel, M. L., and B. B. Hyde. 1963. Protein synthesis by isolated pea nucleoli. J. Cell Biol. 18:41-50.

Bishop, C. T., S. T. Bayley, and G. Setterfield. 1958. Chemical constitution of the primary cell walls of *Avena* coleoptiles. Plant Physiol. 33:283-289.

Biswas, S., and B. B. Biswas. 1966. Characteriza-

tion of ribonucleoprotein particles and protein synthesis in chloroplasts. Indian J. Biochem. **3:** 96-100.

Brachet, J. L. A. 1965. *Acetabularia.* Endeavour **24**(93):155-161.

Brandes, D., D. E. Buetow, F. Bertini, and D. B. Malkoff. 1964. Role of lysosomes in cellular lytic processes. I. Effect of carbon starvation in *Euglena gracilis.* Exp. Molec. Path. **3:**583-609.

Cocking, E. C. 1960. A method for the isolation of plant protoplasts and vacuoles. Nature (London) **187:**927-929.

Das, H. K., and T. Mukherjee. 1964. Protein synthesis in plant mitochondria. III. Characterization of mitochondria and the microsomal fraction of the seedlings of *Vigna sinensis.* Biochim. Biophys. Acta **93:**304-310.

Dashek, W. V., and W. G. Rosen. 1966. Electron microscopical localization of chemical components in the growth zone of lily pollen tubes. Protoplasma **61:**192-204.

de Dvue, C. 1965. The separation and characterization of subcellular particles. Harvey Lect. **59:** 49-87.

Falk, H., U. Lüttge, and J. Weigl. 1966. Research on the physiology of plasmolysed cells. II. Ion uptake, oxygen exchange, transport. Z. Pflanzenphysiol. **54:**446-462.

Flamm, W. G., and M. L. Birnstiel. 1964. The nuclear synthesis of ribosomes in cell cultures. Biochim. Biophys. Acta **87:**101-110.

Frey-Wyssling, A. 1952. Growth of plant cell walls. Sympos. Soc. Exp. Biol. **6:**320-328.

Gahan, P. B., and A. J. Maple. 1966. The behaviour of lysosome-like particles during cell differentiation. J. Exp. Bot. **17:**151-155.

Gibor, A., and M. Izawa. 1963. The DNA content of the chloroplasts of *Acetabularia.* Proc. Nat. Acad. Sci. U. S. A. **50:**1164-1169.

Goffeau, A., and J. Brachet. 1965. Deoxyribonucleic acid–dependent incorporation of amino acids into the proteins of chloroplasts isolated from anucleate *Acetabularia* fragments. Biochim. Biophys. Acta **95:**302-313.

Granick, S. 1963. The plastids; their morphological and chemical differentiation, p. 144-174. *In* M. Locke. Cytodifferentiation and macromolecular synthesis. Academic Press, Inc., New York.

Green, D. E. 1964. The mitochondrion. Sci. Amer. **210:**63-74.

Gropp, A. 1963. Morphologie und Verhalten lebender Zellen. Med. Welt **1:**20-23.

Heber, U. 1963. Ribonucleic acids in the chloroplasts of the leaf cell. [Transl. title.] Planta **59:** 600-616.

Jacobson, A. B., H. Swift, and L. Bogorad. 1963. Cytochemical studies concerning the occurrence

and distribution of RNA in plastids of *Zea mays.* J. Cell Biol. **17:**557-570.

James, W. O., and A. M. Richens. 1963. Energy transport from mitochondria to nuclei. Proc. Roy. Soc. [Biol.] **157:**149-159.

Küntzel, H., and H. Noll. 1967. Mitochondrial and cytoplasmic polysomes from *Neurospora crassa.* Nature (London) **215:**1340-1345.

Lamport, D. T. A. 1963. Oxygen fixation into hydroxyproline of plant cell wall protein. J. Biol. Chem. **238:**1438-1440.

Levitt, J., G. W. Scarth, and R. D. Gibbs. 1936. Water permeability of isolated protoplasts in relation to volume change. Protoplasma **26:**237-248.

Lichtenthaler, H. K., and R. B. Park. 1963. Chemical composition of chloroplast lamellae from spinach. Nature (London) **198:**1070-1072.

Lindenbein, W. 1965. Tetrazolium testing. Proc. Int. Seed Test. Ass. **30:**89-97.

Livingston, L. G. 1964. The nature of plasmodesmata in normal (living) plant tissue. Amer. J. Bot. **51:**950-957.

Lyttleton, J. W. 1962. Isolation of ribosomes from spinach chloroplasts. Exp. Cell Res. **26:**312-317.

Marinos, N. G. 1963. Vacuolation in plant cells. J. Ultrastruc. Res. **9:**177-185.

Martin, P. G. 1966. Variation in the amounts of nucleic acids in the cells of different species of higher plants. Exp. Cell Res. **44:**84-94.

Matile, P., P. J. Balz, E. Semadeni, and M. Jost. 1965. Isolation of spherosomes with lysosome characteristics from seedlings. Z. Naturforsch. [B] **20:**693-698.

Mühlethaler, K. 1967. Ultrastructure and formation of plant cell walls. Ann. Rev. Plant Physiol. **18:**1-24.

Newcomb, E. H. 1963. Cytoplasm–cell wall relationships. Ann. Rev. Plant Physiol. **14:**43-64.

Olson, A. C. 1964. Proteins and plant cell walls; proline to hydroxyproline in tobacco suspension cultures. Plant Physiol. **39:**543-550.

Palade, G. E. 1963. Cell structure, p. 57-74. *In* M. Fishbein. Birth defects. J. B. Lippincott Co., Philadelphia.

Park, R. B., and J. Biggins. 1964. Quantasome; size and composition. Science **144:**1009-1011.

Ruppel, H. G. 1964. Nucleic acids in chloroplasts of *Allium porrum* and *Antirrhinum majus.* Biochim. Biophys. Acta **80:**63-72.

Schmitt, F. O. 1963. The macromolecular assembly, a hierarchical entity in cellular organization. Develop. Biol. **7:**546-559.

Schnepf, E. 1963. Zur Cytologie und Physiologie pflanzlicher Drüsen. I. Über den Fangschleim der Insektivoren. Flora **153:**1-22.

Semadeni, E. G. 1967. Enzymatische Charakterisierung der Lysosomenäquivalente (Sphäroso-men) von Maiskeimlingen. Planta **72**:91-118.

Setterfield, G., and S. T. Bayley. 1961. Structure and physiology of cell walls. Ann. Rev. Plant Physiol. **12**:35-62.

Spencer, D. 1965. Protein synthesis by isolated spinach chloroplasts. Arch. Biochem. **111**:381-390.

Sponsler, O. L., and J. D. Bath. 1942. Molecular structure in protoplasm, p. 41-85. *In* W. Seifriz. The structure of protoplasm. Monogr. Amer. Soc. Plant Physiol. Iowa State University Press, Ames, Iowa.

Steward, F. C., M. O. Mapes, A. E. Kent, and R. D. Holstein. 1964. Growth and development of cultured plant cells. Science **143**:20-27.

Tazawa, M. 1964. Studies on *Nitella* having artificial cell sap. I. Replacement of the cell sap with artificial solutions. Plant Cell Physiol. **5**: 33-43.

Tewari, K. K., and S. G. Wildman. 1966. Chloroplast DNA from tobacco leaves. Science **153**: 1269-1271.

Ts'o, P. O. P. 1962. The ribosomes-ribonucleoprotein particles. Ann. Rev. Plant Physiol. **13**:45-80.

Werz, G. 1964. Elektronenmikroskopische Untersuchungen zur Genese des Golgi-Apparates (Dictyosomen) und ihrer Kernabhängigkeit bei *Acetabularia*. Planta **63**:366-381.

Whaley, W. G., and H. H. Mollenhauer. 1963. The Golgi apparatus and cell plate; a postulate. J. Cell Biol. **17**:216-221.

Wolfe, F. H., and C. M. Kay. 1967. Physicochemical, chemical, and biological studies on wheat embryo ribosomes. Biochemistry (Wash.) **6**: 2853-2860.

Wollgiehn, R., and K. Mothes. 1964. On the incorporation of ^3H thymidine into DNA of chloroplasts of *Nicotiana rustica*. Exp. Cell Res. **35**:52-57.

GENERAL REFERENCES

Edds, M. V., Jr. (Ed.). 1961. Macromolecular complexes. The Ronald Press Co., New York.

Frey-Wyssling, A. 1964. Ultrastructural cell organelles. Proceedings 10th International Botanical Congress, p. 57-68.

Frey-Wyssling, A., and K. Mühlethaler. 1965. Ultrastructural plant cytology. American Elsevier Publishing Co., Inc., New York.

Lehninger, A. L. 1964. The mitochondrion; molecular basis of structure and function. W. A. Benjamin, Inc., New York.

Seifriz, W. 1936. Protoplasm. McGraw-Hill Book Co., New York.

Siegel, S. M. 1962. The plant cell wall. Pergamon Press, Inc., New York.

Sitte, P. 1965. Bau und Feinbau der Pflanzenzelle. Gustav Fischer Verlag, Stuttgart.

Strugger, S. 1952. Praktikum der Zell- und Gewebephysiologie der Pflanze. Springer-Verlag, Berlin.

part II

BIOPHYSICS

chapter 3

Acidity

A. Actual acidity
B. Titratable acidity and buffers
C. pH measurements

ACTUAL ACIDITY

Measurements have revealed that, in nearly all plant cells, the vacuole sap is more acid than the protoplasm. But how are such measurements made, and how is the cell's acidity expressed quantitatively?

According to Bronsted, an acid is a proton donor—a substance capable of releasing a proton (H^+); and a base is a proton acceptor—a substance capable of combining with a proton. Water itself is both a proton donor, since it dissociates H^+,

$$H_2O \rightleftharpoons H^+ + OH^-$$

and a proton acceptor, since these H^+'s do not remain free but combine with a water molecule to form the hydronium ion (H_3O^+):

$$H_2O + H^+ \rightleftharpoons H_3O^+$$

For simplicity, however, the H^+ can be written in this way if we realize that it actually exists in the hydrated (hydronium ion) form. This notation is, in fact, necessary for the sake of uniformity since other ions are also normally hydrated, yet they are always referred to in the nonhydrated form (e.g., Na^+, K^+, etc.).

There are, of course, many different proton donors (acids) and proton acceptors (bases). In an aqueous system such as exists in the plant if an acid is added, it will increase the concentration of protons in aqueous solution; if a base is added, it will combine with the protons, leading to an increase in OH^- concentration, because of dissociation of more and more water molecules to replace the H^+ removed. Therefore, the following quantitative relations hold in aqueous systems:

Acidity $\propto$ [H^+]
Basicity (or alkalinity) $\propto$ [OH^-]

where the square brackets = concentration in gram atoms per liter.

In aqueous solutions:

$$[H^+] \times [OH^-] = K$$

Therefore the H^+ alone reveals both the acidity and alkalinity. Since K in the preceding equation is 10^{-14}, a solution that is *neutral* (i.e., acidity is equal to alkalinity) and therefore has equal concentrations of hydrogen and hydroxyl ions must have [H^+] = 10^{-7} (1/10,000,000) gram atoms per liter.

The range of acidity for the cell sap of plants is from about 10^{-1} to 10^{-7} gram atoms of hydrogen ions per liter, the one extreme being a million times as acid as the other. In order to cover such a large range conveniently, it is necessary to use an exponential method of expressing the values. Consequently, the negative logarithm of the hydrogen ion concentration is used and is called the pH. Thus pH is -log [H$^+$] or log (1/[H$^+$]); conversely [H$^+$] = 1/(antilog pH). Expressed in another way, if [H$^+$] = 10^{-x}, then pH = x.

In terms of pH, then, the range of values for cell sap is from about 1 to 7 (in rare cases 8), the former being a million times as acid as the latter. Since each whole number represents an acidity 10 times as great as the succeeding number, the unit change in pH represents a smaller and smaller change in acidity as neutrality (pH 7) is approached. Thus a change in acidity on advancing from pH 1 to 2 and from pH 6 to 7 in each case represents a reduction in acidity to a tenth (0.1) of the original value; yet the actual decrease in acidity in the first case 0.1 − 0.01 = 0.09 gram atoms of hydrogen ions per liter. In the second case the decrease is only 0.000001 − 0.0000001 = 0.0000009 gram atoms of hydrogen ions per liter. Consequently, the former unit increase in pH represents 100,000 times as great a drop in acidity as the latter. Just as a unit decrease in pH represents a 10 times increase in acidity, so a fraction of a unit is a definite multiple of a higher one (Fig. 3-1).

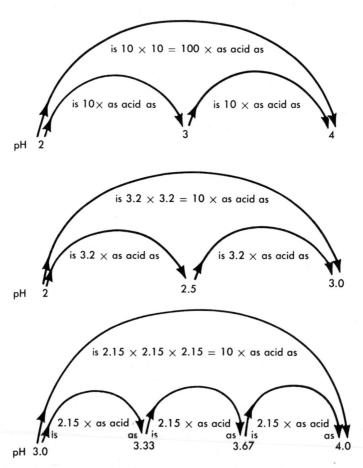

Fig. 3-1. Relation between pH and hydrogen ion concentration.

TITRATABLE ACIDITY
AND BUFFERS

Acidity, as just defined, is a measure of actual H ion concentration. Besides this true acidity, there is the *titratable acidity*, or the normality of the acid as determined by adding just enough base of known normality to bring the pH to the neutral point. The titratable acidity cannot be predicted from the actual H^+ concentration (Table 3-1). Thus the normality (or titratable acidity) of lemon juice (mainly citric acid) is 10 times that of *Begonia* leaf juice (mainly oxalic acid), but the latter has a greater hydrogen ion concentration and therefore a lower pH. The titratable acidity includes not only the actual but also the potential hydrogen ion concentration, that is, the concentration of hydrogen atoms that may change to hydrogen ions if, for instance, some of the hydrogen ions already present are deionized by neutralization or titration with hydroxyl ions. The true acidity is expressed as gram atoms of hydrogen ions, and the titratable acidity as normality of acid. In the case of a strong (completely or nearly completely dissociated into ions) monoprotic acid such as hydrochloric acid, the two are almost identical; but in the case of a weak acid (with only a small fraction of the molecules dissociated) the true acidity is much less than the titratable acidity. Since pH is related to true acidity, a strong acid has a much lower pH than a weak acid of the same normality. Thus, the pH of 1 N hydrochloric acid is 0.10, which is close to the ideal value of 0.00. The pH of 1 N acetic acid is 2.37 because only 0.42% of the acid molecules are dissociated. This means that the 1 N hydrochloric acid is nearly 200 times as acidic as the 1 N acetic acid, although the same amount of alkali would be required to neutralize equal volumes of the two solutions. The strength of an acid, therefore, depends on how much of it is ionized, and this is expressed quantitatively by the ionization constant (Table 3-2).

Since the weak acid has nearly all its titratable hydrogen atoms in the form of potential hydrogen ions, when mixed with its salt it maintains a nearly constant pH, even though large amounts of hydrogen or

Table 3-1. Titratable acidity (normality) and hydrogen ion concentration (expressed as the negative logarithm or pH) of plant juices*

Organ from which sap was expressed	Normality of sap	pH of sap
Lemon fruit	0.95	2.4
Red-black fruits of blackberry	0.23	2.7
Rhubarb petioles	0.22	3.2
Unripe grapes	0.21	3.0
Oxalis leaves	0.16	2.3
Green cooking apple	0.13	3.2
Begonia rex leaves	0.11	2.2
Begonia tuberosa leaves	0.10	2.2
Ripe tomato fruit	0.063	4.4
Ripe Worcester Pearnian apple	0.045	3.9
Celery petioles	0.025	5.2
Root of white beet	0.025	5.8

*Adapted from Thomas, M. 1951. Endeavour **10**(39):160-165.

Table 3-2. Dissociation or ionization constants ($K = \dfrac{[H^+] \times [A^-]}{[HA]}$) of some acids found in plants and their respective pK values (the pH at the middle of their buffering zones)

		K	pK
Inorganic acids			
Phosphoric	1st H to ionize	7.5×10^{-3}	2.1
	2nd H	6.2×10^{-8}	7.2
	3rd H	1.0×10^{-12}	12.0
Carbonic	1st H	0.45×10^{-6}	6.3
Organic acids			
Oxalic	1st H	5.9×10^{-2}	1.2
	2nd H	6.4×10^{-5}	4.2
Malic	1st H	4.0×10^{-4}	3.4
Tartaric	1st H	9.7×10^{-4}	3.0
Citric	1st H	8.7×10^{-4}	3.1
Acetic		1.8×10^{-5}	4.7

hydroxyl ions are added to it. Such a mixture of weak acid plus its salt (or weak base plus its salt) is called a *buffered solution* because of its ability to prevent changes in pH.

The acid dissociates as follows:

$$HA \rightleftharpoons H^+ + A^-$$

Buffering is based on the relation

$$\frac{[H^+] \times [A^-]}{[HA]} = K$$

where K = dissociation constant of the acid (HA).

At equilibrium ($\rightleftharpoons$) there is a constant proportion of undissociated acid molecules and of ions, the proportion depending on the value of K for the particular acid. In the case of a strong acid, for every ten molecules the dissociation may result in:

$$1HA + 9H^+ + 9A^-$$

In the case of a weak acid, it may be:

$$9HA + 1H^+ + 1A^-$$

Addition of a foreign base disturbs the equilibrium by adding OH⁻, which combines with H⁺s. The equilibrium ratio is immediately regained by dissociation of more HA, as a result of which the [H⁺] is nearly unaltered. The pH of a weak acid is therefore buffered against the addition of base because of its large reserve of HA.

But in order for a solution to be buffered against the addition of both H⁺ and OH⁻, it must have a reserve of both A⁻ and HA. This is achieved by mixing a weak acid with its salt. For every ten molecules of each, dissociation may lead to

$$Acid = 9HA + 1H^+ + 1A^-$$
$$Salt = 1NaA + 9Na^+ + 9A^-$$

since salts of even weak acids are highly dissociated. If the two are mixed, a new equilibrium will be produced, but the concentrations of HA and A⁻ will remain high. Consequently, a mixture of equal quantities (of equal normality) of a weak acid and its salt has a pH that tends to remain constant because it is buffered against the addition of both H⁺ (by the high [A⁻]) and OH⁻ (by the potential H⁺ of the high [HA]). A good example is a mixture of equimolar quantities of acetic acid and sodium acetate. A drop of concentrated hydrochloric

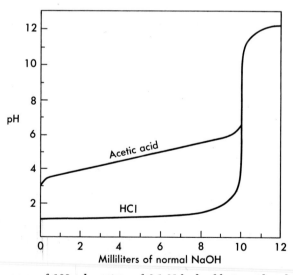

Fig. 3-2. Titration curves of 100 ml portions of 0.1 N hydrochloric acid and 0.1 N acetic acid with N sodium hydroxide. (From Michaelis, L. 1926. Hydrogen ion concentration. The Williams & Wilkins Co., Baltimore.)

acid added to a liter of such a buffered solution has a negligible effect on the pH. In contrast, a drop of concentrated hydrochloric acid added to a liter of unbuffered water lowers its pH about 3.7 units (i.e., increases its acidity about 5000 times).

The buffer range of an acid can be observed from its titration curve because it is that portion of the curve which shows the most gradual rise in pH (Fig. 3-2). A strong acid (0.1 N HCl) also shows a region of gradual rise with only a unit increase in pH, although enough alkali is added to neutralize 90% of it. But this is not as much of a buffering action as it appears to be since the [H$^+$] is reduced by about 0.09 gram atom per liter. The buffer zone of the weak acid is much higher on the pH scale; therefore, although a similar rise in pH occurs, it represents only a fraction of 1% as much of a decrease in [H$^+$] for the same addition of OH$^-$ as in the case of the hydrochloric acid. This is because of the large store of potential hydrogen ions that ionize as more alkali is added.

Each buffer mixture is effective only within a certain range of pH (Table 3-3). The maximum buffering of any mixture is obtained when half the acid (or base) molecules are dissociated that is, when equal quantities of undissociated acid (or base) and hydrogen (or hydroxyl) ions are present. This occurs at an [H$^+$] that is numerically equal to the dissociation constant K of the acid because if an acid HA dissociates into H$^+$ and A$^-$, then:

$$K = \frac{[H^+] \times [A^-]}{[HA]}$$

When the acid is half dissociated:

$$[A^-] = [HA] \text{ and } K = [H^+]$$

Since the dissociation constants of most acids are known, the optimum buffer zone for each can be easily calculated because the midpoint of this zone will occur at the pH equal to $-\log K$ (the so-called pK).

In plant juices several organic and inorganic salts and acids (Table 3-2) are responsible for the buffering action (Small, 1954). Around the neutral point, phosphates (best between 5.5 and 7.5) and bicarbonates are the commonest buffers; on the acid side, organic salts and acids such as malates (apples, peas, asparagus), citrates (tomato, citrus fruits), oxalates (tomato, plum, strawberry), tartrates (grapes), etc. buffer best between pH 3.6 and 6. Usually more than one buffer is present, for example, malate and acetate in onion, malate

Table 3-3. Some buffer solutions and their pH ranges*

	Acidic component	Alkaline component	pH range
1.	Hydrochloric acid	Glycine	1.0– 3.7
2.	Hydrochloric acid	Potassium hydrogen phthalate	2.2– 4.0
3.	Citric acid	Disodium hydrogen phosphate	2.2– 8.0
4.	Acetic acid	Sodium acetate	3.7– 5.6
5.	Potassium hydrogen phthalate	Sodium hydroxide	4.0– 6.2
6.	Potassium dihydrogen phosphate	Sodium hydroxide	5.8– 8.0
7.	Hydrochloric acid	Tris(hydroxymethyl)aminomethane	7.0– 9.0
8.	Diethylbarbituric acid	Sodium diethylbarbiturate	7.0– 9.2
9.	Hydrochloric acid or boric acid	Borax	8.0– 9.2
10.	Glycine	Sodium hydroxide	8.2–10.1
11.	Borax	Sodium hydroxide	9.2–11.0
12.	Sodium bicarbonate	Sodium hydroxide	9.6–11.0
13.	Disodium hydrogen phosphate	Sodium hydroxide	11.0–12.0

*From Altman, P. L., and D. S. Dittmer. 1964. Federation of American Societies for Experimental Biology, Washington, D. C.

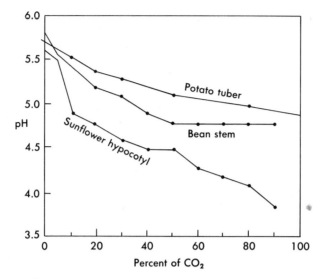

Fig. 3-3. Effect of carbon dioxide on pH of plant juices. (From Small, J. 1954. pH and plants. Baillière, Tindall, Cassell, Ltd., London.)

and oxalate in lettuce. Most of the preceding substances are in the vacuole. The protoplasm is buffered by its proteins, which are known to be effective over a pH range of 4 to 10 (Webb, 1963), although other substances may also have some effect, and metabolic control of pH is probably decisive. Some plant saps are not well buffered, and their pH may be greatly shifted by addition of carbon dioxide (Fig. 3-3). In some cases excess carbon dioxide may produce indirect effects leading to a rise in pH.

Buffer solutions are very important in all preparative procedures, for example, in separating organelles from plant cells, in preparing proteins for enzymatic work, etc. It is a more crucial problem when dealing with plants than with animals since the plant cell consists largely of vacuole sap, which is much more acidic than the protoplasm. If plant cells are broken open without the addition of a buffer solution, the extracted protoplasm will immediately take on the pH of the vacuolar sap because of the much larger volume of the latter. The increased acidity is usually sufficient to destroy the normal properties of both the organelles and the proteins. Many buffer solutions have been developed to prevent such injurious changes. They are usually prepared with pH's of 7 to 8. Phosphate, Tris, and borate buffer solutions are among the best (Good et al., 1966).

pH MEASUREMENTS

There are two main methods of measuring pH: one by the use of indicators (dyes that change color at more or less specific pH's), the other by pH meters. Each has its own virtues, although the modern, line-operated pH meters with glass electrodes are unequaled for precision, speed, and simplicity. The indicator method is still sometimes used, for instance in field work. It is accurate enough for most biological purposes. In fact, it is usually just as *accurate* a measure of biological pH's as the far more *precise* pH meter, because although the pH meter enables measurement of pH to 0.01 units and indicators cannot do better than 0.2 units, the procedure of obtaining the plant juice for measurement is certain to change the pH at least 0.1 to 0.2 units. For determining the pH of individual cells it is the only method available since elec-

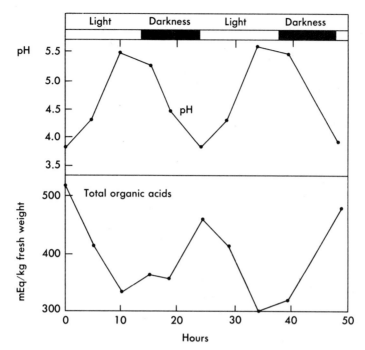

Fig. 3-4. Effect of light and darkness on pH and organic acid content of *Bryophyllum* leaves. (From Vickery, H. B. 1952. Plant Physiol. **27**:9-17.)

trodes cannot be made small enough to insert into most living cells. In the case of large cells, however, microelectrodes have been used. The sap of a single cell of *Nitella flexilis* was in this way found to have a pH of 5.6 (Hirakawa and Yoshimura, 1964). Naturally occurring indicators inside the cell may sometimes be made use of: flavones turn from colorless to yellow above pH 8; anthocyanins are usually pink or red below pH 5 and blue or purple above it.

By the use of these methods the pH's of many juices (mainly from cell vacuoles) have been determined. Fruit juices usually range from pH 2.5 to 4.5 (Table 3-1), although there are some exceptions (e.g., lime, pH 1.7). Juices expressed from other plant parts also show wide variations, although they commonly fall within the range of pH 5 to 6.5. Phloem exudate has the exceptionally high value of 7 to 8. In plants

with acid metabolism (e.g., succulents) the pH may be around 3. Even in the case of a single plant it may vary greatly. In many plants it drops at night and rises during the day. Extreme examples of such diurnal pH changes are found among succulents: in *Opuntia phaeacantha* (a cactus) from pH 5.5 at 4:00 P.M. to 1.4 at 6:45 A.M. Where the change in pH is small, it may at least partly result from increases in carbon dioxide content at night (from respiration), and decreases during the day (because of use in photosynthesis). But such extreme changes as in the *Opuntia* and other succulents result from accumulation of organic acids in the dark and their disappearance in the light (Fig. 3-4).

The pH of plant protoplasm is much more difficult to determine, partly because of the far greater quantity of acid vacuolar sap adjacent to it. It is perhaps because of this and the acid of injury which often arises when protoplasm is damaged, that values as low as pH 5.6 have been recorded for plant protoplasm. Because of these sources of error and because protoplasm of other organisms has usually been found to be around the neutral point, it is highly probable that higher plant protoplasm has a pH around neutrality. Indirect evidence of this is the fact that when protoplasmic proteins are isolated from plants, some of them fail to remain in solution unless the pH is maintained above 6 to 6.5.

Judging from the range of pH's found in the vacuole (1 to 7), the plant is apparently much more tolerant of acidity than of alkalinity. Even the tolerance of externally applied pH is not so great in the alkaline range. The most acidophilic plants grow at soil pH's down to 4 or even 3.5 in extreme cases, and the basophilic plants grow at soil pH's to 8 or 8.4. Thus the tolerance of alkalinity is greater than it appears from vacuole measurements, but it is still less than the tolerance of acidity. This is at least partly associated with indirect effects of pH, for example, the unavailability of some soil nutrients at high pH, rather than a direct damage of protoplasm by pH ions.

Similarly, plants that do poorly in acid soils, (hemp, mustard, asparagus, cucumber, spinach, squash, alfalfa) are, at least in some cases, injured not by the H^+ directly but by the excess iron absorbed at low pH (Wallace and Bhan, 1962).

A highly acid-resistant *Chlorella ellipsoida* has been found that is capable of growth within the entire range of pH 2 to 10 (Kessler and Kramer, 1960). According to Olson (1953), a direct damage by hydroxyl ions does not occur until pH 10.5, exactly the same distance from the neutral point as the pH (3.5) causing direct damage because of hydrogen ions.

The role of pH in many physiological processes will be discussed in later sections (e.g., protein hydration, stomatal movement, enzyme action, etc.).

QUESTIONS

1. What is meant by acidity? By alkalinity?
2. What is the acidity at the neutral point?
3. Is it necessary to measure both?
4. Does an increase in pH represent an increase or decrease in acidity?
5. How much of a change in acidity is represented by one pH unit?
6. How does titratable acidity differ from true acidity?
7. What are the units for each?
8. If two plants change their pH's from 6.5 to 6, does this necessarily mean an equal change in true or titratable acidity?
9. Does a change from 6.5 to 6 represent the same, a greater, or a smaller change in true acidity than a change from 5.5 to 5?
10. If the pH's of a strong and a weak acid are the same, which (if either) will have the greater titratable acidity?
11. What is a buffered solution a buffer against?
12. What does it consist of?
13. Do all buffer solutions cover the same range?
14. What substances in plants produce a buffering effect?
15. What is the acidity of the vacuole?
16. What is the acidity of protoplasm?
17. What pH range will living cells tolerate?
18. How do the night and day pH's of some plants compare?
19. How do the pH's of fruit compare with those of other plant parts?
20. How do the pH's of succulents compare with those of other plants?
21. What is the effect of carbon dioxide on pH?

SPECIFIC REFERENCES

Good, N. E., G. D. Winget, W. Winter, T. N. Connolly, S. Izawa, and R. M. M. Singh. 1966. Hydrogen ion buffers for biological research. Biochemistry (Wash.) 5:467-477.

Hirakawa, S., and H. Yoshimura. 1964. Measurements of the intracellular pH in a single cell of *Nitella flexilis* by means of micro-glass pH electrodes. Jap. J. Physiol. **14**:45-55.

Kessler, E., and H. Kramer. 1960. Physiological research on a highly acid resistant *Chlorella*. [Transl. title.] Arch. Mikrobiol. **37**:245-255.

Olson, C. 1953. The significance of concentration for the rate of ion absorption by higher plants in water culture. IV. The influence of hydrogen ion concentration. Physiol. Plant. **6**:848-858.

Thomas, M. 1951. Vegetable acids in higher plants. Endeavour **10**(39):160-165.

Vickery, H. B. 1952. The behavior of isocitric acid in excised leaves of *Bryophyllum calycinum* during culture in alternating light and darkness. Plant Physiol. **27**:9-17.

Wallace, A., and K. C. Bhan. 1962. Plants that do poorly in acid soils, p. 36-38. *In* A. Wallace. A decade of synthetic chelating agents in inorganic plant nutrition. A. Wallace, Los Angeles.

Webb, J. L. 1963. Enzyme and metabolic inhibitors. Vol. I. General principles of enzyme inhibition. Academic Press, Inc., New York.

GENERAL REFERENCES

Altman, P. L., and D. S. Dittmer. 1964. Biology data book. Federation of American Societies for Experimental Biology, Washington, D. C.

Christensen, H. N. 1964. pH and dissociation. W. B. Saunders Co., Philadelphia.

Clark, W. M. 1922. The determination of hydrogen ions. The Williams & Wilkins Co., Baltimore.

Michaelis, L. 1926. Hydrogen ion concentration. The Williams & Wilkins Co., Baltimore.

Pauling, L. 1947. General chemistry. W. H. Freeman & Co., Publishers, San Francisco.

Small, J. 1954. Modern aspects of pH. Baillière, Tindall, & Cassell, Ltd., London.

Willard, H. H., L. L. Merritt, and J. A. Dean. 1951. Instrumental methods of analysis, p. 179-205. D. Van Nostrand Co., Inc., Princeton, N. J.

chapter 4

Specific surface and adsorption

A. Specific surface and surface energy
B. Polarity and adsorption

The surface of cells, as well as that of their organelles, possesses special properties and plays special roles in the life of the organism. It has, in fact, been suggested (Ling, 1962) that life processes are under the control of specific "sites," which reside at such surfaces. These "cardinal sites," are supposed to be small in number but of such strategic importance that they can control and modulate the behavior of a large number of "fixed sites." They must owe their importance to the presence of specific molecules or ions (adenosine triphosphate [ATP], hormones, drugs, etc.), which may be held, at least temporarily, at these sites by the forces of adsorption. In order to comprehend the properties of living cells, it is, therefore, necessary to understand the nature of surfaces and their ability to adsorb substances.

SPECIFIC SURFACE AND SURFACE ENERGY

The surface of a sphere is given by the formula:

$$S = 4\pi r^2$$

This means that if spheres of different sizes are compared, the surface increases with the square of the radius. The larger the volume of a sphere the larger its surface.

On the other hand, the specific surface is the amount of surface area per unit mass or volume:

$$S_v = \frac{S}{V}$$

where S_v = specific surface (on a volume basis)
S = actual surface area
V = volume

In the case of a sphere:

$$S_v = \frac{4\pi r^2}{4/3\pi r^3} = \frac{3}{r}$$

Or, the specific surface of a sphere varies inversely with its radius. In contrast to actual surface the *larger* the volume of a sphere the *smaller* is its specific surface. This becomes obvious when a potato tuber is compared with its individual cells (Table 4-1). Thus, although the actual tuber surface is a million times that of a single cell, its specific surface is one thousandth that of the cell. This means that the total surface of *all* of the tuber cells is 1000 times that of the tuber itself. Going beyond the cell, the specific surface of its organelles will be still greater since this value increases with the decrease in volume of the body.

The importance of surface and specific surface is a result of the existence of surface energy. This follows from the first law of

Table 4-1. Actual as opposed to specific surface of a potato tuber and its cells

	Diameter	Relative radius	Relative actual surface (S)	Relative actual volume (V)	Relative specific surface (S_v)
Tuber	10 cm	10^3	10^6	10^9	10^{-3}
Cell	0.1 mm (0.01 cm)	1	1	1	1

Table 4-2. Direct relationship between specific surface and plant activity

Surface involved	Active state (large specific surface)	Inactive state (small specific surface)
1. External plant surface		
a. Deciduous trees	Leaves	No leaves
b. Herbaceous stems	Branched and foliate	Spherical (tubers, corms)
2. External cell surface	Meristematic cells (small)	Parenchyma cells (large)
3. Internal protoplasmic surface	Many organelles	Few organelles
4. Internal organelle surface		
a. Mitochondria	Cristae	Few or no cristae
b. Chloroplasts	Lamellae	No lamellae

thermodynamics, which states that energy cannot be created or destroyed but can be converted from one form to another. If a body is cut into smaller pieces, new surfaces are created, and part of the energy used to cut it up is converted into free surface energy. This energy is either in the form of electric charge (solids) or surface tension (liquids). Any other method of creating new surfaces will produce the same result. Consequently, if a potato sprouts and produces a leafy shoot, it is producing new surfaces and new surface energy at the expense of the chemical energy released by respiration. If, for instance, the sprout has 100 times the specific surface of the tuber (and other factors are equal), it is able to absorb oxygen and give off carbon dioxide at 100 times as great a rate per unit volume as can the tuber. In other words, the active, growing form of the plant (the shoot and root systems) is characterized by high specific surface and surface energy; the inactive, nongrowing form of the plant (the tuber bulb, etc.) is characterized by low specific surface and surface energy. This relation has been found to hold at all levels from the macroscopic to the electron microscopic (Table 4-2). Whether or not the high specific surface can be thought of as the cause or the effect of the activity can only be considered after the active and inactive states of the plant are discussed in more detail. There is, of course, a limit to the generality of this relation. Deciduous trees, for instance, reduce their outer surface tremendously when they become dormant. Evergreens do not, but this may partly account for their success in climates having low enough temperatures during their dormant period to reduce their metabolism to a minimum without any reduction in surface.

As shown in Table 4-1, even more important and much larger than the external surface of a plant, is its internal surface. It is not surprising, therefore, that the major metabolism of the plant is controlled by the organelles (Table 2-3). Much of this no doubt occurs at their surfaces (e.g., the surfaces of the cristae in mitochondria and of the lamellae in chloroplasts).

POLARITY AND ADSORPTION

From the second law of thermodynamics it follows that the free surface energy tends to be converted into unavailable energy. In other words, the surface charge will tend to be neutralized by opposite charges, the surface tension to be lowered by other substances. In both cases the reduction in surface energy can be brought about only if molecules or particles that reduce surface energy accumulate at the surface in higher concentration than in the surrounding medium. This concentration at a surface is known as *adsorption*. It is more pronounced at some surfaces than at others, and from some media than from others. In some cases there may even be negative adsorption if this results in a lower free energy of the surface (i.e., if the only available particles have the same sign of charge as the surface or if they raise surface tension).

Even though most substances are not ionized in the plant, there may be an asymmetrical distribution of electrons within the molecule, leading to a polarization of the molecule with an electrically positive and an electrically negative end. Oxygen is more electronegative (i.e., has a greater attraction for electrons) than hydrogen (Table 4-3). Therefore, when the two combine in the water molecule, the electrons of the hydrogen atoms are displaced toward the oxygen atom:

$$:\overset{\cdot}{O}\cdot + \cdot H \qquad :\overset{\cdot\cdot}{O}:- H$$
$$\qquad\qquad \cdot H \qquad \underset{H}{\overset{|}{}} +$$

The oxygen end of the molecule, therefore, becomes negative to the hydrogen end and the molecule is, therefore, a dipole. Carbon and hydrogen are nearly equally electronegative (Table 4-3); therefore, the methane molecule (CH_4) is unpolarized. Water is, therefore, a *polar* substance; methane is *apolar*. But there are all degrees of polarization of molecules from essentially zero, as in methane, to a complete splitting into positive and negative ions, as in salts (Fig. 4-1). The greater the number of apolar groups ($-CH_3$, $-C_6H_5$, etc.) the more apolar is the substance. The greater the number of polar groups ($-OH$, $-COOH$, $-CO$, $-NH_2$, metals, etc.) the more polar is the substance (Table 4-4).

The division of substances into (relatively) polar and (relatively) apolar sub-

Table 4-3. Electronegativity values for some biologically important elements; relative positions as in the periodic table*

H				
2.1				
	C	N	O	F
	2.5	3.0	3.5	4.0
		P	S	Cl
		2.1	2.5	3.0

*From Pauling, L. 1958. General chemistry. W. H. Freeman & Co., Publishers, San Francisco.

Table 4-4. Cohesion forces between organic groups; the larger the forces the more polar the group*

Groups		Molar cohesion energy (kcal/mole)
Aliphatic C:		
Methyl and methylene groups	$-CH_3$ and $=CH_2$	1.78
	$=CH_2$ $=CH-$	0.99
Ether bridge	$-O-$	1.63
Amino group	$-NH_2$	3.53
Carbonyl group	$=CO$	4.27
Aldehyde group	$-CHO$	4.70
Hydroxyl group	$-OH$	7.25
Carboxyl group	$-COOH$	8.97

*From Frey-Wyssling, A., and K. Mühlethaler. 1965. Ultrastructural plant cytology. American Elsevier Publishing Co., Inc., New York.

stances is practically useful since polar substances attract each other and are mutually soluble. The small degree of polarity in apolar substances results in their ready solubility in each other. Polar and apolar substances, on the other hand, are not soluble in each other because the mutual attraction of the molecules of a polar substance results in a repulsion of apolar substances. Even individual chemical groups on large molecules show the same relation. The long protein chains in the aqueous me-

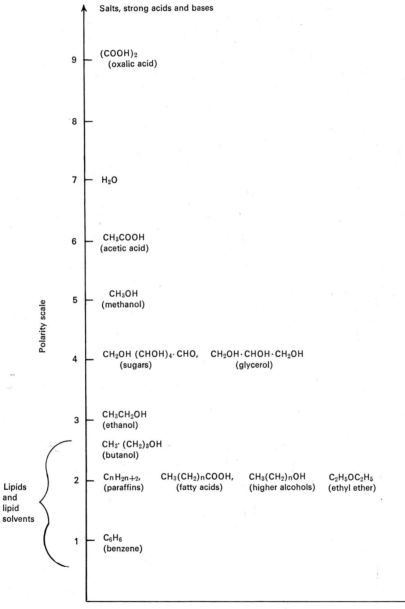

Fig. 4-1. Relative polarity scale of some organic substances calculated from the theoretical values in Table 4-4.

dium of the cell are, therefore, folded in specific patterns, and the polar groups are held in place by other polar groups and by the external polar water molecules that attract them electrically. The apolar groups are pushed together mostly inside the molecule because of the polar nature of the aqueous medium. It is this attraction of polar groups for each other, and the much smaller attraction of apolar groups for each other, that controls the phenomenon of adsorption.

The adsorption of particles at a surface reduces the surface energy in the two following ways:

1. By lowering the surface tension. The surface tension of a liquid depends on the polarity of its molecules. Water, therefore, possesses a high surface tension; alcohols possess much lower surface tensions. But the surface tension of the pure liquid may be very different from that of the impure liquid, that is, solutions of substances in the liquid. Because of the already high surface tension of water, the addition of solutes can do little to raise it but can lower it considerably. Other substances that also consist of polar molecules tend to attract water molecules electrically and can be thought of as being pulled into the body of an aqueous solution (negative adsorption). Salts (and sugars) are polar substances and therefore do not concentrate at the surface. Consequently they do not reduce the surface tension of water. They are therefore said to be surface inactive, although they may actually slightly increase surface tension (Fig. 4-2). Most organic substances are more or less apolar and therefore are not so strongly attracted to the water molecules as the water molecules are to each other. As a result they can be

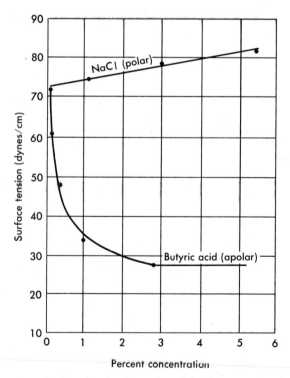

Fig. 4-2. Surface tension of solutions of polar and apolar substances. (Redrawn from Freundlich, H. 1922. Colloid and capillary chemistry. E. P. Dutton & Co., Inc., New York.)

thought of as being "pushed" to the surface, where they reduce the surface tension because of the small molecular forces of attraction between them. They can even form a monomolecular layer on the water, with their polar groups extending down into the polar water layer and their apolar groups away from the water. Such substances are said to be surface active. Proteins may be surface active; they therefore form an adsorbed layer around latex particles in rubber-producing plants. But proteins do possess definite electrical properties and therefore are not nearly as surface active with respect to water as are fatty substances. Consequently, in the case of living protoplasm there is a far greater concentration of lipids than of proteins at the surface. As a result, the surface tension of protoplasm is lowered from 72 dynes/cm (for pure water) to a range of 0.1 to 2 dynes/cm. According to Danielli's calculations (1966), a bimolecular lipid membrane associated with proteins has less surface-free energy than that of any other thickness. He, therefore, concludes (on the basis of the second law of thermodynamics) that biological membranes normally must be bimolecular.

2. Neutralization of surface charge. The phenomenon of staining is caused by this kind of adsorption. Dyes are weak organic (aromatic) acids or bases, consisting mainly of apolar groups. The dissociation of these weak acids and bases is extremely low, and because of their high degree of apolarity they are essentially insoluble in water. But the salts of even weak acids and bases may have a high degree of dissociation (e.g., sodium salts of the acids, chlorides of the bases). Consequently, the salts of these apolar dye acids and bases possess a charge and are, therefore, polar. As one result, the salts are highly soluble in water (a polar solvent); as a second result, these dye ions are strongly adsorbed to oppositely charged surfaces. The salt of a dye base is called a basic dye, the salt of a dye acid, an acidic dye:

1. $R_1OH + HCl \rightleftharpoons R_1Cl + H_2O$
   Dye   Dye salt
   base   (basic dye)

$$R_1Cl \rightleftharpoons R_1^+ + Cl^-$$

2. $R_2H + NaOH \rightleftharpoons R_2Na + H_2O$
   Dye   Dye salt
   acid   (acidic dye)

$$R_2Na \rightleftharpoons R_2^- + Na^+$$

Vital staining is caused by both the polar and apolar properties of dyes. Neutral red is a basic dye (RCl) and, therefore, dissociates as the red (R+) dye ion at low

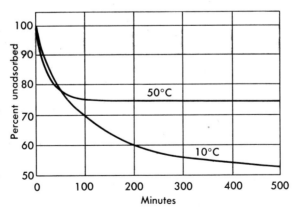

Fig. 4-3. Effect of temperature on the rate of adsorption of Congo red by paper. (Redrawn from Belehradek, J. 1935. Temperature and living matter. Protoplasma-Monographien 8. Gebrüder Borntraeger, Berlin.)

(acid) pH's. At high pH's (about the neutral point) it is converted to the yellow, undissociated base (ROH) since the dye ion (R+) is also in equilibrium with the OH⁻ of the water and must combine with it when the OH⁻ concentration becomes relatively high. Because of this color change it can be used as a pH indicator. The red dye ion is polar and, therefore, highly soluble in water and readily stains (i.e., adsorbs to) oppositely charged negative surfaces. The cell wall is negatively charged, partly because of dissociation of pectic acids to negative pectate ions. Consequently, the cell wall stains in red (acid) solutions of neutral red. In the presence of Ca^{++} this staining does not occur because the doubly charged calcium ions are preferentially adsorbed to the pectate or other negative surfaces.

But the vacuole of most plant cells also contains negatively charged particles, primarily tannins. Yet the vacuole does not stain when the living cell is immersed in a red solution of neutral red. This is because the surface of the living cell consists of apolar lipids, in which the polar NR+ is insoluble. If, however, the pH of the neutral red solution is raised to 7 to 8, it becomes orange in color because of the presence of both the red (NR+) and yellow (NROH) forms. The latter has no charge and is apolar. It, therefore, is soluble in the plasma membranes and diffuses through both. On reaching the vacuole, with its acid pH, the apolar molecule (NROH) is converted to the ion (NR+) and is quickly adsorbed by the tannin particles (T⁻), thus staining the vacuole vitally. It is obvious, then, that vital staining of the cell sap is possible only (1) with basic dyes; (2) if the dye is applied at a high enough pH to produce a sufficient concentration of the apolar, undissociated molecule; and finally (3) strong staining requires the presence of a high concentration of negatively charged (tannin) particles in the vacuole.

Adsorption has proved to be important in the case of uptake of nutrients from the surrounding medium by plant roots. There is strong evidence that the first step in the uptake is adsorption at a specific surface in the roots. It has been found that in weak solutions the uptake is more or less proportional to the external concentration, but that in more concentrated solutions it is almost independent of the external concentration. This is believed to result from saturation of the surface by the adsorbed nutrient at relatively low concentrations (Robertson, 1941).

Temperature has a pronounced effect on adsorption. The amount adsorbed greatly decreases with a rise in temperature. This can readily be understood since the higher the temperature the greater is the kinetic energy of the adsorbed molecules or ions and the more easily they can escape from the surface. However, temperature has the opposite effect on the time required for adsorption equilibrium to be attained; this decreases with a rise in temperature (Fig. 4-3).

QUESTIONS

1. What is meant by specific surface?
2. How does it vary with particle size?
3. What is meant by surface energy?
4. In what two forms does it exist in plants?
5. What relation is there between the extent of plant surface and the activity of the plant?
6. With respect to what is a polar substance polarized?
7. On what property of the elements does the polarity of a substance depend?
8. Is water polar?
9. Are fatty substances polar?
10. Are salts polar?
11. Is there a sharp line of separation between the two kinds?
12. Do polar substances attract or repel each other?

13. Do apolar substances attract or repel each other?
14. What do we know about solubilities of polar and apolar substances in each other?
15. How can the free energy at a surface be decreased?
16. How can the polarity of chemical groups affect a single isolated large molecule?
17. What is the relation between the polarity of a liquid and its surface tension?
18. How can added substances affect the surface tension of a liquid?
19. What are "surface active" substances?
20. What is the surface tension of water? Of protoplasm?
21. What substances can be expected to concentrate at the protoplasm surface?
22. What is a basic dye? An acidic dye?
23. Which dye is suitable for vital staining? Why?
24. Which form of a dye is readily adsorbed?
25. Which form readily penetrates a living cell? A dead cell?
26. Which form concentrates in the vacuole of living cells? Why?
27. Which form concentrates in cell walls? Why?
28. Which cell substances are responsible for each of the preceding two kinds of staining?
29. At what pH will acidic dyes penetrate living cells readily?
30. What relation is there between adsorption and temperature?

SPECIFIC REFERENCES

Belehradek, J. 1935. Temperature and living matter. Protoplasma-Monographien 8. Gebrüder Borntraeger, Berlin.

Danielli, J. F. 1966. On the thickness of lipid membranes. J. Theor. Biol. **12**:439-441.

Ling, G. N. 1962. A physical theory of the living state. Blaisdell Publishing Co., Division Ginn & Co., Waltham, Mass.

Robertson, R. N. 1941. Studies in the metabolism of plant cells. I. Accumulation of chlorides by plant cells and its relation to respiration. Aust. J. Exp. Biol. Med. Sci. **19**:265-278.

GENERAL REFERENCES

Conn, H. J. 1961. Biological stains. The Williams & Wilkins Co., Baltimore.

Danielli, J. F., K. G. A. Parkhurst, and A. C. Riddiford (Eds.). 1964. Recent progress in surface science. Vol. 2. Academic Press, Inc., New York.

Freundlich, H. 1922. Colloid and capillary chemistry. [Transl. by George Barger.] E. P. Dutton & Co., Inc., New York.

Gortner, R. A. 1949. Outlines of biochemistry. John Wiley & Sons, Inc., New York.

Scarth, G. W., and F. E. Lloyd. 1930. Elementary course in general physiology. John Wiley & Sons, Inc., New York.

chapter 5

Colloids

SIGNIFICANCE IN PHYSIOLOGY

In science, as in any human field of activity, there are always fads. A topic will become popular, then later it will be forgotten. The role of colloids in the plant is an excellent example. For a time all of the properties of protoplasm were explained as being basically a result of its colloidal character. With the development of the electron microscope the organelles of protoplasm were discovered, proving that protoplasm is much more than a complex colloidal system. The pendulum has, therefore, swung to the other extreme, and the trend now is to ignore the colloidal properties of protoplasm in particular and of the cell in general. The result is an unfortunate loss of understanding of many cell properties that are still definitely colloidal and therefore explainable in no other way.

For, although the organelles explain many of the most important properties of protoplasm, they fail to explain the properties of the ground protoplasm or hyaloplasm. Thus, in spite of the high water content of protoplasm, it is much more viscous than most ordinary aqueous solutions of the same water content. Furthermore, unlike pure solutions, it possesses some of the properties of solids, for example, elasticity. These properties cannot be understood without a consideration of the colloidal state of protoplasm. Even some of the organelles themselves are in the colloidal size range (see later), and both the cell wall and the vacuole have colloidal properties. It is, therefore, impossible to have a complete understanding of the properties of the cell without and understanding of the properties of colloids.

The most obvious factor in adsorption is the extent of surface—the greater the surface area per unit volume (i.e., the specific surface) the greater is the amount adsorbed per unit volume. Adsorption therefore becomes more pronounced the more finely divided is the adsorbent. It reaches a maximum, in fact, in the *colloidal* range, where particle sizes are below visibility with the optical microscope but above ordinary molecular size (Table 5-1). It is for this reason that adsorption must play such an important role in the living cell and particularly in protoplasm, which consists largely of substances in the colloidal state.

Since the colloidal state depends simply on the particle size, any substance can (at least theoretically) be a colloid. There is,

Table 5-1. Size limits (logarithmic scale) and properties of colloids relative to other dispersions*

Properties	Approximate ranges of diameters (in mμ or nm†)		
	2000 200 20	2	0.2
Name	Suspension (2μ − 200 mμ)	Colloidal dispersion (20 − 5 mμ)	Molecular dispersion (5 − 0.1 mμ)
Visibility	Optical microscope	Ultramicroscope and electron microscope	Near or beyond limits of detection with electron microscope
Separability	Filtrable	Dialyzable	Nondialyzable except with semipermeable membranes
Stability	Unstable	Relatively stable	Stable
Example	Starch suspension	Proteins with mol wt 125,000 (diameter 6.6 mμ‡) and larger	Proteins with mol wt 15,000 (diameter 3.3 mμ‡) and smaller

*Adapted from Scarth, G. W., and F. E. Lloyd. 1930. Elementary course in general physiology. John Wiley & Sons, Inc., New York.

†1 mm = 1000μ (microns) or 1000 μm (micrometer); 1μ = 1000 mμ or 1 μm = 1000 nm (nanometer).

‡See Frey-Wyssling, A., and K. Mühlethaler. 1965. Ultrastructural plant cytology. American Elsevier Publishing Co., Inc., New York.

of course, no sharp line between colloids and noncolloids. Any particles that fall in the range of about 5 mμ to 0.2μ are usually considered to be colloidal. Sometimes the upper limit may be extended to 0.5μ since certain systems with particles of this size are stable. The lower limit includes some molecules since these may be large enough to take on colloidal properties. This is particularly true of proteins.

Colloidal systems really consist of two components, the *disperse phase* (particles) and the *dispersion medium*. Since both of these may be in any of the three states of matter (solid, liquid, and gas), there are nine possible combinations. Of these only one cannot represent a colloid, viz., gas disperse phase in gaseous medium, since all gases are molecularly miscible in all proportions. Of the eight remaining possible combinations, only one, the solid disperse phase in a liquid medium, is important in physiology; and the only liquid medium of

physiologic significance is water. Consequently, the following discussion applies primarily to aqueous colloidal dispersions.

PROPERTIES OF COLLOIDS

There are many ways in which colloidal dispersions differ from molecular dispersions or *true solutions*. All these differences depend on the particle size. Thus the colloidal particles are large enough to scatter light. As a result, colloidal dispersions frequently show a *Tyndall effect;* if a beam of light is shone on the dispersion, the path of the beam can be seen from either side at right angles to the direction of the beam. But such light scattering is not possible when the colloidal particles have a refractive index that does not differ greatly from that of the dispersion medium, for instance, when the particles are highly hydrated as in the case of many proteins.

Another result of their relatively large particle size is the slow rate of diffusion of colloids: as little as 0.001 the rate of small molecules (Table 5-2).

The osmotic effects (see Chapter 6) of colloidal solutions are small, again because of the large particle size (Table 5-3).

The colloidal particles are too large to pass through the pores of many membranes that are easily penetrated by molecularly dispersed salts and sugars. This fact is made use of to free colloidal systems of crystalloids. When confined in such a membrane that is suspended in water, the crystalloids diffuse out, leaving the colloids behind. This process is known as *dialysis*. A similar process including the application of pressure is known as *ultrafiltration*.

Colloids may differ from each other in other ways beside size. Some may be nonspherical, and this may lead to other properties such as gel formation. In spite of their large particle size, some may be electrolytes and, because of their dissociation,

become charged. Proteins supply examples of both of these properties.

STABILITY OF COLLOIDAL DISPERSIONS

Molecular dispersions are stable because of the high kinetic energy of the molecules. When this energy is reduced, for instance when the molecules crystallize, they precipitate. In the same way, colloidal particles aggregate to form larger particles, which will then precipitate. If the colloid does not precipitate, it is stable; and since the particles do not possess the high kinetic energy of molecules, the colloid must owe its stability to some other property that prevents aggregation. There are two main factors that contribute to its stability: *charge* and *hydration*.

The charge of a colloidal particle results either from the capture of an ion or from ionization (dissociation) of the colloid. Colloidal bases (e.g., alkaloids, basic dyes, hydroxides of metals) become positively charged in water. All other colloids are negative in water. Proteins are *amphoteric* (either positively or negatively charged) depending on whether they dissociate as bases or as acids. At some point between, the *isoelectric point*, their net charge is zero, and they are therefore least stable. Some proteins (the albumins) are *isostable;* that is, their own ionization supplies a sufficient net charge. Others (globulins) need other ions and therefore are stable only in

Table 5-2. Relative diffusion rates (or diffusion coefficients) of molecularly and colloidally dispersed substances*

Dispersion	Substance	Diffusion coefficient
Molecular	Nitric acid	2.1
	Sucrose	0.31
Colloidal	Nuclear gold (1.7 mμ)	0.27
	Egg albumin	0.059
	Antitetanus serum	0.0021

*Adapted from Gortner, R. A., 1938. Outlines of biochemistry. John Wiley & Sons, Inc., New York.

Table 5-3. Osmotic pressures (atmospheres) of a molecularly dispersed (dextrose) and a colloidally dispersed (serum albumin) substance

Solution (%)	Dextrose	Bovine serum albumin
5	6.7	0.013
10	13.5	0.032
20	27.0	0.12

Table 5-4. Concentration of salts required to flocculate (precipitate) sols*

Negative sol		Positive sol	
Salt	Millimoles	Salt	Millimoles
KCl	50	KCl	80
$BaCl_2$	0.70	K_2SO_4	0.28
$AlCl_3$	0.09	$K_4Fe(CN)_6$	0.08

*Adapted from Scarth, G. W., and F. E. Lloyd. 1930. Elementary course in general physiology. John Wiley & Sons, Inc., New York.

Cations: $Li^+ > Na^+ > K^+ > NH_4^+ > Rb^+ > Cs^+ > Mg^{++}$

Anions: $SO_4^= > HPO_4^= > Cl^- > Br^- > NO_3^- > I^- > CNS^-$

Fig. 5-1. Order of effectiveness of ions in salting out hydrophilic sols: the lyotropic series. (Redrawn from Scarth, G. W., and F. E. Lloyd. 1930. Elementary course in general physiology. John Wiley & Sons, Inc., New York.)

salt solutions. An unstable colloid may therefore be dispersed by the addition of sufficient ions (i.e., by salts). This is called *peptization*. But if too high a salt concentration is used, the colloid may be discharged and precipitated. This is called *flocculation*. The higher the valence of the oppositely charged ion the smaller is the concentration of salt required to flocculate a colloid (Table 5-4).

The second cause of stability is hydration. Aqueous colloids may be roughly classified into *hydrophilic* and *hydrophobic* colloids, although there is some gradation between the two groups. The hydrophobic colloids are less stable and can be flocculated by low concentrations of electrolytes (Table 5-4). The hydrophilic colloids cannot be flocculated but may be *salted out* by high salt concentrations. The effectiveness of salts varies with the different ions in a very definite order, which is known as the lyotropic (or Hofmeister) series (Fig. 5-1).

Proteins are typically hydrophilic colloids. They cannot be flocculated by low concentrations of salts but can be precipitated by high concentrations, for example, by saturation or half saturation with ammonium sulfate. Some hydrophilic colloids form aqueous sols (i.e., have the physical properties of a liquid); others form more or less rigid gels. Protoplasm itself is intermediate and exhibits the properties of both sols and gels; it shows streaming and brownian movement, as would be expected in a sol, but both of these movements may be stopped by removal of some of its water, converting it into a rigid gel.

The preceding classification of colloids into hydrophilic and hydrophobic systems, although useful, is an oversimplification.

Booij and de Jong (1956) divide them into (1) nonequilibrium state and (2) equilibrium state colloidal sols. Hydrophobic sols, as a rule, are in the nonequilibrium state; and their stability is more apparent than real. Although it may take several years, every sol of this type flocculates spontaneously because a small fraction of the particles will have sufficient kinetic energy to overcome their mutual electric repulsion. Protein molecules, on the other hand, disperse spontaneously and are, therefore, in true solution. Their sol state is a true equilibrium state. Because of their polar nature, proteins attract water and have, therefore, been called hydrophilic colloids, whether sols or gels (see later). Booij and de Jong (1956) have labeled them macromolecular colloids. The hydrophobic sols are of little importance to the biologist. Equilibrium-state colloids are of fundamental importance since many colloids of this kind occur in the cell.

However, the macromolecular colloids are not the only ones in the equilibrium state. A second group (according to Booij and de Jong) are the association colloids. They may consist of double films with a hydrophobic interior, such as those forming the cell membrane (see Chapter 7). These association colloids may be in the sol, or solution form, (analogous to gases); in the coacervate, or isotropic, form (analogous to liquids); or in the colloid crystal (solid) state. According to Booij and de Jong, the coacervate state resembles protoplasm the most. Their state is closely related to flocculation since they arise as a result of a decrease in solubility. The amount of water associated with a protein, for instance, will decrease if it is part of a coacervate, because of the protection of its polar groups by the micromolecule with which it is associated and which may actually form a salt bond with it.

PROPERTIES OF GELS

Since protoplasm has the properties of gels, it is important to know what these are. In spite of the more or less rigid nature of aqueous gels, their water content is usually high, and this ensures just about as rapid diffusion of water-soluble substances through them as through water itself. The following are their main properties.

Swelling or imbibition pressure

Most aqueous gels are highly hydrophilic. Because of the large adsorptive forces they may have imbibition pressures as high as 1000 atm; that is, they may absorb water

against water-removing forces of this value, or they may hold some water against pressures of 1000 atm in a hydraulic press. This is true of seeds because of their gel nature. The first phase of water uptake in the cotyledons and germ axes of pea seeds consists of pure imbibition (Kühne and Kausch, 1965). The maximal degree of swelling is almost independent of temperature. When this point is reached, the protoplasmic gel is prevented by the cell walls from attaining the maximum possible degree of swelling.

Proteins may retain as much as 10 g water per 100 g protein at 100° C. Water that is adsorbed strongly by colloids is sometimes called *bound water*. The amount of such bound water held by tissues may vary with the conditions of growth (Fig. 5-2), and the water is retained against forces of evaporation of well over 1000 atm.

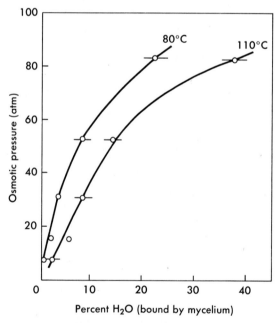

Fig. 5-2. Percent-bound water held by the mold *Aspergillus niger* after growth in media containing 0% to 50% sugar and therefore having osmotic pressures (see Chapter 6) of nearly 0 to 100 atm. The bound water is that held at room temperature in an evacuated desiccator with a relative humidity of almost zero. The values obtained depend on the temperature used to drive off the bound water—less if 80° C than if 110° C. (From Todd, G. W., and J. Levitt. 1951. Plant Physiol. **26**:331-336.)

Hysteresis

If several gelatin gels are made up of different concentrations and then dried, when allowed to reimbibe water, each will tend to regain its original concentration (Fig. 5-3). This is because of the structure of the gel. The colloidal particles are needle shaped and cross each other at definite points, forming a "brush-heap" type of structure. The water is held in the interstices between the *micelles* (particles). Since these micelles tend to maintain a constant position with reference to each other (i.e., they act as though fastened together at the points of contact), the dilute gels will have larger interstices for holding water than will the concentrated gels. This phenomenon may be of fundamental importance in the case of protoplasm. If a plant has grown under low moisture conditions, its protoplasm may conceivably have this tight kind of structure and perhaps be unable to imbibe as much water as if it had grown under conditions of high moisture. This may possibly be one reason why such plants are sometimes permanently set back by early droughts, even though not obviously injured.

Syneresis

Sometimes the ability of a gel to hold water decreases with age, and free water (or solution) is then liberated. This is called syneresis. Syneresis may also conceivably happen in the case of senescent cells. The vacuole would then enlarge at the expense of the protoplasm.

Thixotropy

Some gels may act as sols if stirred or shaken. After this treatment they may pour quite readily. On standing they become gels again and will not pour. This reversible sol $\rightleftharpoons$ gel transformation is called thixotropy. By use of microneedles it is possible to show that protoplasm is thixotropic.

QUESTIONS

1. Is a colloid a particular substance or a state in which the substance occurs?
2. Are the colloidal particles visible?
3. What is the approximate upper size limit?
4. What is the approximate lower size limit?
5. Is there a sharp line between colloids and noncolloids?
6. Can the colloidal particle be a single molecule?
7. Is the specific surface of a colloid large or small?
8. Does it show more or less adsorption than noncolloids?
9. How can one distinguish a colloidal dispersion from a molecular solution?
10. How do the two compare with regard to rate of diffusion?
11. How do the two compare with regard to osmotic pressure?

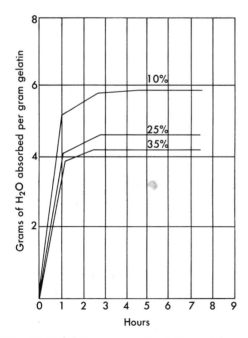

Fig. 5-3. Imbibition curves of gelatin made up of 10%, 25%, and 35% gels, dried to 3% moisture, then allowed to reimbibe water. (From Gortner, R. A. 1938. Outlines of biochemistry. John Wiley & Sons, Inc., New York.)

12. What colloids occur in the cell?
13. Name one property that may give the colloid stability?
14. Name another property that may give the colloid stability?
15. What is flocculation?
16. What is peptization?
17. What is salting out?
18. What is the lyotropic series?
19. Are proteins hydrophilic or hydrophobic (in general)?
20. What causes their charge?
21. Are they acids or bases?
22. What is an amphoteric substance?
23. What is the isoelectric point? How does it compare with the neutral points?
24. Is protoplasm a liquid or a gel?
25. How does the rate of diffusion through gels compare with that through sols?
26. Name three important properties of gels.

SPECIFIC REFERENCES

Booij, H. L., and H. G. B. de Jong. 1956. Biocolloids and their interactions. Protoplasmatologia I(2). Springer-Verlag, Vienna.

Kühne, L., and W. Kausch. 1965. Über das Quellungsmaximum der Kotyledonen und Keimachsen von *Pisum sativum* L. Planta **65**:27-41.

Todd, G. W., and J. Levitt. 1951. Bound water in *Aspergillus niger*. Plant Physiol. **26**:331-336.

GENERAL REFERENCES

Gortner, R. A. 1938. Outlines of biochemistry. John Wiley & Sons, Inc., New York.

Scarth, G. W., and F. E. Lloyd. 1930. Elementary course in general physiology. John Wiley & Sons, Inc., New York.

chapter 6

Diffusion and osmosis

DIFFUSION

Although many of the properties of the cell are dependent on the colloidal nature of its components, others are caused by the free molecular state of the water and the substances dissolved in it. Since these molecules are in constant motion, (thermal agitation) there will be a continuous transfer of molecules from one region to another even without the application of external forces. At equilibrium, when the average free energy of each species of molecules is the same throughout the medium, as many molecules of one kind are transferred in one direction as in the other, and the net effect is zero. If the system is not in equilibrium, more molecules will move in one direction than in the other, and there will be a net transfer of the substance, called *diffusion*. An important question in the physiology of the plant is to what extent diffusion accounts for the continuous transfer of substances into, out of, and within the plant. This question can be answered only if the principles of diffusion are known.

If f is the frictional resistance to molecular movement:

$$\text{Diffusion rate} \propto \frac{1}{f}$$

This explains why the diffusion rate of water vapor is about 1500 times as rapid as that of liquid water. The molecules are so much farther apart in the vapor than in the liquid that the frictional resistance is extremely low. Thus gaseous diffusion is so rapid that no mechanism occurs in the plant for speeding up its movement. Liquid water, on the other hand, diffuses too slowly to keep up with the movement of the water vapor. In order to prevent its tissues from drying up, the land plant has therefore had to develop mechanisms for speeding up the movement of its liquid water and slowing down the movement of its water vapor.

The same principle of diffusion applies to the solutes dissolved in water. This is illustrated by plasmolysis. Hypertonic sucrose solutions take much longer to induce plasmolysis than do calcium chloride solutions of the same tonicity, partly because of their much greater viscosity, partly because of the slower diffusion rate of sucrose than of calcium chloride. But the aqueous solutions in the plant are usually so dilute that their viscosity is nearly that of pure water, and all the solutes are subjected to essentially the same frictional force. Consequently, when considering their movement

51

within the plant, the frictional resistance can usually be neglected, and the diffusion rate will depend on any factor that affects the free energy of the diffusing substance. This is because, from the second law of thermodynamics, molecular movement will always be from a region of higher free energy of the molecules to a region of lower free energy.

Since diffusion depends on molecular movement, it must be quantitatively related to (1) the rate of motion of the average molecule (its thermal agitation) and (2) the number of molecules per unit volume. The free energy of a substance that depends on these two components is called the chemical potential of the substance. Therefore

$$\text{Diffusion} \propto \mu$$

where μ = chemical potential of the diffusing substance. The chemical potential is given by the following equation:

$$\mu = RT\ln a \text{ joules/g}$$

where R = the gas constant (0.082)
T = the temperature (degrees absolute)
a = the activity or "effective" concentration of the substance
ln = the natural logarithm

The two variables, T and a, are measurements of the aforementioned two components. If the temperature is constant, the chemical potential is, therefore, proportional to the activity of the substance. But the activity of a nondissociating substance in the plant is, for all practical purposes, proportional to its concentration (for its dissociating substances, see Chapter 8). Therefore, at constant temperature, it will diffuse from a region of higher concentration to one of lower concentration (Fig. 6-1). The greater this difference the more rapid is the diffusion. However, the rate of diffusion must also depend on two spatial factors. The greater the area across which the substance is diffusing the greater is the number of molecules diffusing per unit time. Conversely, the smaller the distance

between the two concentrations the greater is the number crossing it per unit time. Two of the three factors—the concentration difference and the distance between them—are combined in what is called the *concentration gradient*. This is the concentration difference per unit distance.

The quantitative relation is given by Fick's law of diffusion:

$$\frac{s}{t} = Da \frac{C_1 - C_2}{x}$$

or, expressed as a differential equation:

$$\frac{ds}{dt} = Da \frac{dc}{dx}$$

where s = amount of substance diffusing (moles)
t = time (sec)
D = coefficient of diffusion, or specific diffusion rate
a = area of membrane (cm^2)
C_1 = higher concentration (moles per liter)
C_2 = lower concentration
x = membrane thickness (cm) or distance between C_1 and C_2
$\dfrac{C_1 - C_2}{x}$ or $\dfrac{dc}{dx}$ = concentration gradient

D is a constant for a specific substance diffusing through a specific medium at a standard temperature. When different substances are compared, small molecules are found to diffuse more rapidly than large molecules, that is:

$$D \propto \frac{1}{\text{molecular size}}$$

For small molecules:

$$D \propto \frac{1}{\text{mass}}$$

For large molecules and colloidal particles:

$$D \propto \frac{1}{\text{radius}}$$

It is now obvious why the larger colloidal particles diffuse so much more slowly than the smaller molecularly dispersed substances. In dilute solutions, such as are found in the plant, each solute diffuses independently of the other solutes.

If the partition x (Fig. 6-1) is a structure

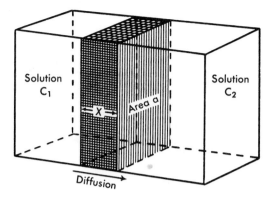

Fig. 6-1. Diffusion of a substance from a region of higher to one of lower concentration. (Redrawn from Scarth, G. W., and F. E. Lloyd. 1930. Elementary course in general physiology. John Wiley & Sons, Inc., New York.)

immersed in the aqueous medium, the concentration gradient in it will depend not only on the two concentrations in the aqueous media but also on the solubility of the diffusing substance in this structure. This solubility in the structure is called the permeability of the structure to the diffusing substance. If the structure is very thin, it is called a membrane. Diffusion of a substance through membranes is, therefore, controlled not only by (1) the temperature, (2) the concentration gradient in the aqueous phase, and (3) the size of the diffusing molecule or particle, but also (4) by the permeability of the membrane to the substance. This will be considered in more detail in Chapter 7.

OSMOSIS AND OSMOTIC PRESSURE
Pressure caused by osmosis

The same laws of diffusion apply to solvent as to the solute. However, because of the much larger quantity of the solvent and because of its liquid state, its diffusion through semipermeable membranes gives rise to hydrostatic pressures. The importance of these pressures has led to the use of the specific term *osmosis,* for diffusion of the solvent, and *osmotic pressure,* for the hydrostatic pressure produced by osmosis

(for a stricter definition see later). Osmosis is sometimes defined as diffusion of solvent across a semipermeable membrane, but the preceding simpler definition is broader and therefore more useful. It permits use of the term, for instance, for the diffusion of water from a solution into a hydrophilic gel, although no semipermeable membrane separates the two systems.

Since pressure is force per unit area ($P = \dfrac{F}{A}$), the osmotic pressure that develops in the cell, because of osmosis, must produce forces of fundamental importance to the plant. These forces give rise to plant *turgor*—the stiffness or rigidity of the plant resulting from the hydrostatic pressure exerted on the cell walls. Plant turgor in its turn results in growth, plant movements, and other plant responses. In order to study these phenomena, osmotic pressure, therefore, must be measured and calculated quantitatively. The qualitative concept of osmotic pressure as the hydrostatic pressure produced by osmosis is therefore inadequate. In the strict quantitative sense, the term *osmotic pressure* is not used for nonequilibrium (i.e., temporary or even steady-state) pressures but only *for the final, equilibrium pressure (above atmospheric) attained as a result of osmosis under specific, standard conditions.* Since forces occur in pairs, osmotic pressure is determined by applying an increasing pressure until the one is found which is just sufficient to stop osmosis. The standard conditions are (1) separation of the solution from pure solvent at atmospheric pressure by a semipermeable membrane and (2) maintenance of the solution and solvent at the same standard temperature (20° C). The osmotic pressure of a solution is, therefore, defined as the extra pressure that must be applied to it in order to stop osmosis from pure water separated from the solution by a semipermeable membrane, when both are at 20° C.

Unfortunately, however, the cell is usu-

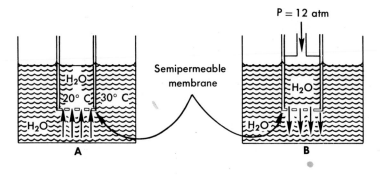

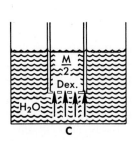

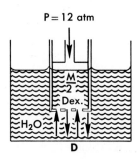

Fig. 6-2. Endosmosis (**A** and **C**) and exosmosis (**B**) produced by differences in **A** temperature, **B** pressure, and **C** solute concentration. There is no osmosis in **D** because the solute effect is exactly counterbalanced by the pressure. The pores in the bottom of the container (or osmometer) indicate a semipermeable membrane.

ally neither under the preceding standard conditions nor in a state of equilibrium. Therefore, the actual forces prevailing in the cell are not those that would result from the defined (equilibrium) osmotic pressure. Furthermore, from Newton's third law of motion, the real forces must exist in pairs, equal in magnitude and opposite in direction. These forces are caused by (1) the pressure exerted by the piston and walls of the osmometer (Fig. 6-2) on the solution, which the physiologist calls wall pressure (P_w); and (2) the hydrostatic pressure exerted by the water on the walls, which the physiologist calls turgor pressure (P_t) since it can convert flexible walls into stiff walls and a flaccid cell into a turgid cell. The important difference between

these three pressures is that osmotic pressure (P_o—the pressure at zero diffusion of water under standard conditions) refers only to the equilibrium pressure. Wall pressure (P_w) and turgor pressure (P_t) refer to the actual pressure under the particular conditions to which the cell is subjected, even though equilibrium has not been attained. At equilibrium under standard conditions, of course the following relation exists:

$$P_o = P_w = P_t$$

In any nonequilibrium or nonstandard state (even in the steady state)

$$P_w = P_t$$

but both differ from P_o.

Table 6-1. Colligative properties of a solution; effect of dissolving 186 g (1 gram molecular weight) of dextrose in 1000 g water, yielding a solution of 1.0 M dextrose

Colligative property	Water	Value for molar dextrose	Effect of dextrose on property
Vapor pressure	17.5 mm Hg at 20° C	17.2 mm Hg	Lowered
Freezing point	0° C	−1.86° C	Lowered
Boiling point	100° C	100.51° C	Raised
Osmotic pressure	0 atm	24 atm	Raised
Water potential	0 joules/g	−2.4 joules/g	Lowered

Energy relations

As in the case of any diffusion process, osmosis will always take place from a region of higher chemical potential of the solvent to one of lower chemical potential. *Endosmosis* (osmosis into a system) occurs if the *external* water is at a higher chemical potential; *exosmosis* (osmosis out of a system) occurs if the *internal* water is at a higher chemical potential. Endosmosis therefore will be induced by any factor that decreases the free energy or chemical potential of the solvent within the system or that increases it outside the system. Exosmosis is induced by either of the reverse changes. There are three such factors (Fig. 6-2). The chemical potential (1) increases with the temperature, since chemical potential is a measure of energy resulting from molecular motion and temperature is simply a measure of the kinetic energy of the molecules; (2) increases with the pressure, since the piston does work on the liquid, raising its free energy; and (3) decreases with increased concentration of solute since the force of attraction between solute and solvent lowers the free energy of the solvent molecules. The effect of the solute on the free energy of the solvent is proved by changes in the colligative properties of solutions (Table 6-1) since these all depend on the free energy of the solvent molecules. For instance, because of the lower free energy of the water molecules in a solution than in the pure solvent, the vapor pressure of the water is lowered and a higher temperature is required to boil the solution than the pure solvent. Similarly, the lower vapor pressure results in a lower diffusion rate (or osmosis) of the water. Theoretically, then, the ideal method for predicting the direction of osmosis between two points is by measuring the chemical potential of the solvent at these two points because the chemical potential measures the net effect of all three factors. Unfortunately, the *actual* chemical potential cannot be measured, but the *change* in relation to pure solvent under standard conditions can be determined.

The experiment illustrated in Fig. 6-2 reveals that a pressure of 12 atm on the 0.5 M sugar solution was required to prevent osmosis from pure water. This must mean that the pressure exactly counteracts the effect of the solute on the free energy of the solvent. Since the solute lowers the free energy of the water molecules (Table 6-1), the pressure must raise it by an equal amount. Therefore, 0.5 mole of sugar per liter of water must lower the free energy of the water by 12 atm. This change in free energy of the water, produced by the solute is called the osmotic potential (π) of the solution. From the foregoing opposing effects of solute and hydrostatic pressure on the free energy of the solvent:

$$\pi = -P_o$$

This means, of course, that only the standard equilibrium pressure, which is called the osmotic pressure, is numerically equal to the osmotic potential (although opposite in sign). In the absence of any pressure, other than atmospheric, the negative value of this theoretical pressure yields the decrease in free energy of the water molecules resulting from the solute. Since solutes always lower the free energy of the solvent, π will always be negative in the plant. Dilution of the solution with solvent will increase the value of π (i.e., will decrease the negative numerical value) until, at infinite dilution, it reaches the maximum value of 0 atm.

Under normal conditions, however, the cell solution is neither at atmospheric pressure nor is it subjected to the full theoretical osmotic pressure since it is turgid but not in equilibrium with pure water. Because of its turgor, the cell contents are subjected to a wall pressure (P_w) somewhere between these two extremes. This wall pressure increases the free energy of the cell's water. The algebraic sum of the preceding two factors (π and P_w) is, therefore, a measure of the net change in chemical potential of the water relative to that of pure water at atmospheric pressure and at the same temperature. This net change in the chemical potential of the water is called the water potential (Ψ), and is, therefore, calculated from:

$$\Psi = \pi + P_w$$

The importance of Ψ lies in its control of the direction (and relative rate) of diffusion. Osmosis will always occur from a region of higher Ψ (smaller negative value) to a region of lower Ψ (larger negative value). Thus, it will occur from cell 1 with a Ψ of -3 atm to cell 2 with a Ψ of -4 atm. Obviously, if the solution is at atmospheric pressure, $P_w = 0$ and $\Psi = \pi$.

The originators of the term *water potential* (Slatyer and Taylor, 1960) define it as "the difference between the partial specific Gibbs' free energy of water in the system

under consideration and of free, pure water at the same temperature."

Since Ψ is always negative (the maximum value being zero for pure water) and since P_w is positive in a turgid cell (Fig. 6-3, *A*) and cannot exceed the numerical value of π, Ψ will be negative in a turgid cell or will have a maximum value of zero. But if a cell is allowed to lose water until it is flaccid (Fig. 6-3, *B*), the whole cell will collapse (unlike the separation of protoplast from wall, which occurs on plasmolysis, Fig. 2-2), and the positive osmotic force will be converted to a negative force or tension. Therefore, P_w will have a negative energy value. Since both π and P_w are negative in the flaccid cell, the water potential must be lower (i.e., more negative) than π alone.

Osmotic quantities have, in the past, been measured in atmospheres. The recent trend, however, is to measure them in *bars* (a unit of the metric system). Fortunately, the values are interchangeable since 1 atm (approximately 15 lb/inch^2) = 1.01 bars. Before the recent adoption of water potential terminology, various other names were given for the water potential—suction pressure (SP), diffusion pressure deficit (DPD), osmotic equivalent (E), etc. These terms were used as positive values for what are now known to be negative water potentials.

In the preceding equations the free energy of the water is measured in terms of a unit of pressure. This is possible because of the twofold significance of pressure:

1. $P = \dfrac{F}{A}$ (pressure = force per unit area)

2. $P = \dfrac{W}{\Delta V}$ (pressure = work done per unit change in volume)

But this requires the use of units of pressure (bars) in place of the units of energy (ergs or joules). The one can be readily converted into the other because there is a very simple relation between the two:

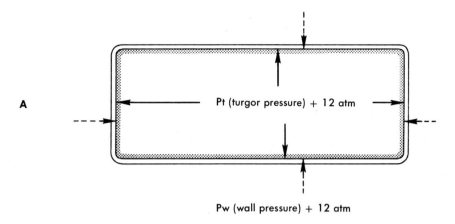

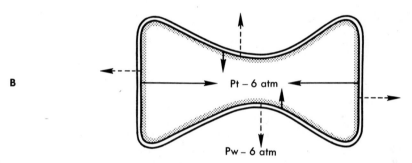

Fig. 6-3. Pressures in **A,** a turgid, and **B,** a flaccid cell. The + sign means (1) an increase in free energy of the water and (2) the direction of the forces in a turgid cell. The − sign means (1) a decrease in free energy of the water and (2) the direction of the forces in a flaccid cell. The extracted sap from the turgid cell would have an osmotic potential of −12 bars if the cell is at full turgor (i.e., in equilibrium with pure water).

$$\Psi \text{ joules/g} = \frac{\Psi \text{ bars}}{10}$$

Another quantity, the *matric potential,* is sometimes included as a theoretical value for the effect of hydrophilic colloids (as opposed to the osmotic potential caused by solutes) on the water potential. It must be realized, however, that there is no difference thermodynamically as to the nature of the effect of the colloid and the solute. Both lower the water potential by virtue of their attraction for the water molecules. Furthermore, the methods normally used to measure the osmotic potential (see later) include any effect resulting from the hydrophilic colloids present. Therefore, in practice, what is called the osmotic potential is really a combination of the two quantities and no distinction between the two can be made. Recently, attempts have been made to measure this matric potential separately (Wiebe, 1966; Boyer, 1967). But as long as the method for measuring osmotic po-

Table 6-2. Approximate ranges of calculated osmotic pressures of plant saps of different environmental groups*

Plant group	Osmotic pressure (atm or bars)
Summer ephemerals	8–42
Succulents and winter ephemerals	4–24
Xerophytes	14–57
Hydrophytes	8–13
Air leaves	18–21
Water leaves	8–9
Epiphytes	3–6
Halophytes	30–115
Parasites	14–17
Hosts	11–14

*From Altman, P. L., and D. S. Dittmer (Eds.). 1964. Biology data book. Federation of American Societies for Experimental Biology, Washington, D. C.

tential combines the effects of both solute and colloid, there is no need to include matric potential as a separate quantity in the equation for Ψ.

From Table 6-2 it is obvious that very large osmotic pressures have been found in plant cells, and since these theoretical pressures may be converted into actual turgor pressures of nearly the same numerical values, the question arises as to the ability of cell walls to withstand pressures of the order of 100 atm (i.e., 1500 lb/inch²). It must be realized, however, that the dimensions of a cell are very small, and therefore the actual force acting on the two opposite wall faces of a cell is normally far too small to exceed the elastic limit of the mature cell wall. In the case of growing cells, on the other hand, the walls are so soft that this elastic limit may be readily exceeded and the result is cell enlargement (see Chapter 17).

It is necessary to distinguish between two pressures, both of which are included in the colligative properties of aqueous solutions, and both of which, therefore, depend on the water potential (Ψ). A decrease in Ψ is accompanied by a decrease in vapor pressure (Table 6-1). However, vapor pressures are too small (usually 0.02 to 0.04 atm) to account for appreciable forces in plant cells. In the liquid state, however, the water molecules are about 1500 times as close together as in the vapor state when both are at the same vapor pressure. The *hydrostatic* pressures produced by a decrease in Ψ (followed by osmosis), therefore, may be expected to attain values 1500 times the change in vapor pressure. Thus, 0.5 M dextrose solution has a vapor pressure only 0.0004 atm less than that of pure water, yet if it is surrounded by a semipermeable membrane and immersed in water, osmosis may lead to a maximum pressure of 12 atm (Fig. 6-2).

It is therefore of fundamental importance not to confuse the small vapor pressure (which may also be called the "diffusion pressure") of water with the much larger hydrostatic pressures, although the latter result from the former. This very confusion gave rise in the earlier literature to terms such as *diffusion pressure deficit* (DPD), which was a measure of potential *hydrostatic* pressure and not the much smaller vapor (or true diffusion) pressure difference.

Significance of osmotic quantities

The water potential of a cell enables the prediction of the direction of osmosis— from a high potential (or low negative value) to a low potential (or high negative value). There is, therefore, a gradient from the highest value in the water-absorbing zone of the plant, the root, to the lowest value in the water evaporating part of the plant, the leaves (Table 6-3). Besides the direction of diffusion of water between two systems, the water potential difference between them also indicates the relative rate of diffusion. This is true as long as relatively small differences are involved (e.g., less than either 150 bars or 15 joules/g). Unfortunately, the preceding equation relating chemical potential to osmotic quantities does not hold for values above these. In such cases the vapor pressure difference must be used, for example, for the diffusion of water from the leaves into the normally dry air (the water potential of the air at 60% relative humidity at normal temperatures is about −600 atm).

The three pressures play specific roles in the physiology of the cell (Table 6-4). The osmotic pressure of a solution depends on the presence of solutes. Measured or calcu-

Table 6-3. Water potential of inner cortex in different organs of ivy[*]

Organ	Water potential (atm)
Petiole	−8.4
Stem (225 cm high)	−5.0
Stem (35 cm high)	−2.9
Root (older portion)	−2.4
Root (absorption zone)	−1.6

*Adapted from Molz, F. J. 1926. Amer. J. Bot. **13**:433-501.

lated values for plant or cell sap are indirect measures of cell sap concentration. Cell sap concentration, measured in moles, will not parallel osmotic pressure values exactly since solutes differ in their osmotic pressures per mole (see later). It is therefore sometimes expressed in *osmolarity*—the molarity of a nearly ideal solution (e.g., dextrose) having the same osmotic pressure as the unknown solution. Osmotic potentials of cell saps, on the other hand, may reveal the relative degree of hydration of the protoplasm. Since both the vacuole and protoplasm are subjected to the same wall pressure, their water potentials relative to each other are solely the result of their osmotic potentials (using the term as including matric potential, see earlier). Consequently, if the solute concentration of the vacuole increases, the osmotic potential must decrease. Water must then move into the vacuole from the protoplasm, and the latter will be partially dehydrated. Rapidly metabolizing and growing cells are char-

Table 6-4. Significance of osmotic quantities

Quantity	Symbol	Sign in turgid cells	Meaning	Significance
Osmotic pressure (of physicist)	P	+	Calculated force per unit area required to stop osmosis	1. Theoretical maximum wall and turgor pressure 2. Measure of solute concentration
Osmotic potential	π	−	Water potential resulting from solutes (and colloids if matric potential included)	Degree of hydration of protoplasm
Turgor pressure	P_t	+	Instantaneous actual hydrostatic pressure exerted on cell wall	Stretches cell wall and may lead to cell enlargement if walls are sufficiently plastic
Wall pressure	P_w	+	Instantaneous back pressure by wall on protoplast	Increases free energy of water molecules
Water potential	ψ	−	Chemical potential of water in system minus that of pure water under standard conditions	Indicates direction and relative rate of diffusion

Table 6-5. Calculated osmotic pressures of plant saps in some Canadian evergreens in relation to the season*

Month	Osmotic pressures (atm)		
	Linnaea borealis	*Picea glauca*	*Pyrola rotundifolia*
Oct.	19.6	17.1	
Dec.	25.0	20.3	24.6
Jan.		20.0	
Feb.			23.9
March	25.6	24.9	
April		20.1	17.2
May	14.3	21.0	
June		19.7	12.6

*Adapted from Altman, P. L., and D. S. Dittmer (Eds.). 1964. Biology data book. Federation of American Societies for Experimental Biology, Washington, D. C.

Table 6-6. Diurnal changes in water potential and in calculated osmotic pressures of the cell sap*

Time	Water potential (atm)	Time	Osmotic pressure (atm)
5:15 A.M. (sunrise)	−7.5	6:00 A.M.	12.5
7:00	−9.5		
9:00	−12.0	10:00	15.3
11:00	−15.0		
1:00 P.M.	−17.5	2:00 P.M.	17.4
3:00	−17.5		
4:30	−15.0	5:00	16.5
7:30	−15.0	8:00	16.3

*Adapted from Herrick, E. M. 1933. Amer. J. Bot. **20:**18-34.

acterized by low cell sap concentrations and therefore high protoplasmic hydration. High cell sap concentrations and, therefore, low protoplasmic hydration are commonly found in inactive, dormant cells, for example, during the winter (Table 6-5). Osmotic pressure is also related to the ecology of the plant (Table 6-2), although there is, of course, considerable overlap between ecological groups. The highest values are definitely found among the halophytes, the lowest among the epiphytes, hydrophytes, and succulents. Osmotic pressure values may also reflect the balance between transpiration and absorption, as can be seen by the diurnal changes (Table 6-6), although changes in water potential are more striking.

Because it reflects the water status of the plant in all these cases, Walter (1963) has used the osmotic potential as an indirect measure of what he calls the "hydrature" of the plant, by analogy with temperature, which he describes as the "status" of the heat in a system. Hydrature, according to Walter, is the relative water vapor pressure (or relative humidity) of an aqueous system, independent of external factors, (i.e., at constant temperature and external pres-

sure). It is therefore proportional to osmotic potential.

The wall pressure raises the water potential of the water in the cell. This increases the escaping tendency of the water molecules and may lead to diffusion out of the cell. But the resulting force is insufficient to squeeze the water through the extremely fine "pores" of the plasma membrane. Artificial pressures of 100 to 150 atm may be required to produce this flow. Nevertheless, wall pressure may be thought of as producing a flow in another way—the flow of water up the xylem (see Chapter 8). Turgor pressure, on the other hand, may lead to wall stretching and therefore cell enlargement (see Chapter 17).

The cells of land plants are not normally at equilibrium but are continuously losing and gaining water. The actual turgor and wall pressures therefore may be much less than the osmotic pressure of the cell sap because of loss of water to the environment. Even when immersed in pure water (Fig. 6-4), the cell's turgor pressure can never fully attain the osmotic pressure of the cell sap at zero turgor pressure because any development of turgor results from endosmosis, which dilutes the cell sap and lowers its

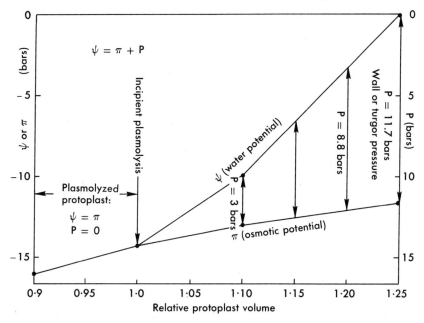

Fig. 6-4. Changes in osmotic quantities of cells of a sunflower petiole during deplasmolysis and final expansion in water. All three quantities increase, but the increase in Ψ for any one increase in cell volume always equals the sum of the increases in π and P. Therefore, at any one point $\Psi = \pi + P$; for example:

Protoplast volume	Ψ	π	P
0.9	−16	−16	0
1.1	−10	−13	3
1.15	−6.7	−12.7	6
1.25	0	−11.7	11.7

(Redrawn from Clark, J. A. 1956. Ph.D. Thesis. University of Missouri, Columbia, Mo.)

osmotic pressure (or raises its osmotic potential). As endosmosis occurs, all three values change, but the equation holds for any instantaneous set of values. These relations are shown in Fig. 6-4.

Starting with a sunflower leaf protoplast plasmolyzed to 0.9 of the cell wall volume, the osmotic potential is −16 bars and so is the water potential (Ψ) since wall pressure is zero. When the protoplast has expanded sufficiently just to fill the cell wall, it has an osmotic potential and water potential of −14.3 bars, and wall pressure is still zero. At a cell volume of 1.1, the osmotic potential has decreased still further (to −13.0 bars) because of dilution of the cell contents by the entering water. The wall pressure has risen from zero to 3 bars and the Ψ is −10 bars (osmotic potential plus wall pressure). At full turgor (i.e., in equilibrium with pure water) the Ψ is zero and the osmotic potential and wall pressure are −11.7 bars and +11.7 bars, respectively (where + simply means above atmospheric pressure).

Determination of osmotic quantities

Osmotic potential. In most cases the osmotic quantities of cells are not determined directly. The values are much more easily obtained by comparing the cell with solutions of known osmotic values. It is therefore convenient to be able to calculate the osmotic potential of a solution from its concentration. This can be done by use of the following equation,

$$\pi = -CRT$$

where π = osmotic potential (bars)
C = concentration (moles per liter)
R = gas constant (0.082)
T = temperature (degrees absolute or Kelvin)

If a solution is reasonably dilute and consists of monomolecular solute particles, this equation gives values that are sufficiently accurate for most physiological work. If, however, the molecules dissociate into smaller particles as in the case of electrolytes (salts, acids, bases), the osmotic potential will be numerically larger than calculated by an amount equal to the excess of particles over the expected number of molecules. In the case of sodium chloride, for instance, the right side of the equation has to be multipled by 1.8 and in the case of calcium chloride, by 2.4. This factor is known as the *isotonic coefficient*, and the equation becomes

$$\pi = -iCRT$$

where i = isotonic coefficient (average number of particles per molecule). This equation is not accurate at high concentrations because of the hydration of the particles; and the error cannot be calculated for, because the degree of hydration is usually not known accurately enough and is not constant. Accurate osmotic potentials can, however, be calculated from two colligative properties of solutions: (1) the freezing point lowering and (2) the vapor pressure, both of which values may be obtained for many solutions from published tables (e.g., *Handbook of Chemistry and Physics* edited by C. D. Hodgman and others).

1. $\pi = -12.06 \ \Delta$ bars

where Δ = freezing point lowering of the solution (compared to pure water)

2. $\pi = -RT \ln \dfrac{p_o}{p}$

where $\ln$ = the natural logarithm
p_o = vapor pressure of pure water
p = vapor pressure of the solution at the same temperature as p_o

The preceding calculations must be applied to determine the exact values for the solu-

Table 6-7. Osmotic pressures of solutions of several substances, calculated from freezing point lowerings of the solutions listed in International Critical Tables; calculations from Lewis's formula

Molarity (weight molar)	Osmotic pressure (atm or bars)				
	$CaCl_2$	NaCl	KNO_3	Sucrose	Glucose
0.01	0.7	0.5	0.5		
0.10	6.2	4.5	4.3	2.6	2.4
0.20	12.3	8.9	8.2	5.1	4.8
0.30	18.5	13.2	11.9	7.7	7.2
0.40	25.3	17.6	15.4	10.2	9.6
0.50	32.2	21.9	18.8	12.9	12.1
0.60	39.2	26.2	22.0	15.5	14.5
0.70	47.2	30.5	25.1	18.3	17.0
0.80	56.2	34.9	27.9	21.1	19.4
0.90	65.4	39.2	30.6	23.9	21.8
1.00	75.0	43.6	33.2	26.9	24.3
2.00	194	88.8		56.9	
3.00	346	138.2		92.9	
4.00				137	
5.00				187	

tions used in measuring the osmotic quantities of cells and tissues (Table 6-7). These solutions may be used for two different kinds of measurement.

Osmotic potential of the cell at zero turgor. The two simplest methods are (1) the minimum length method and (2) the incipient plasmolysis method. Since the stretching or contraction of living cells is caused by turgor pressure, once turgor pressure is zero, no further cell wall contraction can occur (when cells are in aqueous solutions). Consequently, the most dilute solution that is capable of reducing the cell (or tissue in the case of uniform strips) to minimum size has the same osmotic potential as the cell sap (i.e., is isotonic or isosmotic with it). In the second method the point is determined by going a little beyond it and observing for incipient (i.e., beginnings of) plasmolysis. If the merest trace of plasmolysis can be detected, the two methods give essentially identical results.

Osmotic potential of the cell at its momentary state. If the juice of the tissue or cell is extracted, its freezing point lowering can be measured and the osmotic potential calculated as just described. The value will, of course, vary depending on the degree of turgor or wilting of the plant. However, as long as the plant material is not wilted, the values obtained by this method should agree within about 10% with the values obtained for the cell at zero turgor.

Water potential. Since no diffusion of water can occur between systems at the same water potentials, the simplest method is to find the solution of known water potential in which the living cells show neither endosmosis nor exosmosis, that is, in which they neither swell nor contract. This can be done by measurements of cell or tissue strip length.

Turgor and wall pressure. Although these values have been measured directly in the case of large algal cells, and by an ingenious method (Virgin, 1955) in the case of higher plants, they are usually calculated from the osmotic potential and the water potential, using the transposed form of the fundamental equation:

$$P_w = \Psi - \pi$$

Many ingenious methods for measuring water potentials and osmotic potentials have been developed (Levitt, 1964). Recently, the precision of the vapor pressure method has been greatly increased in the hands of many investigators. These methods are constantly being improved, but their accuracy is still rather low because of many complicating factors that are difficult to control or to correct for. The most recent improvements probably yield results as accurate as by any of the classical methods.

QUESTIONS

1. What kind of movement is diffusion caused by?
2. If diffusion stops (because of equilibrium), does molecular movement stop?
3. If different substances are compared under the same conditions, will all diffuse at the same rate?
4. Does diffusion depend in any way on the medium?
5. What difference must exist in order for diffusion to occur?
6. What effect does temperature have on diffusion?
7. What three factors are included in Fick's law of diffusion?
8. What property of a substance can always be related to diffusion, including concentration difference, temperature, etc.?
9. What effect does a solute have on the free energy of the solvent molecules?
10. What is meant by osmosis?
11. What is meant by osmotic pressure?
12. What is the effect of pressure on the free energy of water molecules?
13. If a solution is separated from water by a semipermeable membrane, what happens?

14. How long will this process continue?
15. What are the real pressures in a living cell?
16. What is the effect of each pressure?
17. How do these pressures compare quantitatively?
18. How can these pressures be equal when the cell enlarges?
19. What quantity gives a quantitative measure of the ability of the cell's water to diffuse?
20. Does this quantity vary directly or inversely with this ability to diffuse?
21. How is this quantity related to π and P?
22. When a cell is in equilibrium with pure water, what is its Ψ?
23. What is the osmotic potential of 0.33 M dextrose?
24. What is the osmotic potential of 0.33 M sodium chloride?
25. At what temperature would a cell with an osmotic potential of -24 bars freeze?
26. If a cell is at incipient plasmolysis, what cell quantity does the external solution reveal?
27. If a cell remains unchanged in size on transfer to a solution, what cell quantity does the solution give?
28. How do you measure P_w? How do you measure P_t?
29. What osmotic quantity can be calculated from the freezing point lowering of the plant juice?

SPECIFIC REFERENCES

Altman, P. L., and D. S. Dittmer (Eds.). 1964. Biology data book. Federation of American Societies for Experimental Biology, Washington, D. C.

Boyer, J. S. 1967. Matric potentials of leaves. Plant Physiol. **42**:213-217.

Clark, A. J. 1956. An investigation of the drought hardening of the soybean plant. Ph.D. Thesis. University of Missouri, Columbia, Mo.

Herrick, E. M. 1933. Seasonal and diurnal variations in the osmotic values and suction tension values in the aerial portions of *Ambrosia trifida.* Amer. J. Bot. **20**:18-34.

Molz, F. J. 1926. A study of suction force by the simplified method. Amer. J. Bot. **13**:433-501.

Scarth, G. W., and F. E. Lloyd. 1930. Elementary course in general physiology. John Wiley & Sons, Inc., New York.

Slatyer, R. O., and S. A. Taylor. 1960. Terminology in plant and soil-water relations. Nature (London) **187**:922-924.

Virgin, H. 1955. A new method for the determination of the turgor of plant tissues. Physiol. Plant. **8**:954-962.

Walter, H. 1963. Zur Klärung des spezifischen Wasserzustandes im Plasma und in der Zellwand bei der höheren Pflanze und seine Bestimmung. Ber. Deutsch. Bot. Ges. **76**:40-71.

Wiebe, H. H. 1966. Matric potential of several plant tissues and biocolloids. Plant Physiol. **41**:1439-1442.

GENERAL REFERENCES

Crafts, A. S., H. B. Currier, and C. R. Stocking. 1949. Water in the physiology of the plant. Chronica Botanica Co., Waltham, Mass.

Dainty, J. 1963. Water relations of plant cells. Adv. Bot. Res. **1**:279-326.

Harris, J. A. 1934. Physico-chemical properties of plant saps in relation to phytogeography. University of Minnesota Press, Minneapolis.

Levitt, J. 1964. Osmotic pressure measuring devices. *In* D. W. Newman. Instrumental methods of experimental biology. The Macmillan Co., New York.

Slatyer, R. O. 1962. Internal water relations of higher plants. Ann. Rev. Plant Physiol. **13**:351-378.

chapter 7

Permeability

A. Diffusion through cell membranes
B. Quantitative relations
C. Mechanism of control by membranes

DIFFUSION THROUGH CELL MEMBRANES

The rate at which solutes diffuse into a cell will depend not only on their concentrations inside and outside the cell but also on the permeability of the cell to them. Thus, from Fick's law of diffusion it follows that if the concentration difference, area, and distance between the two measured concentrations are kept constant and at a standard temperature, D (Fick's constant) will be a measure of the rate of movement specific for each substance investigated. D is called the specific diffusivity of the substance. However, D will be constant for a specific substance only when it diffuses through a specific medium, for example, water. Yet this specific diffusivity of a substance in water will not give the rate of its diffusion into a living cell, even though the water content of the cell may be 85% to 95%. This is because the protoplast is surrounded by an apolar structure of very low water content called a membrane. Diffusion through membranes depends on their permeability. In its simplest sense, *the permeability of a structure to a substance may be defined as the specific diffusivity of the substance in the structure.* Permeability, therefore, varies directly with

the solubility of the substance in the structure (as well as inversely with its molecular size as in the case of diffusion in general). Most water-soluble substances have low solubilities in these apolar membranes, and therefore their rates of diffusion into or out of the cell are usually low. When the rate of diffusion of a substance into a cell is measured, the specific diffusivity (D) then becomes a measurement of the permeability of the cell or of its membranes to the substances. It is, then, commonly referred to as the permeability constant of the cell. Measurements are available for the permeability of the cell to many substances (Stadelmann, 1962).

The same principles apply to the solvent as to the solute, and the membrane, therefore, may have its permeability measured to water as well as to the solute dissolved in it. Yet the statement that the cell is semipermeable may seem to imply that it offers no resistance to the movement of water but completely prevents the movement of solutes. Both of these conclusions are incorrect. Thus, if a collapsed but still living cell is allowed to expand in water, the cell wall expands ahead of the protoplasm (Fig. 7-1), showing that it is more permeable to the water than is the protoplast. Direct

65

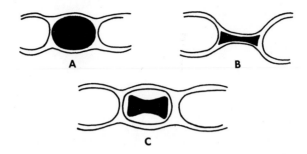

Fig. 7-1. Expansion of a collapsed (i.e., dried) living cell: **A,** in water; **B,** in air; **C,** immediately after transfer from air to water. Final stage as in **A.** (Redrawn from Iljin, W. S. 1930. Protoplasma **10**:379-414.)

measurements on *Nitella* cells have indicated that the wall is 3.6 times as permeable to water as is the protoplasm (Kamiya et al., 1962).

That the living cells are nearly ideally semipermeable with respect to their solute has been shown recently by Scholander et al. (1965, see Chapter 9). By placing all except the cut end of shoots in a bomb and leading in nitrogen gas under pressure, they were able to force as much as 90% of the cell water out of the cut end. This exuded water was almost completely free of solutes because the freezing point was only −0.01° C. Yet there are many substances to which the living cell is highly permeable, although most of these do not normally occur in them. With respect to such substances the cell is differentially permeable rather than semipermeable.

There is direct experimental evidence for the conclusion that the cell's permeability is controlled at the protoplasmic surface. Since the cell wall is a hydrophilic gel, diffusion of solutes through it is relatively unimpeded. This is evident from the fact that hypertonic aqueous solutions of all nontoxic solutes cause a rapid separation of the protoplast from the cell wall in the process of plasmolysis of a living cell, as long as the protoplast is less permeable to the solute than to the water. Similarly, if dyes to which the protoplasm is normally impermeable (or nearly so) are injected into protoplasm, they diffuse freely and rapidly throughout the layer. Finally, there is nothing to prevent free diffusion of solutes throughout the cell sap once they have entered the vacuole. But the same dyes that diffuse readily throughout the protoplasm fail to leave it either by diffusing out through the wall or into the vacuole. Thus the only impediments to free diffusion throughout the cell are the two protoplasmic surfaces, the so-called plasma membranes. The inner one is sometimes called the *tonoplast*, and the outer one, the *plasmalemma*. Although the membranes are too thin to be seen under the optical microscope, they are both readily observed in electron micrographs (Fig. 2-6) and are about 7.5 mμ thick (Sitte, 1958, see Chapter 2). These surface membranes differ from the main body of the protoplasm in their highly lipid (fatty) nature. As a result, they slow up and in many cases completely prevent the passage of solutes in either direction. It is, thus, the lipids of plasma membranes that are responsible for the semipermeability of the living cells. Differences between the permeabilities of the tonoplast and the plasmalemma have often been sought without success. According to recent results, however, the permeability to urea of the plasmalemma of *Nitella* is 30 times that of the tonoplast (Dainty and Ginzburg, 1964*b*).

It has recently been suggested that the

main body of the cytoplasm layer is also responsible for the semipermeability of the cell, but all the evidence is against this. Although the endoplasmic reticulum (ER) forms a system of internal membranes that traverse the cytoplasm, they do not actually compartmentalize the protoplasm. On the other hand, the organelles are themselves surrounded by differentially permeable membranes, but this will not affect movement around them through the main body of the cytoplasm (the groundplasm). Even bodies within an organelle may possess osmotic properties because of the possession of semipermeable membranes, for example, the grana of chloroplasts (Gross and Packer, 1967).

Electron micrographs of the plasma membranes of cells have led to the suggestion that the whole concept of cell permeability must be changed, mainly because these micrographs indicate the presence of visible "pores" in the membranes or connections from the ER to the membrane. It is suggested that substances may therefore enter the cell freely through these pores or through the ER. More recent observations, however, indicate that this earlier interpretation of the electron micrographs was erroneous, and the ER does not connect with the plasma membrane (Frey-Wyssling and Mühlethaler, 1965, see Chapter 2). As to the structure of the membrane, this will be considered later.

QUANTITATIVE RELATIONS

In most cases only relative permeability is measured. This is done by using a single cell or kind of cell and a standard concentration gradient. In this way the area, concentration gradient, and membrane thickness in Fick's law are kept constant. Therefore:

$$D = K \frac{s}{t} \quad \text{or} \quad D \propto \frac{s}{t}$$

$$\text{where } K = \frac{x}{a(C_1 - C_2)}$$

D = specific diffusivity of the substance in the cell, or the cell's permeability to the substance
s = amount of substance diffusing (moles)
t = time (sec)

Relative permeability is then measured simply by $\frac{s}{t}$, the amount of substance entering the cell per unit time.

There are three main methods in common use for measuring cell permeability:

1. The deplasmolysis method. The cells are plasmolyzed with a hypertonic solution of the substance. If the solute penetrates, deplasmolysis occurs at a rate proportional to the permeability of the cell to the substance, provided that the other diffusion factors in Fick's equation are kept constant.

2. The chemical analysis method. The cells are immersed in a solution of the substance, and after a standard time the cell sap is removed and analyzed for the solute. This method is most easily used in the case of large-celled algae, (e.g., species of *Chara, Nitella, Valonia,* etc.), which contain as much as a milliliter or more of sap per cell. This sap can be removed for analysis by means of a hypodermic needle or syringe.

3. The isotope method (Brooks, 1951). If the substance used in method 2 is radioactive, the sensitivity of the method is increased by several orders of magnitude, and it has even been possible to determine the rates of penetration into the protoplasm and vacuole separately. Deuterated and tritiated water (DHO and THO) have been used to measure water permeability.

In the case of water a distinction has been made between the values obtained by methods 1 and 3. Method 1 measures hydraulic conductivity or osmotic permeability; method 2 measures diffusional permeability. Theoretically, the two should be identical, and direct measurements with artificial phospholipid membranes show that they are (Cass and Finkelstein, 1967). In the case of cells or tissues, however, differences between the two values have been obtained. However, the difference probably results from experimental error, and diffusional permeability values are commonly a gross underestimate (Kohn and Dainty, 1966). Recent measurements have, indeed, shown that the osmotic and diffusional permeability constants of the cell are identical (Gutknecht, 1967).

By means of such methods it can be shown that gases penetrate freely, and all small molecules penetrate rapidly. This holds true as long as the molecular weight is not greater than 50 to 60 (e.g., water, ethyl alcohol, ethylene glycol). Electrolytes appear to be exceptions. Although in many cases their molecules are small, they penetrate slowly or not at all. But it must be remembered that electrolytes dissociate into ions that are highly hydrated and therefore form rather large particles (larger than the undissociated molecules) since each ion must diffuse together with its shell of water molecules. The charge on the ions may perhaps also impede penetration. As a result, weak acids and bases penetrate more rapidly than strong acids and bases because the former consist mainly of undissociated and therefore unhydrated (and uncharged) molecules. The solubility of the undissociated electrolytes in the lipid membrane is, therefore, greater than that of the ions.

In the case of nonelectrolytes each substance penetrates at a rate essentially independent of the presence of other substances. In the case of electrolytes this is far from true. Monovalent cations penetrate much more rapidly from single salt solutions than when a salt of a divalent cation is also present. In fact, single salt solutions may be toxic, although the two-salt solutions are quite harmless. Thus a solution of potassium chloride or sodium chloride (or other salts of potassium and sodium) may cause swelling of the cytoplasm after rapid penetration, accompanied by vacuole contraction because of the transfer of water from the vacuole to the swelling cytoplasm (Fig. 7-2). This may eventually lead to injury or death. If, however, one tenth as much calcium chloride (on a molar basis) is added to the potassium chloride or sodium chloride solution, no such extreme swelling occurs; and the cells may remain alive for days. This is because the Ca^{++} reduces or prevents the penetration of the K^+ or Na^+. Such effects of ions on each other are known as *antagonism*. The nontoxic solution of the two salts is called a *balanced solution*. It is interesting to note that potassium chloride and sodium chloride penetrate the cytoplasm rather readily in single salt solutions but do not seem to enter the vacuole as readily. This is sometimes taken as evidence of a difference in permeability between the outer and inner plasma membranes, but it may simply be caused by the great increase in water imbibed by the protoplasmic colloids. Conversely, it has also been suggested (Collander, 1959) that the plasma membrane is actually completely impermeable to the ions. From this view, ion antagonism is not a lowering of the normal permeability of the cell but is a protection of the membrane from injury by the unbalanced, single salt solution, which penetrates only after injuring the membrane.

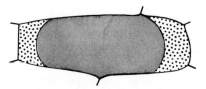

Fig. 7-2. Penetration of potassium into cytoplasm indicated by *vacuole contraction*. This is caused by the increased hydrophily of the cytoplasm, causing movement of water from the vacuole (normally almost filling the cell, see Fig. 2-1) to the cytoplasm (normally an almost invisibly thin layer).

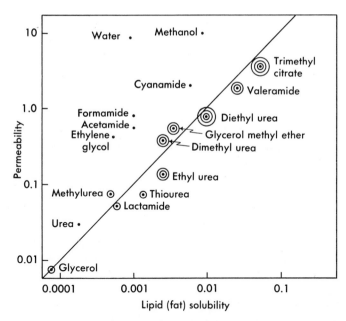

Fig. 7-3. Permeability of *Chara* cells to various organic nonelectrolytes. Molecular size is indicated by the number of circles. (From Collander, R. 1949. Physiol. Plant. 2:300-311.)

Ion antagonism can be explained on the basis of the colloidal properties of the membrane (Booij and de Jong, 1954, see Chapter 5). Ion antagonism occurs between two cations if the concentration necessary to reverse the negative charge of the colloid is more than 10 times as high for one ion as for the other. This antagonism between two ions may occur for one biological colloid and not for another. Phospholipids show the greatest spread of activity between ions, and therefore, antagonism is well developed in the case of these substances. They are the only biological colloids in which antagonism is readily observed between Na$^+$ and Ca^{++}. The existence of such antagonism is, therefore, evidence that cell permeability is controlled by phospholipids. There are also other reasons why phospholipids are ideal compounds for biological membranes. They consist of both apolar (lipid-soluble) and polar (water-soluble) groups in the same molecule. Surface-active molecules of this type can form structures of colloidal size

called micelles, in which the lipid groups form a lipid phase, separated from the surrounding aqueous medium by the water-soluble groups. This applies to both plasma membranes.

The permeability of cells to large molecules (i.e., with molecular weights above 50 to 60) varies from the highest permeability (nearly instantaneous penetration) to zero permeability, depending on the substance. Those large molecules that are lipid (fat) soluble penetrate rapidly, and the more lipid soluble they are the more rapid is the penetration (Fig. 7-3). This relation, explains the accumulation of acids in the vacuole. Those commonly found in the vacuole (oxalic, malic, citric acids) are lipid insoluble and, therefore, cannot leak back into the cytoplasm (Collander, 1959). Lipid-soluble acids (acetic, lactic, pyruvic acids) are never found in the vacuole. Even the same substance will penetrate more rapidly under conditions that increase its lipid solubility. The permeability of cells to organic acids, for instance, increases with a decrease in pH because of the decrease in dissociation of the acid and the consequent increase in concentration of the lipid-soluble molecular form.

MECHANISM OF CONTROL BY MEMBRANES

These results, taken in conjunction with the aforementioned rapid penetration of small molecules that are not lipid soluble, have led to the suggestion that the semipermeable membrane is a lipid sieve; that is, it consists of relatively large fatty particles separated from each other by much smaller aqueous pores (Fig. 7-4). According to the *lipid-sieve theory* of permeability, lipid-soluble molecules of any size pass readily through the fatty particles, and only the smallest lipid-insoluble molecules pass through the aqueous pores.

This concept is simplified and does not agree with the most popular current diagram of the membrane, proposed by Davson and Danielli (1943). They proposed that the plasma membrane consists of a bimolecular leaflet of phospholipids. The nonpolar portions are pictured as oriented within the leaflet, perpendicular to the plane of the membrane; the polar portions must therefore point outward on both sides. The leaflet is supposed to be sandwiched between two layers of protein (Fig. 7-5). Although their concept was based primarily

on permeability results, it has since been supported by electron microscopic and x-ray diffraction data. In further support of their concept, Danielli (1966) has calculated that a bimolecular membrane has less surface-free energy than that of any other thickness and, therefore, is the most stable and is to be expected from the laws of thermodynamics.

There are, however, objections to the Davson and Danielli concept and many facts with which it does not agree (Korn, 1966). Furthermore, some recent electron micrographs have led to a picture in complete conformity with the lipid-sieve theory —a layer of lipid globules stabilized by rows of proteins between them (Sjöstrand, 1963). Recent improved electron micrograph methods have actually revealed a surface sculpturing of the membrane (Frey-Wyssling and Mühlethaler, 1965, see Chapter 2), which may be interpreted according to the lipid-sieve theory. It has also been suggested (Kavanau, 1965) that the membrane is dynamic in living cells, and, therefore, the "holes" may conceivably form and re-

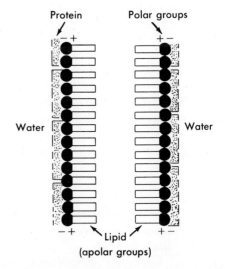

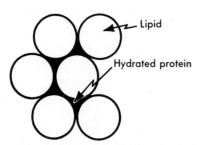

Fig. 7-4. Simplified diagram of the possible submicroscopic structure of the plasma membrane. (Redrawn from Scarth, G. W., and F. E. Lloyd. 1930. Elementary course in general physiology. John Wiley & Sons, Inc., New York.)

Fig. 7-5. Structure of plasma membrane. (From Davson, H., and J. F. Danielli. 1943. The permeability of natural membranes. Cambridge University Press, New York.)

form. Recent evidence (Gutknecht, 1967), however, indicates that no actual "water-filled pores" occur in the membrane, as has been postulated on the basis of erroneous differences between osmotic and diffusional permeability to water (see earlier).

Many attempts have been made to produce artificial membranes having the properties of the plasma membrane. Surprising success has been achieved in some cases. Thus, a bimolecular lipid membrane consisting of phosphatidylcholine, cholesterol, and n-decane has been found to possess a permeability to water well within the range of most living cells (Hanoi and Haydon, 1966). Similar results have been obtained with artificial phospholipid membranes (Cass and Finkelstein, 1967). In spite of this similarity to the plasma membrane of cells, no evidence could be obtained for aqueous pores in them.

The lipid-sieve theory has proved to be a very useful explanation of permeability phenomena, and the relative rates of penetration of different substances can usually be explained in this way. But there are exceptions. Urea (molecular weight 60, lipid insoluble) penetrates some cells more rapidly than does glycerol (molecular weight 92, lipid insoluble) in accordance with theory. Yet there are cells in which the relation is reversed. Such cells usually have low permeabilities to other amides beside urea and are therefore called amidophobic. Differences in relative permeabilities of these two substances may even occur in the same cell at different times (Mayer, 1965).

Modern workers have ascribed many other functions to the plasma membrane beside the control of cell permeability. It is believed to function in protein synthesis, energy transfer, active transport of substances (see Chapter 8), and sometimes in other ways (Korn, 1966). Obviously, such properties are possible only if proteins and other substances are included in the term *membrane*. The lipids alone could not account for these properties.

Many substances when applied to cells can greatly alter permeability. Some of these effects are readily explainable by the lipid-sieve theory. Anesthetics such as ether and chloroform are lipid solvents and therefore would be expected to alter cell permeability because of their solubility in the lipid plasma membrane. This would increase the lipid phase and presumably decrease the cell's permeability to polar substances. If sufficient anesthetic is available to dissolve away some of the lipid membrane, the effect would be the opposite. Thus, such substances may be expected to produce opposite effects depending on the concentrations used. On the other hand, azide (a respiratory inhibitor) would not be expected to affect permeability, yet it decreases cell permeability to glycerol and sucrose (Burström, 1962).

Cells of different plants differ in permeability to a substance. Thus, the values for permeability to water usually range from 1×10^{-4} to 15×10^{-4} cm/sec in different cells (Stadelmann, 1963). Even the permeability of a specific cell to a specific substance may not be constant. Besides changes induced by the ions themselves (antagonism, see earlier), changes may occur, for instance, with the season. The permeability to water increases during fall and winter in many plants (Levitt, 1956). However, this is true only of polar substances, which according to the lipid-sieve theory, "leak" through the smaller aqueous phase of the membrane, the proportion of which is presumably readily altered. Since the plasma membrane is made up largely of lipids, even doubling or trebling the aqueous phase will not change the lipid proportion appreciably, and the cell will be highly permeable to lipid-soluble substances under any conditions.

The fact that the cell is differentially rather than perfectly semipermeable has been clearly shown by the surprisingly large amounts of substances that can be leached from living leaves, both naturally by rain and artificially. Some of this material, of course, comes from dead cells, and the amount leached from the living cells is probably less than 1% of the total. As mentioned earlier, pressure can also remove very small amounts of substances.

QUESTIONS

1. What is the relation between diffusion and permeability?
2. Is the cell perfectly semipermeable?
3. Is the cell wall differentially permeable?
4. Is the protoplasm as freely permeable to water as is the wall?
5. Where is the property of semipermeability located?
6. What is the name of the structure involved?
7. What is the structure's chemical nature?
8. How permeable is the cell to gases?
9. How permeable is the cell to small molecules?
10. How can you prove that a solute does enter a cell?
11. Is the cell as permeable to all substances of small molecular size?
12. How would the permeability to ions relate to the lyotropic series?
13. Does the rate of penetration of a substance depend on the presence or absence of other substances?
14. What is antagonism?
15. What is a balanced solution?
16. Is the cell impermeable to all large molecules?
17. What is the name of the theory that attempts to explain permeability to any substance?
18. Is the permeability of a specific cell to a specific substance constant? Explain.

SPECIFIC REFERENCES

Brooks, S. C. 1951. Penetration of radioactive isotopes, P^{32}, Na^{24}, and K^{42} into *Nitella*. J. Cell. Comp. Physiol. **38**:83-93.

Burström, H. 1962. Influence of azide on the permeability of *Rhoeo* cells. Indian J. Plant Physiol. **5**:88-96.

Cass, A., and A. Finkelstein. 1967. Water permeability of thin lipid membranes. J. Gen. Physiol. **50**:1765-1784.

Dainty, J., and B. Z. Ginzburg. 1964*a*. The measurement of hydraulic conductivity (osmotic permeability to water) of internodal characean cells by means of transcellular osmosis. Biochim. Biophys. Acta **79**:102-111.

Dainty, J., and B. Z. Ginzburg. 1964*b*. The permeability of the cell membranes of *Nitella translucens* to urea, and the effect of high concentrations of sucrose on this permeability. Biochim. Biophys. Acta **79**:112-121.

Danielli, J. F. 1966. On the thickness of lipid membranes. J. Theor. Biol. **12**:439-441.

Gross, E. L., and L. Packer. 1967. Ion transport and conformational changes in spinach chloroplast grana. Arch. Biochem. **121**:779-789.

Gutknecht, J. 1967. Membranes of *Valonia ventricosa;* apparent absence of water-filled pores. Science **158**:787-788.

Hanoi, T., and D. H. Haydon. 1966. The permeability to water of bimolecular lipid membranes. J. Theor. Biol. **11**:370-382.

Iljin, W. S. 1930. Die Ursachen der Resistenz von Pflanzenzellen gegen Austrocknen. Protoplasma **10**:379-414.

Kamiya, N., M. Tazawa, and T. Takata. 1962. Water permeability of the cell wall in *Nitella*. Plant Cell Physiol. **3**:285-292.

Kohn, P. G., and J. Dainty. 1966. The measurement of permeability to water in disks of storage tissue. J. Exp. Bot. **17**:809-821.

Korn, E. D. 1966. Structure of biological membranes. Science **153**:1491-1498.

Levitt, J. 1956. The hardiness of plants. Academic Press, Inc., New York.

Mayer, E. 1965. On pore permeability of protoplasm. Protoplasma **60**:159-161.

Sjöstrand, F. 1963. A new repeat structural element of mitochondrial and certain cytoplasmic membranes. Nature (London) **199**:1262-1264.

Stadelmann, E. J. 1963. Comparison and calculation of permeability constant of water. [Transl. title.] Protoplasma **57**:660-718.

GENERAL REFERENCES

Collander, R. 1949. The permeability of plant protoplasts to small molecules. Physiol. Plant. **2**: 300-311.

Collander, R. 1959. Cell membranes; their resistance to penetration and their capacity for transport, p. 3-104. *In* F. C. Steward. Plant physiology. Vol. II. Academic Press, Inc., New York.

Davson, H., and J. F. Danielli. 1943. The permeability of natural membranes. Cambridge University Press, New York.

Kavanau, J. L. 1965. Structure and function in biological membranes. Holden-Day, Inc., San Francisco.

Scarth, G. W., and F. E. Lloyd. 1930. Elementary course in general physiology. John Wiley & Sons, Inc., New York.

Stadelmann, E. J. 1962. Permeability, p. 493-528. *In* R. A. Lewin. Physiology and biochemistry of algae. Academic Press, Inc., New York.

Strugger, S. 1949. Praktikum der Zell- und Gewebsphysiologie der Pflanze. Springer-Verlag, Berlin.

chapter **8**

Absorption

A. Solute absorption
   1. Passive absorption
   2. Active absorption
B. Water absorption
C. "Negative" absorption, or loss

With the help of the physical principles of diffusion, it should now be possible to attack the transfer of substances in the plant. There are three links in the chain of movement:

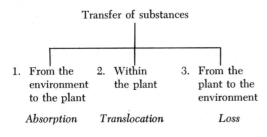

The first of these, absorption, is the entrance of substances into the plant. It is convenient to exclude gas absorption since the mechanism is the same as for gas loss (see Chapter 11).

SOLUTE ABSORPTION

Permeability, by definition, controls entrance into any system only if it is caused by diffusion. Thus differential permeability of the cell completely prevents some substances from diffusing in while permitting others to enter at rates that depend on the permeability of the cell to them. This is true not only in the laboratory but under natural conditions as well. Weed killers, for instance, when sprayed onto plants will enter the cells and produce the damage much more effectively if they are lipid soluble. This is why the lipid-soluble ester is often more effective than the lipid-insoluble salt of the weed killer.

However, substances that cannot diffuse into the cell may also enter. Colloidal substances such as proteins (e.g., ribonuclease) have been shown to enter cells, perhaps by first being converted to lipoproteins. Microscopically visible particles that are too large to diffuse and are lipid insoluble may be coated by lipid, and if brought in contact with the protoplast surface of some plant cells, may then pass through it into the protoplasm. Such plant cells resemble animal cells that ingest (eat) particles by *phagocytosis*. Recent evidence indicates that certain plant cells may ingest solution by a similar method called *pinocytosis* ("drinking")—an infolding of the membrane until it encircles the liquid by pinching off the fold and reconstituting the membrane (Frey-Wyssling and Mühlethaler, 1965, see Chapter 2).

When absorption is simply caused by diffusion, the cell does not participate actively in the process, which is then called *passive absorption*. Phagocytosis and pinocytosis differ from diffusion by the active participation of the cell in the absorption process; that is, the cell uses up energy and does work. Under normal conditions, however,

74

the plant protoplast is surrounded by a cell wall that does not permit visible particles to come in contact with the protoplast surface; nor would it favor pinocytosis, which has been shown only in the case of free tapetal cells suspended in natural fluid and free protoplasts devoid of cell walls (Cocking, 1966). There are, however, other methods by means of which the cell can actively control the absorption process. The main problem, then, is to find out what portion of the absorption of free molecules and ions is passive and what portion is active.

Passive absorption

If passive absorption is a simple diffusion process, the maximum absorption of a substance will be reached when its concentrations inside and outside the cell are equal. Yet substances are commonly "accumulated" by plants as a result of passive absorption; that is, they attain a higher concentration inside than outside the cell. Furthermore, some substances are accumulated in the sense of attaining a larger quantity per unit cell volume, even if not accumulated in the dissolved state. Passive absorption may result in both of these kinds of accumulation by one of the following three methods.

Entering molecules ionize. The gradient of molecular concentration can be maintained by the ionization of entering molecules. If the membrane is impermeable to the ions, absorption of the molecules will continue without loss of the ions, which will then accumulate. The end result will be an accumulation of the substance inside the cell, as in vital staining (Fig. 8-1). The cell is permeable to the neutral red (NROH) molecule because it is a lipid-soluble substance, but is impermeable to the ion (NR+), which is lipid insoluble and large (molecular weight 289). The dye molecule is essentially undissociated at pH 8 and above, mainly dissociated at pH 5.5 and below. The two forms are in approximately equal concentrations at pH 7. If the pH

of the staining solution is about 7 to 8, the dye molecules continue to enter the vacuole and immediately on entering, dissociate because of the acidity of the vacuole. This maintains the molecular concentration gradient, and the dye accumulates as the ion, which cannot leak out. Accumulation stops when the concentration of undissociated molecules is the same inside and outside the vacuole.

Entering solute particles become insoluble either by precipitation or by adsorption on colloidal particles. The concentration gradient in this way will be maintained, and the substance will accumulate in the undissolved form. This occurs when cells containing colloidal tannin accumulate a vital stain in the vacuole; accumulation is then more intense than by the first method.

Entering ions are accumulated electrostatically. As just indicated, absorption is passive if the cell does not participate actively in the process. The substance must, therefore, move in passively if (1) it is at a higher level of free energy outside than inside the cell, and (2) the cell is permeable to it. The rate of absorption will be quantitatively related to these two factors. If the cell is sufficiently permeable to the substance, only the free energy levels need be considered. In the case of a substance in the molecular state, its free energy is caused by the thermal agitation of the molecules and is, therefore, given quantitatively by the chemical potential of the substance. The rate of movement (or diffusion) is proportional to the chemical potential gradient. But ions possess electrical as well as thermal energy. As a result, there is an electrical potential difference between the two sides of a cell membrane, usually of 10 to 100 mv, with the inside of the cell negative to the outside (Curran, 1963). This is caused by the very high electrical resistance of the lipid plasma membrane. If the membrane is 100 Å thick, the potential difference across it must be $10^4 - 10^5$ v/cm, an extremely high value. Thus, there are two

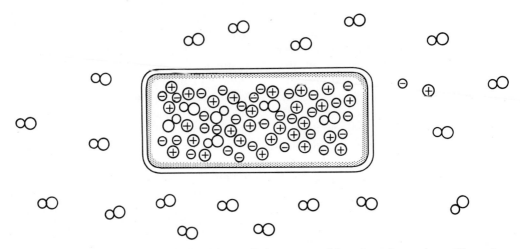

Fig. 8-1. Accumulation of neutral red by a cell from a very dilute dye solution (e.g., 25 ppm) resulting from passive absorption in the molecular (◯◯) form followed by dissociation (⊖⊕). Accumulation continues until the concentration of *undissociated molecules* inside the cell equals that outside it. At this point, total dye (dissociated plus undissociated) is many times as concentrated inside as outside the cell.

"forces" tending to move an ion (Dainty, 1962):(1) that resulting from its chemical potential gradient and (2) that resulting from its electrical potential gradient. Their movement will therefore depend on the algebraic sum of these two quantities, just as the movement of water molecules was seen to depend on the algebraic sum of osmotic potential and pressure. The algebraic sum of the forces acting on ions is called the electrochemical potential gradient:

Electrochemical potential gradient = Chemical potential gradient + Electrical potential gradient

It is therefore possible for a cell to be in equilibrium with its surrounding solution, although the ion concentrations (or chemical potentials) differ. A species of cation, for instance, can be at equilibrium, although it is at a higher concentration inside than outside the cell because the cell contents have a negative electric potential relative to the external solution. But the sum total of these two potentials for the ion—its electrochemical potential—must be the same inside and outside the cell at equilibrium. In other words, the outwardly directed concentration gradient must be equal to the inwardly directed electrical gradient if electrochemical equilibrium prevails. This is analogous to the condition for the cell's water (Fig. 8-2). At equilibrium the cell's osmotic potential may be lower (or its concentration may be greater) than that of the surrounding solution, provided that it is counterbalanced by wall pressure; but the water potential must be the same inside and outside the cell.

A chemical potential gradient between a cell and its surrounding solution is readily produced by a concentration difference between the two. The electrical potential gradient is somewhat more complicated (Dainty, 1962), and only the simplest aspects will be considered here. When a cell is immersed in a solution of electrolyte, each ion tends to move according to its own concentration gradient. However, because of the electrical attraction between oppositely charged ions, they must enter either by (1) exchanging for others of the same

Water

$$\frac{\text{Water potential}}{\text{inside cell}} = \frac{\text{water potential}}{\text{outside cell}}$$

$$\psi_i = \psi_o$$

π_i may be greater or less than π_o

if P_i is less or greater than P_o

where ψ = water potential
π = osmotic potential
P = wall pressure

Ions

$$\frac{\text{Electrochemical}}{\text{potential inside}} = \frac{\text{electrochemical}}{\text{potential outside}}$$
cell   cell

$$\bar{\mu}_i = \bar{\mu}_o$$

μ_i may be greater or less than μ_o

if E_i is less or greater than E_o

$\bar{\mu}$ = electrochemical potential
μ = chemical potential
E = electric potential

Fig. 8-2. Conditions for diffusion equilibrium.

sign or else (2) moving in pairs (Fig. 8-3). The large, negatively charged protein ions (anions) cannot pass through the plasma membrane, whereas the inorganic ions, both of the same sign (anions) and of the opposite sign (cations), can pass, although very slowly. The protein anions inside the cell must be electrically balanced by enough cations to supply an equal number of opposite charges. If now the cell is bathed in a solution containing the same kind and concentration of cation as is inside the cell, plus an anion not found inside the cell (or present there in lower concentration), the new anion will tend to diffuse in. But electrical balance must be maintained because although the more mobile ion of a pair may move slightly ahead of the other, thus converting its thermal energy into a small electric potential *(diffusion potential)*, it cannot move completely out of the sphere of influence of the other ion unless work is done on it. Such a separation would result in a net increase in free energy—an impossibility according to the second law of thermodynamics. Consequently, for each anion that diffuses in, a cation of equal charge tends to be dragged in against its

concentration gradient. Simple diffusion equilibrium therefore cannot prevail, and the cation will reach a concentration inside the cell greater than that outside the cell. On the other hand, the anion diffusing into the cell will reach a concentration less than that outside the cell because of the equal number of ion pairs moving in and out of the cell. This electrochemical equilibrium, which is controlled by electrical as well as diffusion phenomena, is known as the *Donnan equilibrium*. If the cell is in a solution of a salt MA that dissociates into M^+ and A^-, the Donnan equilibrium can be represented quantitatively in simple form:

$$\frac{[M^+]_i}{[M^+]_o} = \frac{[A^-]_o}{[A^-]_i}$$

where $[M^+]_i$ = concentration of diffusible cation inside the cell

$[M^+]_o$ = concentration of diffusible cation outside the cell

$[A^-]$ refers similarly to the diffusible anion

An accumulation of M^+ inside the cell can occur only because of the presence of a nondiffusible ion of opposite charge (R^- in Fig. 8-3), and it must be accompanied by a corresponding negative accumulation of anion.

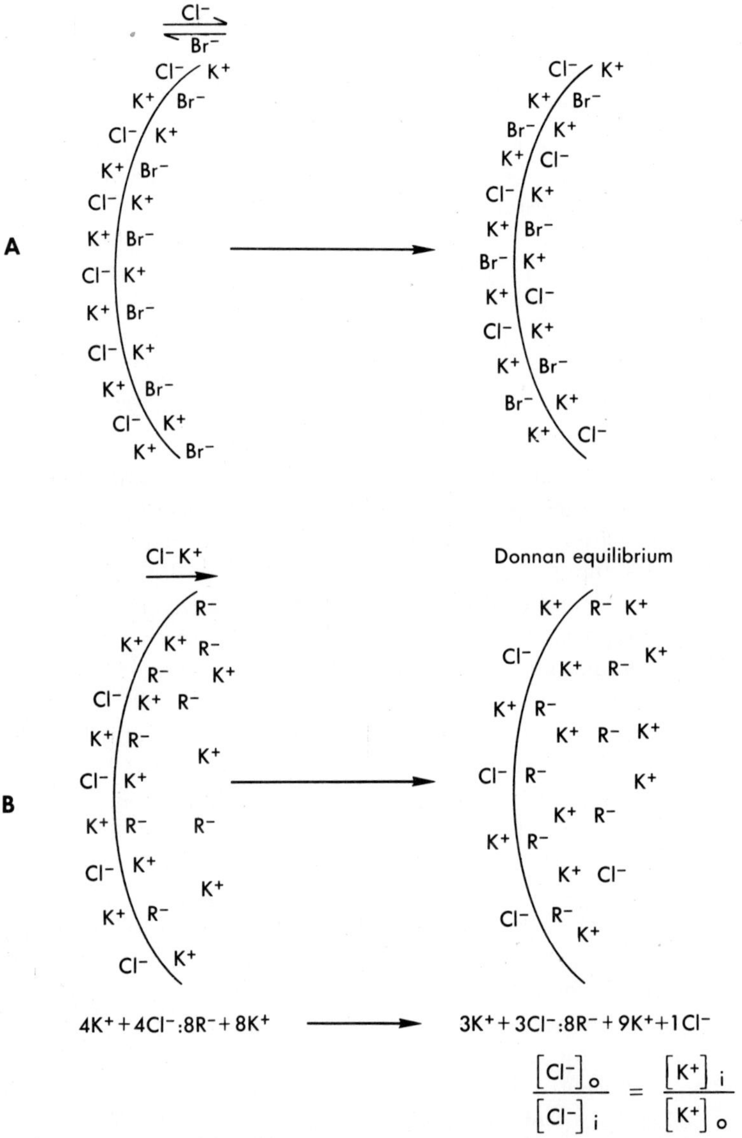

Fig. 8-3. Passive absorption of ions and Donnan equilibrium. **A,** Ion exchange. There is diffusion of Cl^- into the cell without disturbing the electrical neutrality because of a simultaneous and equal diffusion of Br^- out of the cell. **B,** Ion pair diffusion. Ion exchange is prevented because of the presence of nondiffusible ions (R^-) inside the cell. At equilibrium the ratio of Cl^- concentration outside to that inside the cell $\dfrac{[Cl^-]_o}{[Cl^-]_i}$ equals the ratio of K^+ concentration inside to that outside the cell $\dfrac{[K^+]_i}{[K^+]_o}$.

The actual electrical potential gradient associated with accumulation by the Donnan equilibrium can be calculated from the Nernst equation, which in its simplest form is given as

$$E = -59 \log \frac{[M^+]_i}{[M^+]_o}$$

where E = the electric potential difference between the cell and its external solution (mv). Other ions present in the same solution can be treated in the same way (Higinbotham et al., 1967). It is therefore possible to determine from the preceding equation whether or not a specific ion is at electrochemical equilibrium inside and outside the cell.

Because of this Donnan equilibrium, K^+ ions may accumulate in algal cells to 30 times their concentration in the external solution, although at electrochemical equilibrium inside and outside the cells. (Dainty, 1962). The actual accumulation will, of course, vary with the cell and the external concentration.

Active absorption

Passive absorption is therefore unable to lead to an accumulation of solutes in the molecular state since molecular diffusion can lead to no more than an equal concentration inside and outside the cell. As a result of the additional electrical energy of ions, it can, however, lead to their accumulation because of the presence of nondiffusible protein ions inside the cell. If, however, the accumulation is greater than can be accounted for by the Nernst equation, the excess must be caused by some other source of energy—specifically the metabolic energy of the cell. When this condition prevails, the absorption that occurs at the expense of metabolic energy is called active. Unfortunately, it is difficult to apply the Nernst equation to higher plant cells since this requires measurement of the electrical potential between the inside and the outside of a cell (Higinbotham et al., 1967).

In practice the existence of active absorption for any one substance can be demonstrated by any one of the following four criteria:

1. If the rate of absorption is too rapid to be explained by passive absorption, that is, by the permeability of the cell and the chemical or electrochemical potential gradient of the substance
2. If the steady state electrochemical (or in the case of uncharged particles, the chemical) potential is higher inside than outside the cell
3. If the absorption is quantitatively related to the expenditure of metabolic energy
4. If the *mechanism* of absorption can be demonstrated to depend on cell activity

The following discussion gives some of the evidence for all four of these criteria.

1. Since this kind of active absorption is not against a diffusion gradient but simply involves a more rapid rate of absorption than can be accounted for by diffusion, it has been called "facilitated diffusion" (see Hogben and Adrian, 1963). Many substances are readily absorbed by living cells that are almost impermeable to them, according to standard tests—substances with molecular (or hydrated) weights greater than 50 to 60 but insoluble in lipids. This includes many of the substances most important to the plant, for example, sugars and many inorganic salts. That these substances do, indeed, fail to penetrate cells in sections of tissue is shown by the fact that one can plasmolyze the cells in solutions (e.g., of dextrose or calcium chloride), and no deplasmolysis will occur even over a period of several days. On the other hand, that they are absorbed by the living cells of the unsectioned plant can be shown by feeding them to the roots or even by immersing tissue slices (e.g., potato tuber, carrot root, etc.) in these solutions and bubbling air through the solutions. In periods of 24 to 48 hr very considerable amounts will go

into the living cells. This absorption is therefore something more than the result of diffusion and cell permeability.

2. Free ions are usually accumulated by living (and particularly by actively growing) cells until they are present in much higher concentrations in the cell sap than in the surrounding medium (Table 8-1). Although electrochemical potentials were not measured in these cases, the K^+ concentration inside the cell is much higher than the accumulation found by modern methods at electrochemical equilibrium, at least in the case of *Nitella* in pond water. Furthermore, both anions and cations are accumulated, whereas the nondiffusible protein anions in the cell can account only for the accumulation of the oppositely charged cations. Finally, the Donnan equilibrium would lead to a greater accumulation of divalent and trivalent ions than of monovalent ions, whereas the opposite occurs. Consequently, at least some of the ions must be at higher electrochemical potentials inside than outside the cell. Direct measurements of electrochemical potential have shown that this does not necessarily hold for all ions and under all conditions. Thus, when Cl^- ions occurred at a higher potential inside the *Nitella* cell, Na^+ ions were at a lower potential inside than outside the cell (Dainty, 1962). It has therefore been concluded that the cell has inwardly directed "pumps" for some ions and outwardly directed pumps for others. The latter would then lead to negative active absorption (excretion). In the case of pea and oat roots, none of the measured ions (K^+, Na^+, Mg^{++}, Ca^{++}, NO_3^-, $H_2PO_4^-$, $SO_4^=$) was in equilibrium, and none appeared to move passively (Higinbotham et al., 1967). As in the alga *Nitella*, both inwardly (for the anions) and outwardly (for Na^+, Mg^{++}, and Ca^{++}) directed pumps appeared to occur. In the case of sunflower, the anions were found to be actively absorbed, the cations passively absorbed (Bowling, 1966).

Table 8-1. Ratio of sap concentration to concentration in the surrounding medium*

Ion	*Valonia macrophysa* (in sea water)	*Nitella clavata* (in pond water)
Cl	1.03	100.5
Na	0.18	46.1
K	41.6	1065.1
Ca	Very small	13.2
Mg	Very small	10.5
SO₄	0	25.8

*Adapted from Osterhout, W. J. V. 1936. Bot. Rev. 2:283-315.

Animal physiologists use the term *active transport* for those cases where the ion is moved in against an electrochemical potential gradient (Hogben and Adrian, 1963). The plant physiologist has only recently (Etherton and Higinbotham, 1960) begun to measure this gradient because it is so difficult to do in cells of higher plants. In any case the preceding definition of active transport is too narrow for the plant physiologist since the plant is normally not at equilibrium with its environment, nor is the *steady state* between active absorption and passive loss usually achieved. Under normal conditions of plant growth, ions can therefore be actively absorbed without necessarily attaining a higher electrochemical potential inside than outside the cells. Thus NO_3^- and NH_4^+ ions are metabolized inside the cell and cannot be expected to attain either equilibrium or the steady state with the external solution, even though absorbed actively. Ions that are not metabolized are translocated from the root to other parts of the plant, and therefore they, too, cannot possibly attain equilibrium or the steady state with the external medium. The plant physiologist must therefore usually determine whether absorption by the intact plant is active or passive by use of the remaining two criteria.

3. The main source of metabolic energy is the respiratory process. It has been shown

Table 8-2. Absorption of potassium bromide from 0.00075 N solution by carrot discs[*]

Percent O_2	Relative respiration rate	Relative K absorption	Relative Br absorption
2.7	43	22	42
12.2	78	96	86
20.8	100	100	100
43.4	106	117	118

[*]Adapted from Steward, F. C., W. E. Berry, and T. C. Broyer. 1936. Ann. Bot. **50**:345-366.

conclusively that the rate of absorption is frequently proportional to the respiratory rate (Table 8-2). Respiratory inhibitors prevent this kind of absorption. In fact, if the cells are deprived of oxygen, not only are they unable to accumulate ions but they will actually lose much of what they had previously accumulated. This shows that respiratory energy is needed not only to accumulate the ions but also to maintain them in the cells. Thus the plasma membrane may be visualized as a microsieve, through which the accumulated ions are constantly leaking out. Only if the living cell has a pumping system, capable of pumping the ions back into itself as rapidly as they are leaking out, can these high concentrations be maintained.

4. The final method depends on a knowledge of the mechanism of active absorption. If this mechanism could be shown to depend on metabolic activity, this would be final and definitive proof of active absorption. Many suggestions have been made of a "biological pump" that is able to force the ions into or out of the cell. Lundegårdh proposed the first such mechanism, but attempts to support it by direct experimental evidence have failed. Others have proposed that the cell produces "carriers" with different "sites," each of which can specifically capture a particular ion or group of ions. This would account for the selectivity of the process. By selectivity is meant the ability of the cell to absorb one ion in preference to another that is chemically closely related (e.g., K^+ as opposed to Na^+). Such selectivity cannot possibly be accounted for by passive processes. Just what the chemical nature of the carriers is, or where they are located, has not been determined. (Recent evidence has implicated RNA and ATP, see later chapters.)

A mechanism dependent on metabolic energy may be readily visualized. It was mentioned earlier that microscopically visible particles coated with lipid may enter the plasma membrane and therefore the protoplasm. A lipid-coated "carrier" inside the cell (e.g., the mitochondrion) could similarly move out of the internal protoplasm, protruding through the membrane far enough to capture the external ions. Since such particles are too large to diffuse appreciably, they would have to be moved by the cell (e.g., by cytoplasmic streaming), both into the membrane and subsequently back into the cell. Such a mechanism is, unfortunately, difficult to prove. If proved, it would supply direct and unequivocal evidence of active absorption since the cytoplasmic streaming depends on respiratory energy. The formation of "sites" on the "carriers" and the release of the captured ions from them into the protoplasm or vacuole are also conceivable only at the expense of respiratory energy.

All the preceding four methods of investigation have pointed to the requirement of metabolic energy for the normal absorption of ions by cells. Furthermore, it has become obvious that active ion absorption is associated with protoplasmic membranes in general. Thus, the plasma membranes

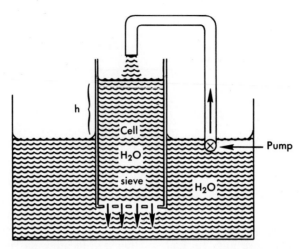

Fig. 8-4. Steady state maintenance of water at a higher free energy level (height **h**) by continuous pumping. The rate of energy expenditure varies directly with the permeability of the sieve.

(specifically the plasmalemma) lead to ion accumulation in the protoplasm as a whole; the mitochondrial membranes lead to ion accumulation in the mitochondrion; and the chloroplast membranes lead to ion accumulation in the chloroplast.

It follows that in order for cells to grow and multiply (and in many cases even to stay alive), they must possess the ability to absorb actively. Although most of the experimental work has used electrolytes, there is some evidence that this is also true of nonelectrolytes such as sugars (Street and Lowe, 1950). On the other hand, the living cell cannot actively absorb substances that are highly lipid soluble since it would require too much energy to do so; these substances would "leak" out through the plasma membrane so rapidly that it would be like keeping a large-pored sieve full by pumping the water back as fast as it leaked out (Fig. 8-4).

Although ions are normally absorbed actively, this does not eliminate passive processes. Cations are adsorbed passively on the root surface and may subsequently be transported actively across the plasma membrane. Similarly, the root contains "free space" into which solutes can diffuse read-

ily. This space accounts for about 10% to 15% of the root volume (Ingelsten and Hylmö, 1961) and consists essentially of the cell walls (Pitman, 1965). Diffusion into this free space must precede active absorption at the protoplasmic surface. It is also probable that beside the active absorption of ions, some passive absorption by the protoplasm and vacuole may also occur, particularly from high external concentrations. This question will be considered later (see Chapter 11).

WATER ABSORPTION

Unlike solutes, absorption of water is purely a passive process, and the cell does not control it by direct expenditure of energy. The reason for this becomes clear from Fig. 8-4. The water potential in the cell can be kept above that of its environment only by the cell continuously pumping water into itself as rapidly as it leaks out through the semipermeable membrane. Since this membrane is highly permeable to water and the specific surface of the cell is considerable, large amounts of energy would have to be expended continuously by the cell. Calculations (Levitt, 1953) reveal that the cell does not release enough

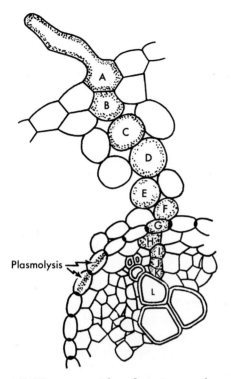

Fig. 8-5. Water potential gradient in roots from a root hair (cell **A**) to a vessel (cell **L**). Water enters by moving along this gradient. Plasmolysis of two endodermal cells is shown to illustrate the strong adhesion of the protoplasts to the Casparian strips. (Redrawn from Priestley, J. H. 1920. New Phytol. **19**:189-200.)

energy to maintain an appreciable increase in this water potential.

Unfortunately, there is some difference in terminology among investigators. "Active" and "passive" water absorption have been used in the past for water uptake resulting from root pressure and transpirational pull, respectively (see later). These terms will not be adopted in this text since (1) it is believed that the same terminology should be used for solutes as for solvents and (2) "active" implies direct participation of metabolic energy.

As in the case of solutes, according to the laws of diffusion, water should enter a cell from any medium in which the chemical potential of the water is higher than that in the cell. Thus, in order for water to diffuse from the root medium through the root epidermis, cortex, endodermis, pericycle, and finally into the vessels, it must follow a water potential gradient (Fig. 8-5). The living cells would affect such water movement by maintaining this gradient. The net water potential gradient would be that between the vessels and the external medium. A hydrostatic (osmotically induced) pressure develops as a result of this difference and is known as the *root pressure* because all the living cells between the xylem vessels of the root and the soil solution act as a single semipermeable membrane. The root xylem is therefore an osmometer immersed in the soil solution, and the maximum osmotic pressure it can develop depends on the concentration of the vessel sap as well as that of the soil solution:

$$P_r = \pi_s - \pi_v$$

where P_r = maximum root pressure
π_s = osmotic potential of soil solution
π_v = osmotic potential of vessel sap

In most cases the vessel sap is very dilute, so that even if the soil solution is pure water, pressures of 1 atm or less are common (Table 8-3). Values as high as 5 to 6 atm have, however, been recorded.

But how can this root pressure be maintained? Since the vessels are nonliving cells, they do not possess semipermeable membranes. It might therefore be expected that the solutes in the vessels would diffuse through the walls into the walls of adjacent cells, and from cell wall to cell wall until they leaked out of the roots into the surrounding medium. This leakage is believed to be prevented by a hollow cylinder of cells, one row thick, surrounding the central stele (in which the vessels are located) and known as the endodermis (Fig. 8-5, *G*). The endodermal cells have lateral walls that are impermeable to water because of impregnation with fatty substances. Sometimes the fatty substances form a ribbon

Table 8-3. Osmotic pressure of tracheal (i.e., xylem) sap[*]

Species	Origin of sap	Osmotic pressure (atm)
Cotoneaster frigida	Centrifuged from stems	0.48–1.08
Cucurbita pepo	Stump exudate	1.9
Fagus sylvatica	Centrifuged from stems	0.26–1.23
Gossypium sp.	Stump exudate (low-salt plants)	0.92
	Stump exudate (high-salt plants)	3.00
Impatiens balsamina	Stump exudate	0.36
Lycopersicon esculentum	Stump exudate	1.5 –2.4
Salix babylonica	Centrifuged from stems	0.41–1.14
Ulmus procera (syn. campestris)	Centrifuged from stems	1.13–3.52
Xanthium strumarium	Stump exudate	0.67
Zea mays	Stump exudate	1.46

[*]From Altman, P. L., and D. S. Dittmer (Eds.). 1966. Federation of American Societies for Experimental Biology, Washington, D. C.

around the cell wall known as the *Casparian strip*. These endodermal cells are alive. Therefore, in order to pass through them, sap solutes would have to penetrate the semipermeable protoplasm since they cannot leak through the lateral fatty cell wall unless they are fat soluble. Even when these cells are plasmolyzed, the protoplasm remains attached to the lateral walls (Fig. 8-5).

A second factor that would oppose maintenance of root pressure is dilution of the vessel sap by the water entering the vessels. This would have to be opposed by more rapid absorption of solutes than of water, in order to ensure a higher concentration in the vessels than in the root medium; or there might also be an active movement of solutes (e.g., a secretion of salts or sugars) into the vessels from the adjacent living cells. Thus active absorption or active secretion *of solutes* must accompany absorption of water if root pressure is to be maintained. Experiments clearly demonstrate the need of oxygen for maximum water absorption by roots (Rosene, 1950). This is not surprising in view of the need for active absorption of solutes in order to maintain root pressure.

The movement of water into a cell may therefore be said to involve *osmotic work*. But it is important to understand the true meaning of this term in order to avoid the misconception that this implies work done on the water molecules by the plant. Osmotic work does not involve a "water pump." The work is performed by the plant in pumping *solute* molecules into itself. This lowers the free energy of the water molecules in the plant, leading to a passive diffusion, or osmosis, of water into the plant. The osmotic work is, therefore, equal to the work done by the water molecules in diffusing into the plant, and it cannot be greater than the work done by the plant in actively absorbing its solutes. The amount of energy involved in osmotic work is small—of the order of 1% of the total expended by the living cell (Robertson, 1941).

Some aspects of root pressure remain to be explained. Although a water potential gradient does normally exist between the external medium and the vessel sap (i.e., the vessel sap does have a lower osmotic potential than the soil solution) and although a cell-to-cell gradient has been found from the epidermis to the endodermis, yet the gradient apparently reverses itself from the endodermis to the vessel sap. It must be realized, however, that all measurements of these quantities are probably meaningless since the steady state between water absorption by roots and water loss by the plant as a whole is maintained in the intact plant under normal conditions. Any sectioning of the plant for experimental determinations destroys this steady state and therefore changes the conditions of the cells.

"NEGATIVE" ABSORPTION, OR LOSS

It has long been known that roots not only absorb substances from the soil but they also give off substances to the soil. This follows from the mechanism of absorption. If passive absorption occurs, it must be from a high electrochemical or chemical potential to a lower one. Consequently, those substances at a higher potential outside the cell will tend to move in; those at a higher potential inside the cell will tend to move out. If active absorption occurs, there must again be a two-way movement. If the movement is carrier mediated, absorption of cations from the soil must be at the expense of H^+ given off into the soil, and the soil may be expected to become more acid. Similarly, if anions are absorbed via carriers, they must be exchanged for other anions (e.g., HCO_3^-), and the soil may be expected to increase in bicarbonates.

But many other substances may also be given off into the root medium, including plant metabolites: sugars, amides, organic acids (Slankis et al., 1964). The quantities, however, are small, requiring labeling with C^{14} for detection. Thus, as many as thirty-five radioactive substances have been detected in the medium in which *Pinus strobus* seedlings were grown aseptically (Street, 1966). These substances that were given off by the roots included a wide variety of metabolites: alkaloids, vitamins, nucleotides, flavones, sugars, auxins, amino acids, organic acids. According to Street, the loss from roots is probably of minor significance in the physiology of higher plants. This release of substances from the plant into its environment is known as *excretion*, or *secretion*. Substances may also be given off from the top of the plant. The excretion of droplets of very dilute solution from a highly turgid plant into a saturated atmosphere is called *guttation*. Nectaries secrete sugar solutions. Insectivorous plants secrete solutions of digestive enzymes. Some halophytes excrete salt. Considerable quantities of both organic and inorganic metabolites may be leached from plants by rain, dew, and mist (Tukey, 1966). As in the case of roots, only a very small fraction of the plant's solutes can be removed in this way. Conversely, dew can be absorbed by leaves, but only in relatively small quantities. The term *secretion* has also been used for the active movement of solutes from the protoplasm into the vacuole of the same cell, and it therefore may imply either absorption or loss of substances.

QUESTIONS

1. Can a cell absorb substances to which it is impermeable?
2. Give some examples.
3. If absorption is caused by diffusion, what will the concentration inside the cell be at equilibrium?
4. If there is more of a substance per unit volume of cell than per unit volume of external solution, can this still be caused by diffusion?
5. How can this be explained?
6. When can you be sure that the substances inside the cell are not there solely as a result of diffusion?
7. What is meant by passive absorption?
8. Can an ion diffuse across a membrane from a region of higher concentration to one of lower concentration (for the same ion) without any other ions moving?
9. What two possibilities are there for such movement?
10. In a nonliving system at equilibrium can a higher concentration of an ion occur on one side of a membrane?

11. What is the name of this phenomenon?
12. What must be present inside the membrane to give rise to this phenomenon?
13. Can it explain the higher concentration of ions inside the cell than outside?
14. If it can, explain why. If it cannot, explain why not.
15. What is meant by active absorption?
16. With what process can we associate absorption of ions by cells against an electrochemical potential gradient?
17. Which ions are absorbed actively—cations or anions?
18. What is the source of the energy for the process?
19. What evidence is there that this is the source of energy?
20. Can lipid soluble substances be actively absorbed?
21. Can water be actively absorbed?
22. What is root pressure?
23. How can it be measured?
24. How is the concentration of the vessel sap maintained?
25. What is the relation of water absorption to oxygen?

SPECIFIC REFERENCES

Altman, P. L., and D. S. Dittmer (Eds.). 1966. Environmental biology. Federation of American Societies for Experimental Biology, Washington, D. C.

Bowling, D. J. F. 1966. Active transport of ions across sunflower roots. Planta 69:377-382.

Cocking, E. C. 1966. Electron microscopic studies on isolated plant protoplasts. Z. Naturforsch. [B] 21b:581-584.

Curran, P. F. 1963. The biophysical nature of biological membranes. *In* R. H. Wasserman. The transfer of calcium and strontium across biological membranes. Academic Press, Inc., New York.

Etherton, B., and N. Higinbotham. 1960. Transmembrane potential measurements of cells of higher plants as related to salt uptake. Science 131:409-410.

Higinbotham, N., B. Etherton, and R. J. Foster. 1967. Mineral ion contents and cell transmembrane electropotentials of pea and oat seedling tissue. Plant Physiol. 42:37-46.

Hogben, C., and M. Adrian. 1963. General aspects of ion transport. *In* R. H. Wasserman. The transfer of calcium and strontium across biological membranes. Academic Press, Inc., New York.

Ingelsten, B., and B. Hylmö. 1961. Apparent free space and surface film determined by a centrifugation method. Physiol. Plant. 14:157-170.

Levitt, J. 1953. Further remarks on the thermodynamics of active (non-osmotic) water absorption. Physiol. Plant. 6:240-252.

Pitman, M. G. 1965. The location of Donnan free space in disks of beetroot tissue. Aust. J. Biol. Sci. 18:547-553.

Priestley, J. H. 1920. The mechanism of root pressure. New Phytol. 19:189-200.

Robertson, R. W. 1941. Studies in the metabolism of plant cells. I. Accumulation of chlorides by plant cells and its relation to respiration. Aust. J. Exp. Biol. Med. Sci. 19:265-278.

Rosene, H. F. 1950. The effect of anoxia on water exchange and oxygen consumption of onion root tissues. J. Cell. Comp. Physiol. 35:179-193.

Slankis, V., V. C. Runeckles, and G. Krotkov. 1964. Metabolites liberated by roots of white pine (*Pinus strobus* L.) seedlings. Physiol. Plant. 17:301-313.

Steward, F. C., W. E. Berry, and T. C. Broyer. 1936. The absorption and accumulation of solutes by living plant cells. Ann. Bot. 50:345-366.

Street, H. E., and J. S. Lowe. 1950. The carbohydrate nutrition of tomato roots. Part II. The mechanism of sucrose absorption by excised roots. Ann. Bot. 14:307-329.

Street, H. E. 1966. The physiology of root growth. Ann. Rev. Plant Physiol. 17:315-344.

Tukey, H. B., Jr. 1966. Leaching and metabolites from above-ground plant parts and its implications. Bull. Torrey Bot. Club 93:385-401.

GENERAL REFERENCES

Dainty, J. 1962. Ion transport and electrical potentials in plants cells. Ann. Rev. Plant Physiol. 13:379-402.

Lundegårdh, H. 1945. Absorption, transport, and exudation of inorganic ions by the roots. Ark. Bot. 32A(12):1-139.

Osterhout, W. J. V. 1936. The absorption of electrolytes in large plant cells. I. Bot. Rev. 2:283-315.

Sutcliffe, J. F. 1962. Mineral salts absorption in plants. Pergamon Press, Inc., New York.

chapter 9

Ascent of sap

A. Bulk (or mass) flow
B. Root pressure as opposed to transpirational pull

BULK (OR MASS) FLOW

After the water has entered the roots it rises to all parts of the plant. This process is known as the ascent of sap. Speeds of as high as 75 cm/min have been recorded (Huber, 1932). Calculations from Fick's law reveal that diffusion can account for only an infinitesimal fraction of this movement (Levitt, 1956). Consequently, another kind of movement must be involved.

When a liquid diffuses, it moves as a result of the thermal agitation of the individual molecules. Diffusion can therefore occur under conditions of constant and uniform hydrostatic pressure. When, however, the liquid is exposed to a greater pressure in one region than in another, movement will occur from the region of higher to that of lower pressure. This movement is called *bulk flow* by physicists, or mass flow by many physiologists.

The bulk flow of a liquid can be represented by:

$$F = kp$$

where F = flux or flow (e.g., cm^3/hr)
p = pressure difference between the two ends of the flowing system (bars)
k = conductivity of the system in which the liquid is flowing (cm^3/hr/bar)

As in the case of electricity, the conductivity factor in water flow (k) is in- versely proportional to resistance to flow (r):

$$k \propto \frac{1}{r}$$

Both the similarity to and the difference from diffusion is obvious if the pressure is expressed as a gradient $\left(\frac{p}{l}\right)$:

$$F = \frac{kp}{l}$$

where l = length of the path of flow or the distance between the two points differing in pressure by p bars.

It is obvious from the preceding that bulk flow differs from diffusion flow in being proportional to the pressure gradient instead of the concentration gradient. Therefore, in order for the sap to flow through the tree, the pressure gradient $\frac{p}{l}$ must be sufficient to overcome the frictional resistance to movement. But when water flows through any tube, the frictional resistance is that between the flowing water and a stationary film of water held at the inner surface of the tube. In other words, the frictional resistance is the viscosity of the water. The relation between flow or flux (F) through a horizontal tube and viscosity is given by Poiseuille's law:

$$F = \frac{pr^2}{8ln}$$

where F = flux

      p = pressure difference at two ends of the tube

      r = tube diameter

      l = tube length

      n = viscosity (0.01 cgs units for water)

Poiseuille's law is simply related to the first equation since it can be rearranged as follows:

$$F = \frac{r^2}{8n} \times \frac{p}{l}$$

It is now obvious that $\frac{r^2}{8n}$ is simply the constant k in the first equation and is therefore a measure of the conductivity of the tube for a specific liquid. Therefore, for any one pressure gradient, the rate of flow is directly proportional to the square of the radius of the tube through which the liquid flows, and inversely proportional to the viscosity of the flowing liquid.

Poiseuille's law can also be rearranged to permit calculation of the pressure necessary to cause a flow under any one set of conditions:

$$p = \frac{8Fln}{r^2}$$

But Poiseuille's law applies to flow through a horizontal tube. Flow up a tree requires an additional pressure to overcome the force of gravity, that is, 1 atm. for every 30 feet of vertical tube length. In the case of sap flowing at the maximum measured rate (75 cm/min) through a tree 30 m (i.e., nearly 100 feet) tall, the necessary pressure for commonly found vessel diameters will be found in Table 9-1.

How do these theoretical values compare with experimentally determined values for trees? Early investigators found that the pressure required to force a flow through a horizontal log is about equal to the pressure required to support a stationary column in the vertical log. Recent measured values have yielded a gradient of 0.1 atm/m (Scholander et al., 1965). Although the rates of flow and the diameters of the vessels through which the major flow occurred were not measured in these cases, the values certainly fall within the calculated range. Consequently, it can be concluded that the ascent of sap follows the laws of bulk flow.

The secret of success of vines is revealed by Poiseuille's law. In spite of the small cross section of their stems, they are able to transport water rapidly to great heights. This results from their unusually large vessel diameters, permitting maximum flow rates. Tube diameter is particularly important to them since the viscosity of liquids is directly proportional to flux velocity. In the case of trees with slow flux velocities, viscosity is so low that xylem elements of small diameter are fully adequate.

Bulk flow as described by Poiseuille's law is similar in some respects to diffusion as

Table 9-1. Pressures required to overcome gravity and frictional resistance to flow in vessels of different diameters (calculated from Poiseuille's law)

Vessel diameter (mm)	Pressure (bars)			
	To overcome gravity	To overcome frictional resistance	Total	$\frac{p}{l}$ (bar/m)
1.0	3	0.12	3.12	0.1
0.2	3	3	6	0.2
0.1	3	12	15	0.5

described by Fick's law. Thus, if we use the symbols for Poiseuille's law, Fick's law becomes:

$$F_d \propto r^2 \frac{C_1 - C_2}{l}$$

Poiseuille's law becomes:

$$F_b \propto r^2 \frac{p_1 - p_2}{l}$$

where F_d = flow caused by diffusion
F_b = bulk flow

In both cases the flux or movement is proportional to the gradient—pressure gradient in the case of Poiseuille's law, concentration gradient in the case of Fick's law. In both cases the flux or movement is proportional to the cross-sectional area and inversely proportional to the resistance to flow. In the case of diffusion the resistance factor is inversely related to the constant D. Because of these similarities, it may sometimes be difficult to distinguish between

the two. For this reason it has been suggested that when water diffuses into a cell, the portion moving through the "microtubes" of the semipermeable membrane is actually a bulk flow (Ray, 1960). More recent results, however, oppose this concept of water-filled pores in the membrane (see Chapter 7).

In general, however, there are three differences between the two kinds of movement (Fig. 9-1):

1. Bulk flow is always associated with a measurable hydrostatic pressure gradient, and diffusion is not.
2. If solutes are present, they move by bulk flow at the same rate and in the same direction as the solvent, but in the case of diffusion they do not.
3. Bulk flow usually involves much more rapid movement than that caused by diffusion since the whole column of

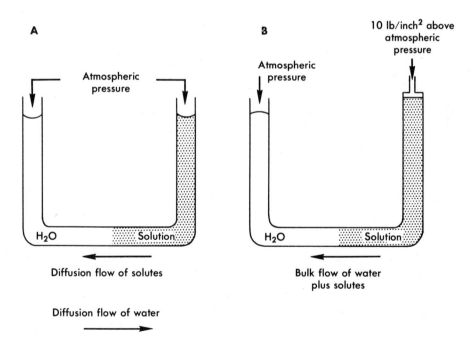

Fig. 9-1. Diffusion flow as opposed to bulk flow. **A,** Diffusion flow: $F \propto \dfrac{\Psi \text{ solution}}{x}$, where x is distance between water and solution. Water and solutes move independently. **B,** Bulk flow: $F \propto p/l$, where 1 is distance between applied pressure (p) and liquid at atmospheric pressure. Water and solute move together and flow is independent of any diffusion within the tube.

water moves at once as a single body. The speed of movement in bulk flow is limited only by the frictional resistance or viscosity of the water and the impressed pressure gradient.

ROOT PRESSURE AS OPPOSED TO TRANSPIRATIONAL PULL

The bulk flow may result either from (1) a push—a larger positive (hydrostatic) force at the base than at the top of the plant, or (2) a pull—a smaller negative force (or tension) at the base than at the top of the plant. The former is caused by root pressure (a rise in pressure at the bottom of the plant); the latter is caused by transpirational pull (a fall in pressure at the top of the plant). In either case, water is flowing against the force of gravity, and 1 atm of pressure difference is required per 30 feet, just to support the column of water. Consequently, for a 300-foot tree a pressure of 10 atm is just sufficient to support the column of water; and in order to overcome frictional resistance (see earlier), according to the available experimental evidence, about another 10 atm is needed if the water is flowing through the tree at a normal rate.

Since the maximum root pressure (P_r) is the excess of the osmotic potential of the soil solution over that of the root:

$$P_r = \pi_s - \pi_v$$

where P_r = pressure difference between base of column and top of column (which is at atmospheric pressure)

π_s = osmotic potential of soil solution

π_v = osmotic potential of vessel sap

Therefore, the flow of water caused by root pressure (F_r) is

$$F_r = k\,(\pi_s - \pi_v)$$

where k is the water conductivity of the root (g/hr/bar, Brouwer, 1953).

Since measurements of the root pressure of trees usually yield values of only 1 to 2 atm, the ascent of sap in tall trees cannot be caused by root pressure. This does not, of course, detract from the importance of root pressure. In grapevines, for instance, the vessels are emptied during winter and are refilled in spring because of a root pressure of 2 to 5 atm (Scholander et al., 1955). At this time there is a gradient of about 0.1 bar/m in the stem. In opposition to the grapevine, however, most trees keep their vessels filled with water throughout the winter and, in fact, have their highest water contents at this time (Gibbs, 1935). This difference is undoubtedly related to the unusually wide vessels in the grapevine.

The most conclusive evidence that the ascent of sap is not commonly the result of root pressure is the fact that tensions or stresses instead of positive forces or pres-

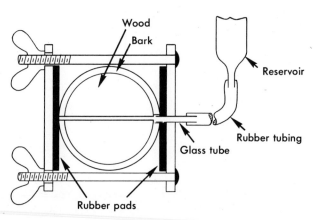

Fig. 9-2. Apparatus used for tree injection. (From Thomas, L. A., and W. A. Roach. 1934. J. Pomol. Hort. Sci. **12**:151-166.)

sures usually exist in the xylem (the water-transporting tissues) under normal growing conditions. This is shown by the absorption of water through bore holes in tree trunks. If such holes are connected by a syphon to an elevated bottle of water (Fig. 9-2), as much as 1 gallon can be absorbed over a 24 hr period by an average-sized apple tree on a warm summer day. On the other hand, when root pressure is operative, sap exudes from such bore holes. The only time such positive forces occur in the xylem is when the leaves are not on the plant (in the case of deciduous plants) or when the plant is in a saturated or near-saturated condition. In the latter case water may actually be exuded from specialized structures in the leaves called hydathodes. This exudation is known as *guttation,* and it may be stopped by reducing the root pressure, for example, by watering the soil with a sugar or salt solution, decreasing the osmotic potential of the root medium. The guttation fluid is not, however, pure water but may contain sugars, amino acids, inorganic elements, and a variety of vitamins (Goatley and Lewis, 1966).

That the stress or tension increases during the day can be shown by the use of a sensitive instrument known as a *dendrograph.* This measures the diameter of the tree trunk and reveals a definite contraction during the day and an expansion at night (Fig. 9-3).

Thus the ascent of sap is usually associated with a pull from above rather than a push from below. If a capillary tube is held vertically in a beaker of water, a simi-

lar pull occurs, resulting in the rise of water in the tube to well above the surface of the water in the beaker. It is easy to calculate whether the rise in the plant can be explained by a capillary pull in the vessels:

$$h = \frac{2S}{rdg}$$

where h = height (cm) supported in the capillary tube
S = surface tension of liquid (dynes/cm)
r = radius of tube (cm)
d = density of liquid (g/cm^2)
g = acceleration caused by gravity (980 cm/sec^2)

A vessel diameter of 0.1 mm (0.01 cm) would account for only a 30 cm rise. This explains why plants like the grapevine, in which the vessels empty in the fall, cannot refill by capillarity and therefore must depend on root pressure to refill them. But in most plants the water system remains unbroken (at least in part, see later) throughout the life of the plant. Consequently, it is not in the vessels themselves that the capillary pull occurs. Since this is a surface tension phenomenon, the pull must occur at the water surface, and since the water system of a plant is continuous from the lowest roots to the highest leaves, the upper surface of the water column is in the leaves. Therefore, it must, be at the outer surface of the leaf (mesophyll) cells that are in contact with the intercellular spaces, that is, in the microcapillaries of their walls. Since these microcapillaries are so fine that they cannot be seen with the ordinary optical microscope, their diameters must be less than 0.1μ. Using this value in the pre-

June 27, 1921

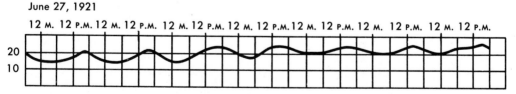

Fig. 9-3. Daily expansion and contraction of a redwood tree, amplified ten times. **M,** noon, **P.M.,** midnight. (Redrawn from MacDougal, D. T., and F. Shreve. 1924. Carnegie Institute of Washington, Pub. No. 350. Washington, D. C.)

ceding equation for capillary height, we find that these microcapillaries are capable of exerting sufficient pull to support a water column 300 m high—3 times the height of the tallest trees!

Unfortunately, physicists have not determined within what limits the preceding equation holds. Some go so far as to state that the rise is basically a push from below by atmospheric pressure and therefore cannot exceed 30 feet. However, this concept is easily shown to be incorrect since the capillary rise occurs just as readily in a vacuum. In the case of microcapillaries, 200 Å (i.e., 0.02μ) in radius, recent measurements by Fedyakin (Derjaguin, 1966) have shown that the calculated value is double the real value. But this still leaves a capillary stress (or transpirational pull) of more than enough to account for the rise to the top of the tallest trees. The classical Askenasy experiment has shown that the rise in purely physical systems can be 2 to 3 times that caused by atmospheric pressure. However, such experiments are difficult to perform and have not succeeded in duplicating the rise of hundreds of feet in tall trees. The theoretical concept is shown in Fig. 9-4. The following three stages in the process may be distinguished: (1) transpiration (the evaporation of water) from the surface of the microcapillaries; (2) capillary rise of water threads resulting from the force of adhesion between the water and the cell wall (i.e., the force of imbibition by the cell wall colloids); and (3) bulk flow—the whole column of water moves en masse because of the force of cohesion between the water molecules. These three stages, of course, occur almost simultaneously. This concept is usually known as Dixon's Cohesion Theory of the ascent of sap, although it was proposed independently by Dixon and Joly in 1894 and by Askenasy in 1895.

Some points are difficult (Greenidge, 1958), although not impossible to explain by Dixon's Cohesion Theory. Even when it is stationary, a column of water 100 m high and 1 inch in diameter is subjected to a downward force of 150 lb because of the gravitational pressure of 10 atm, and if the preceding reasoning is correct, this is counteracted by an upward pull caused by surface tension pull equal to this. The column is therefore being subjected to a powerful force tending to break it. The question is whether a liquid such as water has sufficient

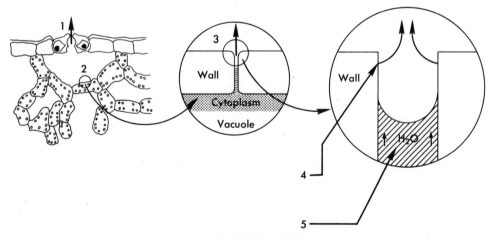

Fig. 9-4. The rise of sap as a result of transpirational pull. *1,* Transpiration (stoma indicated by arrow); *2,* evaporation from microcapillaries into intercellular space; *3,* partial drying of wall surface; *4,* water from film attracted by imbibitional force of partially dried wall; *5,* whole water thread rising because of cohesive force of water.

tensile strength to resist the pull without rupture, and whether the force of attraction between the water and the vessel wall is sufficient to prevent separation from the wall. In other words, are the forces of cohesion and adhesion adequate?

Physical chemists have shown experimentally that pure water has a cohesive force capable of resisting a stress of hundreds of atmospheres (theoretically of the order of 1000 atm). But the water in the vessels is not pure. In addition to solids, gases are present in solution. The reduction in pressure because of the capillary pull may therefore reduce the solubility of these gases until they separate from the liquid, causing rupture of the column. Under conditions of excessive tension, vessels have in fact been found to become gas filled. By means of an apparatus that detects vibrations in the plant, it has recently been possible to demonstrate such "cavitation" in the xylem of many plants (Milburn and Johnson, 1966). But since there are many columns of vessels side by side, it is unnecessary for all of them to be continuous, and therefore the temporary filling of some vessels with gas may not be injurious. When the tension is relieved, by rain or simply at night, the gases go back into solution and the column becomes continuous again. When the diameter of the vessel is very large (e.g., 1 mm), as in grapevine, the bubbles of gas are carried along with the stream and have little or no effect on the water movement (Scholander et al., 1955). Cavitation is actually prevented in the case of gymnosperms even when their xylem sap is frozen, presumably because of their bordered pits (Hammel, 1967).

Many investigators have attempted to measure the stresses in the xylem using ingenious indirect methods, and a wide range of values has been reported (Loomis et al., 1960). Scholander and co-workers (1965) have recently developed an elegant method called the "pressure-bomb" technique (Fig. 9-5). They sever a shoot, al-

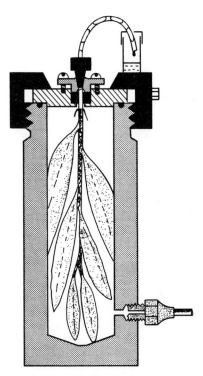

Fig. 9-5. Pressure bomb for the determination of the negative pressure of xylem sap. (From Scholander, P. F., et al. 1966. Plant Physiol. **41**:529-532.)

lowing the tension in the vessels to pull the liquid back, and immediately enclose all except the cut surface in the bomb. They then measure the counterpressure (by means of nitrogen gas) that must be exerted on the shoot to force the liquid back to the cut surface. They discovered stresses as high as 80 atm (Fig. 9-6).

Because of its relation to transpiration (or evaporation), the capillary pull is sometimes called *transpirational pull*. Thus the whole column is pulled upward en masse as though it were a solid wire. This pull, of course, would be still greater than the forces needed simply to support the column since the moving column must also overcome the frictional resistance to movement. Since the column of water moves en masse, this means that the transpirational pull is

94   *Biophysics*

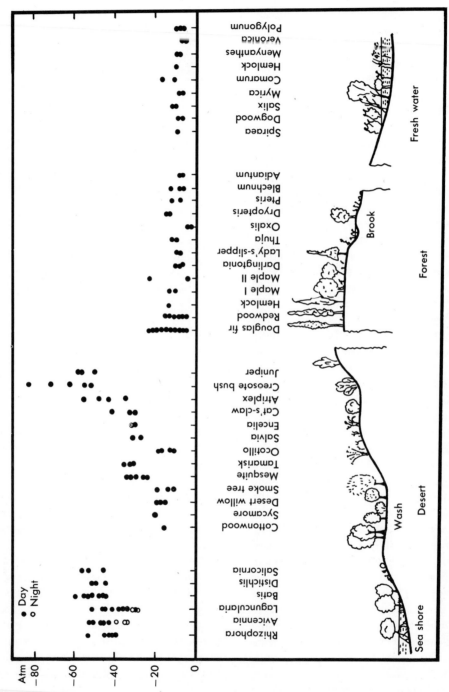

Fig. 9-6. Sap stresses in a variety of flowering plants, conifers, and ferns. Most measurements were taken during the daytime in strong sunlight. (From Scholander, P. F., et al. 1965. Science **148**:339-346. Copyright 1965 by the American Association for the Advancement of Science.)

transmitted from the microcapillaries in the mesophyll cell walls to the protoplasm and vacuole, to the adjacent cells, and to the nearest vessel. It is transmitted down the vessels through the leaf blade, petiole, stem, and all the way to the roots with the possible exception of the plasma membrane, across which diffusion occurs according to the classical concept. As mentioned earlier it has recently been suggested, however, that bulk flow may be possible even across these membranes (Ray, 1960). The pull is also thought to move water along the cell walls, bypassing the protoplasts. Direct evidence of this has been produced for cell walls in the leaves (Gaff et al., 1964), but indirect evidence (Weatherley, 1963) indicates that it cannot account for much of the flow through roots. From the base of the vessels in the roots the pull is transmitted through the adjacent cells to the epidermal cells and even to the medium surrounding these cells, that is, solution or soil particles. Thus the transpirational pull is responsible not only for the movement of water within the plant but also for absorption of water from the root medium. When the transpirational pull is pronounced, it follows that the absorption of water is greatly enhanced because even if no bulk flow occurs across the membranes, diffusion would be directly involved only over an infinitesimally small fraction of the total distance. Indirectly, of course, diffusion is involved since transpiration is itself a diffusion process (see Chapter 11).

The mass movement of water under tension in the plant is rapid, as can be shown by cutting the stem of a wilted plant under water. Recovery occurs more rapidly than if the roots of the intact plant are watered. This illustrates the fact that although nearly the whole path of water movement is through dead vessels, the few living cells in the root offer a greater resistance to movement than do the vessels. This is no doubt because of the retarding effect of the semipermeable plasma membranes, which are considerably less permeable to water than are the cell walls. The resistance to flow in the vessels is simply caused by the low frictional resistance or viscosity of the water itself.

The high tensions in the vessels and tracheids point to the need of the thick walls always found in these cells. In the absence of such thick walls, the cells would collapse, closing the cavity and stopping the flow of water. Since water movement occurs only in a relatively narrow zone of the sapwood (the heartwood vessels are blocked by growths known as tyloses and by resins), it is here that the tension is localized. This means that when tensions are high, the vapor pressure of the water in the sapwood is greatly reduced below that in the heartwood. As a result, there is evidence that water distills from the heartwood to the sapwood on a hot summer day (Reynolds, 1939).

QUESTIONS

1. What is meant by bulk flow?
2. To what is the rate of bulk flow proportional?
3. How does bulk flow differ from diffusion flow?
4. Can the ascent of sap be caused by diffusion? Why?
5. Can it be generally caused by root pressure? At any time?
6. How tall a tree could be supplied with water by a root pressure of 5 atm?
7. What physiological process, other than ascent of sap, is caused by root pressure?
8. What would be the root pressure on a normal sunny day?
9. How does a tree trunk change during a summer day?
10. What is the instrument that measures this?
11. Is the water pushed up or pulled up a tree on a sunny summer day?
12. What is the pull called?
13. What theory explains this pull?

14. What forces are primarily responsible for the rise?
15. Is the role of capillarity a result of the xylem capillaries?
16. What is the relation of water absorption to the ascent of sap?
17. Is the cohesion of water adequate to account for the rise of water to the top of trees?
18. Is there any complicating factor under natural conditions?
19. Where is the greatest resistance to the flow of liquid water in the plant?
20. Are the cohesive tensions on the water throughout the wood?
21. Do vessels collapse under tension? Why?

SPECIFIC REFERENCES

Brouwer, R. 1953. Water absorption by the roots of *Vicia faba* at various transpiration strengths. Proc. Kon. Nederl. Akad. Wet. [Biol. Med.] **56**:106-136.

Derjaguin, B. V. 1966. Recent research into the properties of water in thin films and in microcapillaries. Sympos. Soc. Expr. Biol. **19**:55-60.

Gaff, D. F., T. C. Chambers, and K. Markus. 1964. Studies of extrafascicular movement of water in the leaf. Aust. J. Biol. Sci. **17**:581-586.

Gibbs, R. D. 1935. Studies of wood. Canad. J. Res. **12**:715-787.

Goatley, J. L., and R. W. Lewis. 1966. Composition of guttation fluid from rye, wheat, and barley seedlings. Plant Physiol. **41**:373-375.

Greenidge, K. N. H. 1958. Moisture movement in plants. Scientia **93**:191-195.

Hammel, H. T. 1967. Freezing of xylem sap without cavitation. Plant Physiol. **42**:55-66.

Huber, B. 1932. Beobachtung und Messung pflanzlicher Saftströme. Ber. Deutsch. Bot. Ges. **50**: 89-109.

Levitt, J. 1956. The physical nature of transpirational pull. Plant Physiol. **31**:248-251.

Loomis, W. E., R. Santamaria, and R. S. Gage. 1960. Cohesion of water in plants. Plant Physiol. **35**:300-306.

MacDougal, D. T., and F. Shreve. 1924. Growth in trees and massive organs of plants. Carnegie Institute of Washington, Pub. No. 350. Washington, D. C.

Milburn, J. A., and R. P. C. Johnson. 1966. The conduction of sap. II. Detection of vibrations produced by sap cavitation in *Ricinus* xylem. Planta **69**:43-52.

Ray, P. M. 1960. On the theory of osmotic water movement. Plant Physiol. **35**:783-795.

Reynolds, E. S. 1939. Tree temperatures and thermostasy. Ann. Mo. Bot. Gard. **26**:165-255.

Scholander, P. F., W. E. Love, and J. Kanwisher. 1955. The rise of sap in tall grapevines. Plant Physiol. **30**:93-104.

Scholander, P. F., H. T. Hammel, E. D. Bradstreet, and E. A. Hemmingsen. 1965. Sap pressure in vascular plants. Science **148**:339-346.

Scholander, P. F., E. D. Bradstreet, H. T. Hammel, and E. A. Hemmingsen. 1966. Sap concentrations in halophytes and some other plants. Plant Physiol. **41**:529-532.

Thomas, L. A., and W. A. Roach. 1934. Injection of fruit trees; preliminary experiments with artificial manures. J. Pomol. Hort. Sci. **12**:151-166.

Weatherley, P. E. 1963. The pathway of water movement across the root cortex and leaf mesophyll of transpiring plants. *In* A. J. Rutter and F. W. Whitehead. The water relations of plants. John Wiley & Sons, Inc., New York.

GENERAL REFERENCES

Dixon, H. H. 1924. The transpiration stream. University of London Press, London.

Kramer, P. J. 1949. Plant and soil water relationships. McGraw-Hill Book Co., New York.

chapter 10

Translocation of solutes

A. Paths of movement
B. Mechanisms of movement

Although the aim of physiology is to explain all living processes in terms of physics and chemistry, this goal in most cases has not yet been reached. As a result, it is frequently necessary to give a number of possible explanations (hypotheses) and to attempt to evaluate each on the basis of physical and chemical theory, together with the available physiological evidence. Sometimes it is possible in this way to exclude all the hypotheses except one. This has been done in the case of the ascent of sap. Thus, although several explanations have been proposed from time to time for the phenomenon, all except the cohesion theory have been conclusively disproved. In an elementary course there is no time to discuss these disproved theories, and the cohesion theory is accepted, not because it is conclusively proved, but because it, and it alone, agrees with the known facts. In the case of the translocation of solutes, however, the evidence does not permit acceptance of any one hypothesis. This has led to more controversy than in almost any other field of plant physiology (see General References). The development of many new techniques has produced a large body of new information and has led to the hope that the uncertainties will soon be resolved. But a clear understanding of the problem requires a consideration of both the old and the new results, both the old and the new hypotheses.

PATHS OF MOVEMENT

There are two principal translocatory systems in the plant, the xylem vessels and the phloem sieve tubes. These two systems are quite different, both morphologically and physiologically (Table 10-1).

The structure of the vessels is very simple and has long been known since they are dead when mature and consist only of cell walls, which can be adequately studied under the optical microscope in dead, permanently mounted sections. The sieve tubes, however, are much more difficult to study. Because they are so long and narrow, it is difficult to cut thin enough sections of living tissue without cutting across every sieve tube and therefore killing them. Improved techniques, however, have succeeded in revealing some of the properties of the living sieve tubes; for example, it is possible to plasmolyze them in the same way as other living plant cells (Esau, 1966). Since they are alive when functional, they must be carefully fixed, then observed under the electron microscope in order to determine the protoplasmic structure, including that of the organelles. This, too,

Table 10-1. Comparison of the two translocatory systems

Vessels or tracheids	Sieve tubes
A. Anatomical comparisons	
1. Occur in xylem tissue	1. Occur in phloem tissue
2. Have open ends	2. Sieve plates at ends (and sometimes at sides)
3. Long, narrow cells (usually wider than sieve tubes)	3. Long, very narrow cells
4. Thick, rigid walls	4. Thin, more extensible walls
B. Physiological comparisons	
1. Dead when mature and functional	1. Alive when mature and functional
2. Permeable to both solute and solvent	2. Semipermeable
3. Low sap concentration	3. High sap concentration (osmotic pressure of 15 to 34 atm)
4. No turgor pressure	4. High turgor pressures (in turgid plants)
5. Partially collapsed when functioning (except when root pressure is operating)	5. Distended by pressure when functioning
6. Absorb water or air when cut (except when root pressure is operating)	6. Exude when cut (if functioning)
7. Translocate both solvent and solutes	7. Translocate only solutes
8. Speeds up to 75 cm/min	8. Speeds up to 5 cm/min

Fig. 10-1. A cucurbit sieve element with slime-impregnated filaments running from sieve plate to sieve plate; the cut end is at the bottom. (From Crafts, A. S. 1961. Translocation in plants. Holt, Rinehart & Winston, Inc., New York.)

gives rise to difficulties, and some investigators disagree as to the details of protoplasmic structure. No nucleus is found in the mature sieve tube, and many investigators have produced evidence that the tonoplast also disappears (Buvat, 1963). Mitochondria, however, do remain, although having fewer internal membranes than in parenchyma cells. Ribosomes and dictyosomes are lacking (Murmanis and Evert, 1966). However, it is the sieve plates that have been most intensely studied, and the connecting strands across the pores have been clearly demonstrated (Fig. 10-1) by some although not by all observers (Esau, 1966). At least in some cases these appear to consist of endoplasmic reticulum (ER), which fills the pores. More recent observations indicate that beside the ER, strands derived from slime bodies traverse the pores, which are lined by the plasmalemma (Murmanis and Evert, 1966).

Pure vessel sap is much more easily obtained than sieve-tube sap. All that is necessary is to cut a length of stem (cut surfaces at both ends) and stand it on the shoulder of a specially constricted centrifuge tube. On centrifuging the stem the vessel sap is thrown out of the vessels into the bottom of the centrifuge tube. A relatively simple method of obtaining sieve-tube sap is also used, although there may be complications. The sieve tubes are normally near full turgor because of the close proximity

Table 10-2. Analysis of tracheal (vessel) sap of the pear*

Element or ion	Parts per million in sap	
	Nov. 10	May 10
Ca	16.6	84.7
Mg	0.8	23.5
K	23.6	59.6
Fe	1.0	2.1
SO₄	8.3	31.8
Cl	3.2	4.5
PO₄	10.6	25.2

*Adapted from Anderssen, F. G. 1929. Plant Physiol· 4:459-476.

Table 10-3. Analysis of phloem exudate of *Robinia pseudo-acacia* during August*

Element or substance	Milligrams per cubic centimeter
Ca	0.72
Mg	0.38
K	0.95
Li	Trace
Reducing sugars	0.51
Sucrose	199.94
Total sugars	200.45
Nitrate nitrogen	0.135
Total nitrogen	0.56

*Adapted from Moose, C. A. 1938. Plant Physiol. **13**:365-380.

of the vessels with their sap of high water potential. Thus in a transpiring plant, although the vessels are slightly collapsed by the tension, the sieve tubes may still be distended by their turgor pressure. As a result, a cut across the stele causes exudation ("bleeding") from the sieve tubes, and the exudate may be collected for analysis. Of course, a high rate of transpiration and relatively low rate of absorption by the plant may result in a sufficiently severe tension on the vessels to reduce the sieve-tube turgor to zero. Under these conditions no exudation is possible. Exudation will also occur from the vessels after some time because of root pressure. Consequently, the phloem exudate must be collected immediately after the cut is made in order to prevent contamination with vessel sap. In spite of such precautions, the phloem exudate may be contaminated with substances from other cells when obtained from a cut surface. A far more elegant method of obtaining pure sieve-tube exudate therefore, has been adopted by recent investigators. The aphid is able to insert its proboscis between the cells of a plant until it reaches a sieve tube, which it then penetrates. If the anesthetized insect is separated from its proboscis, this now acts as a micropipette from which pure sieve-tube sap continues to exude for days. With modern methods of chromatography this sieve-tube sap has been quantitatively analyzed.

Although the ascent of water was long ago conclusively shown to occur in the xylem vessels and tracheids, the path of solute movement has been the subject of considerable controversy. According to the classical concept, mineral substances are carried up the xylem vessels by the ascending sap, but organic substances move downward in the pholem sieve tubes. There is much indirect evidence for this view. Analysis of the xylem sap (obtained by centrifuging) reveals the presence of inorganic salts (Table 10-2). On the other hand, sieve-tube exudate consists mainly of sugars (Table 10-3); and the concentration of total solutes may be as high as 15% to 30%, of which 50% to 90% may be sucrose.

Many recent investigations have identified several larger moleculed sugars than the disaccharide (see Chapter 14) sucrose in the phloem exudate, for example, raffinose (a trisaccharide), stachyose (a tetrasaccharide), and verbascose (a pentasaccharide). But sucrose is still the commonest sugar in the phloem of most plants and the one usually found in the largest quantity. Hexoses (glucose and fructose) do not occur in the sieve tubes. Most of the rest is organic nitrogen (amino acids and amides), although calcium, magnesium and potassium may be present in significant amounts (Table 10-3).

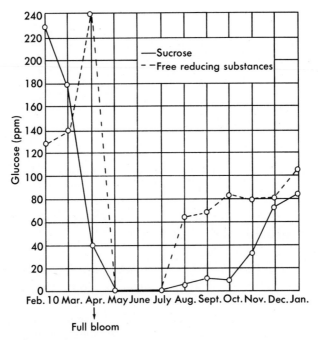

Fig. 10-2. Free reducing substances and sucrose in tracheal sap of pear branches. (From Anderssen, F. G. 1929. Plant Physiol. 4:459-476.)

Nevertheless, it cannot be concluded that the only translocatory system in which organic substances occur is the phloem. Sugars may reach a relatively high concentration in the xylem, at least during spring (Fig. 10-2). Furthermore, although it is true that organic nitrogen is translocated downward and even upward (Joy and Antcliff, 1966) in the sieve tubes of many herbaceous plants, in the case of trees this occurs primarily, if not solely, upward in the vessels (Bollard, 1957, Barnes, 1963). The organic nitrogen compounds of major importance in the translocatory stream through the xylem of fifty-three species of trees are allantoic acid, arginine, asparagine, aspartic acid, citrulline, glutamic acid, and glutamine. In seven species of *Pinus*, glutamine accounted for 73% to 88% of the nitrogen.

Even the early workers soon began to question the classical concept that organic substances move in the phloem, inorganic in the xylem. Many of the preceding anal-

yses showed that this could not always be true, because if sugars are found in the xylem sap, they must be translocated in this system; and if inorganic ions are found in the phloem sap, they must be translocated there. Some even suggested that all movement occurred only in the xylem, others that all occurred only in the phloem. To test these concepts it was not sufficient to analyze the two saps; it was necessary actually to show movement.

One method of determining the path of movement was to interrupt the flow by cutting across one or other of the two translocatory systems, leaving only the other intact. *Girdling* or *ringing*, for instance, is the removal of a ring of bark (containing phloem), leaving only the xylem intact. The general effects of ringing have long been known (e.g., the accumulation of carbohydrates above the ring), and such effects have been explained by the movement of sugars in the phloem. In other

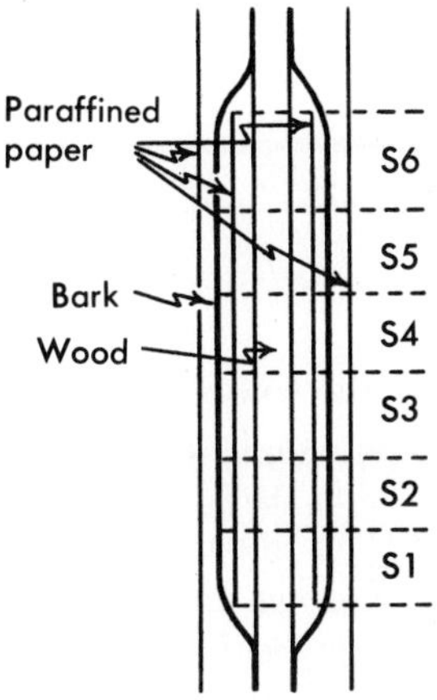

	Willow (Gain of K, 5 hr after feeding)			Geranium (Gain of PO$_4$, 6 hr after feeding)	
	Bark	Wood		Bark	Wood
				270	860
			S8	9.0	112
	53	47	S7	0.5	120
S6	11.6	119	S6	0.6	132
S5	0.9	122	S5	0.8	138
S4	0.7	112	S4	<0.3	147
S3	<0.3	98	S3	<0.5	137
S2	<0.3	108	S2	<0.3	152
S1	20	113	S1	11.1	131
	84	58		316	442

Paraffined paper

Bark

Wood

Fig. 10-3. Upward movement of radioisotopes fed to the roots of willow and geranium (5 mEq per liter). Bark is stripped from the wood 1½ hr before feeding. (Redrawn from Stout, P. R., and D. R. Hoagland. 1939. Amer. J. Bot. 26:320-324.)

cases the two systems were separated from each other in order to supply substances only to one of the two systems. More recently, the movement of solutes has been followed mainly by the use of tagged atoms, for example, radioisotopes. The great advantage of this method, of course, is the ability to follow quickly and quantitatively the movement of even small amounts of introduced substances by measuring their radiation. With the aid of such substances Stout and Hoagland (1939) conclusively showed that inorganic substances move up the plant in the xylem (Fig. 10-3). These results also revealed one source of difficulties: Whenever bark and wood were left in normal contact, the substances moved readily into the phloem from the xylem. The substances in this way reached relatively high concentrations in the phloem both above and below the region of separation of bark from wood, sometimes even higher than in the adjacent xylem. Yet the extremely low concentrations in the separated phloem revealed that no movement occurred up or down the phloem other than that caused by simple diffusion.

Even these experiments must be accepted with some reservation. As in all such techniques, injury to the bark is difficult to preclude completely, and their results do not prove that movement of these substances cannot also occur in the phloem. In fact, later results conclusively showed that they can. Thus Biddulph and Markle (1944) injected radiophosphorus into leaves of cotton plants having bark and wood separated as in Stout and Hoagland's experiment. In this case movement was downward instead of upward. They found it moved only via the phloem and at rates in excess of 21 cm/hr (Table 10-4). As in Stout and Hoagland's experiments, the radioisotope was able to move from one tissue to the other when the two were left in contact. These results also seem to disprove any suggestion of injury

Table 10-4. Downward movement of radiophosphorus injected into a leaf of a cotton plant; leaf strip immersed in solution for 5 min, then in water; bark stripped as in Fig. 10-3*

	Micrograms PO$_4$ in stripped stem after 3 hr		Micrograms PO$_4$ in unstripped stem after 1 hr	
	Bark	Wood	Bark	Wood
	0.904		0.444	
Bark ⎡	0.684	0.004	0.160	0.055
separated ⎨	0.544	0.000	0.103	0.063
from ⎪	0.120	0.019	0.055	0.018
wood ⎣	0.123	0.002	0.026	0.007
		0.160		0.152

*Adapted from Biddulph, O., and J. Markle. 1944. Amer. J. Bot. **31**:65-70.

to the phloem in Stout and Hoagland's experiments because if separation of the phloem from the xylem fails to prevent downward movement in the phloem, it is hardly likely to prevent upward movement. But it is not yet possible to generalize safely since not all inorganic substances have been investigated. Manganese, for instance, has been reported to be translocated mainly in the phloem (Klimovitskaya, 1963).

Without the help of radioisotopes, Mason, Phillis, and Maskell had earlier produced evidence that nitrogen, phosphorus, potassium, sulfur, magnesium, and chlorine are phloem mobile in a downward direction but that calcium is not (in opposition to the results of Moose, Table 10-3). More recent results have confirmed this immobility of calcium in the phloem (Arisz, 1952). But others have shown that at least in some plants there can be a very slow movement of calcium in the phloem. Thus we are forced to admit that inorganic substances may move both in the xylem and the phloem. However, the available evidence favors the view that upward movement of these substances is primarily, if not solely, in the xylem; downward movement under normal conditions is probably solely in the phloem. Since by far the greatest part of the movement of these substances is in the upward direction, this would assign to the xylem the main path of movement of inorganic substances, whereas the phloem is normally almost the sole path of movement of organic substances (except for the upward movement in trees [1] of organic nitrogen and [2] of sugars during spring) and the path of downward movement of all solutes. The classical concept is therefore reasonably near the truth.

MECHANISMS OF MOVEMENT

As to the mechanism of movement in the phloem, two principal theories have been proposed: (1) the mass, bulk, or pressure flow theory (for arguments in its favor see Crafts, 1951, 1961, and Esau, 1966); and (2) the cytoplasmic streaming theory (for the earlier defense of this theory see Curtis and Clark, 1950). A third concept, "activated (or facilitated) diffusion," is simply giving a name to an unknown process. It is based on two facts: (1) the movement usually follows a diffusion gradient and (2) the rate is much too rapid (40,000 times in some cases) to be accounted for by diffusion. No attempt is made to explain the "activation" mechanism.

The first of these was proposed by Münch in 1926. He thinks of translocation in the plant as a kind of circulation analogous to that of the bloodstream, with the mesophyll cells in the leaves acting as a kind of heart. Water and salts move up the xylem, organic substances move down the phloem, and an osmotic mechanism controls the latter (Fig. 10-4). As a result of

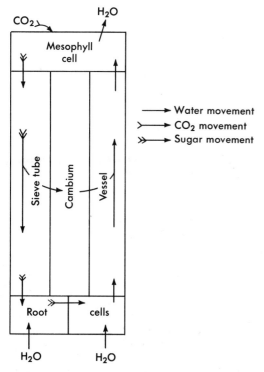

CO₂ and H₂O labels appear at the top, with Mesophyll cell, Sieve tube, Cambium, Vessel, Root, and cells labeled.

Legend:
→ Water movement
> → CO₂ movement
» → Sugar movement

Fig. 10-4. Diagrammatic representation of the pressure flow theory. (Redrawn from Münch, E. 1931. Die Stoffbewegungen in der Pflanze. Gustav Fischer Verlag, Stuttgart.)

photosynthesis, the cell sap concentration of the mesophyll cells at the top of this circulatory system is maintained high, even though some solutes are being exported down the phloem. Because of the upward movement of water in the xylem, the water content is maintained at a high level. These two factors combine to yield a high turgor pressure in the mesophyll cells. Small pores known as pits occur in the cell wall through which cytoplasmic strands, known as plasmodesmata, connect these living cells with each other. These connections occur from mesophyll cell to mesophyll cell and ultimately to the sieve tubes. Through these connections the turgor pressure succeeds in forcing some of the cell solution into and down the sieve tubes. The loss of solutes from the mesophyll cells would then be compensated for by newly synthesized or-

ganic substances, the loss of water by movement from the xylem. Excess water forced into the phloem would be "squeezed" laterally through the cambium into the xylem at the same time as the solutes are conducted downward. Movement in the phloem would then always have to be from a region of high to one of low turgor pressure. It could, therefore, occur in an upward direction from storage tissues. But at any one time all substances would have to move in the phloem in the same direction. Movement would also depend on the existence of open plasmodesmata between all living cells. There are many other factors that have to be taken into account in evaluating the Münch "Druckströn" theory. The strongest experimental evidence in its favor is the existence of exudation pressure in the phloem. The Münch mass-flow or pressure-flow theory gives a complete, logical, and apparently reasonable concept of translocation in the phloem and therefore has been accepted by many physiologists as "the mechanism."

The second theory was suggested by De Vries and later championed by Curtis (1935). According to this concept, movement is caused by a combination of diffusion and cytoplasmic streaming. Diffusion would occur from sieve tube to sieve tube across the intervening cell wall. Within the sieve tube the substances that have diffused in from the cell above would be carried downward by the cytoplasmic stream; those that diffused in from the cell below would be carried upward by the same cyclic stream. Thus this theory accounts for the known movement of substances along concentration gradients and for the apparent fact that two substances can be simultaneously translocated in opposite directions (Mason and Phillis, 1937; Palmquist, 1939). By use of C^{14} and P^{32}, Chen (1951) also produced strong evidence of the simultaneous movement of carbohydrates and phosphates in opposite directions. But in view of the results of Stout and Hoagland

(Fig. 10-3), the P^{32} may have moved up in the vessels and laterally to the sieve tubes. More direct evidence for the phenomenon has been produced by Biddulph. His more recent (1965) results, however, indicate that the bidirectional movement occurs only in separate phloem bundles. The most conclusive evidence of simultaneous movement of substances in opposite directions was recently produced using the aphid method of obtaining sieve-tube sap (Eschrich, 1967). The leaf below the aphid was provided with fluorescein, that above with a C^{14} compound. The honeydew mostly contained both tracers. In opposition to such results there is evidence from other sources that some organic substances can be translocated only in the same direction as sugars, for example, the weed killer 2,4-D and viruses. Several other objections have been raised against the cytoplasmic streaming theory, although some at least are no longer valid. Thus it was stated that no streaming occurs in sieve tubes. But careful technique has now succeeded in demonstrating that it does occur (Thaine, 1964). Perhaps the most difficult point to reconcile with this theory is the fact that the phloem exudate is apparently cell sap and not cytoplasm, yet it contains high concentrations of the translocated substances (Table 10-3). This would seem to indicate that translocation is via the vacuole rather than the cytoplasm. Electron micrography, however, may have eliminated this objection because if the tonoplast really does disappear in the mature sieve tube, cytoplasm and vacuole would become one system ("myctoplasm," Esau, 1966). Thaine, in fact, has observed in living cells what he calls "transcellular strands" through which streaming occurs. This streaming is therefore in the center of the sieve tube and actually appears to traverse the sieve-tube pores, continuing from cell to cell. According to these observations, the diffusion part of the cytoplasmic streaming theory would be eliminated, and translocation would be caused by streaming not only within each cell but also from cell to cell. However, the opposing argument states that the observed rates of streaming are far too slow to account for the measured translocation rates.

A quantitative comparison, however, no longer upholds this objection. Streaming rates as high as $107\mu/\sec$ at $27°$ C have been recorded (Barr and Broyer, 1964). This is 6.5 mm/min or about one-eighth the maximum rate of translocation found in the phloem (5 cm/min). Since the phloem is highly specialized for translocation and possesses both ample respirable material and an active ATP–forming system, it is not unreasonable to postulate a streaming rate 8 times that in nonspecialized cells. However, it is not clear whether the "transcellular strands" would be capable of a cyclosis since they seem to traverse the pores; yet a one-way flow would seem inconceivable since all the sieve tubes would become emptied of protoplasm.

One possible solution of the problem can be visualized. The transcellular streaming strands appear to consist primarily of slime, that is, of polysaccharides. Presumably, these can be synthesized reversibly from sucrose. Therefore, as the strands stream from cell to cell, they may be converted in each sieve tube to sucrose, which can then move laterally. If there is no lateral movement, slime is resynthesized and translocated further. In other words, translocation of carbohydrates may be in the insoluble state, as a part of the moving strands, rather than as a solution. Since the sieve strands are highly hydrated, they would also carry with them whatever solutes are present in the sieve tubes.

It is now obvious that both Münch's pressure-flow theory and De Vries' cytoplasmic streaming theory as originally stated are incorrect. The heartlike pump visualized by Münch does not exist in the mesophyll cells since the turgor pressure gradient is from sieve tube to mesophyll cell instead of the reverse (Arisz, 1952);

furthermore, the concentration of sucrose in the sieve tubes is 20 times that in the meso-phyll cells of the leaf (Wanner, 1953). This reverse concentration gradient also elimi-nates Münch's concept of a closed cycle of flow that continues up the xylem into the mesophyll cells and from the mesophyll cells down the phloem.

On the other hand, many lines of evi-dence indicate that the flow through the sieve tubes is dependent on metabolism. The earliest evidence was the fact that flow cannot occur across a thin portion of phloem killed by scalding or cooled to near freezing. Since some, although not all of the cytoplasmic structures have been found in mature sieve tubes (Buvat, 1963), they therefore must be capable of some metab-olism. Hexoses from the mesophyll cells do move into the sieve tube but are completely converted to sucrose by what must be a metabolic process. As a result, the osmotic concentration of the sieve tube soon exceeds that of the mesophyll cells supplying it. Labeling has proved that some of the sucrose that moves down the sieve tubes is respired to carbon dioxide (Canny, 1962). It is even possible to inhibit sieve tube translocation reversibly by a localized ap-plication of cyanide, a metabolic inhibitor (Willenbrink, 1966). Early investigators recognized the role of metabolism by calling the movement "activated diffusion." How-ever, one aspect of the mass-flow theory remains unchallenged. Because of the large pits in the sieve plates, the sieve tubes form a continuous system under pressure. For this reason the plant must have a mecha-nism for rapidly sealing off the connecting holes between cells. This is accomplished by the formation of callose plugs as soon as the sieve tubes are wounded or when they become nonfunctional, for example, in the fall.

In short, the translocation of solutes in the phloem cannot be a simple cycling such as visualized by the two major theories, but it must include many processes associated with the life of the cells, for example: (1) active absorption of sugars and other sub-stances by the sieve tubes and (2) enzy-matically controlled syntheses of larger moleculed sugars from smaller ones. Respi-ratory energy is required for both of these and perhaps for other associated processes. It is this participation of numerous physio-logical processes that makes it so difficult to determine the exact mechanism.

In the light of modern evidence, those who cling to the Münch hypothesis must modify it as follows: The initiating turgor pressure in the mesophyll cells, the flow across plasmodesmata between mesophyll cells and sieve tubes, and the complete cycle through the mesophyll cells and root paren-chyma would all have to be eliminated. The initiating force would now have to be the turgor pressure in the sieve tubes them-selves, which would follow a gradient with the maximum pressure at the top of the sieve tube adjacent to the source of supply in the mesophyll cells. But the development of this pressure would depend not only on the photosynthetic production of sugars in the mesophyll cells but also on a presum-ably respiratory release of energy that would enable the active absorption of hexoses from the mesophyll cells and their immediate enzymatic conversion to sucrose and higher sugars. Even if this profound modification of the Münch hypothesis should prove correct, it would be only a partial explanation of the complex mecha-nism of translocation in the phloem. Finally, if the sieve pores are filled with strands of ER, the whole basis for the pressure-flow theory is destroyed.

On the other hand, those who wish to champion the cytoplasmic streaming theory would also have to make modifications in the original concept, for example, elimina-tion of diffusion as a component if transcell-ular streaming occurs, and the discovery of some mechanism to speed up the flow in the intact plant. But the accumulating evi-dence is leading more and more investigators to the suggestion (Wooding, 1967) that the fibrils found in the sieve tube may have a similar function to that postulated for fibrils in streaming protoplasm—they could produce a rapid flow in the sieve tube and could also keep the sieve tube contents thoroughly mixed.

QUESTIONS

1. What are the two translocatory tissues?
2. Which cells in each are primarily involved?
3. Are these cells dead or alive?
4. Are they normally distended or partially collapsed?
5. Are they semipermeable or freely permeable?
6. Are they thin walled or thick walled?
7. Do they have high or low sap concentrations?
8. Do they have high or low turgor pressures?
9. Do they exude or absorb on being cut?
10. How can one obtain vessel sap for analysis?
11. How can one obtain sieve-tube sap for analysis?
12. What is the major constituent of each sap?
13. Does vessel sap ever contain organic substances?
14. How can one interrupt movement in one system without affecting the other?
15. In what direction are inorganic substances mainly translocated?
16. In what direction are organic substances mainly translocated?
17. What evidence is there for this (or these) direction(s) of movement?
18. What are isotopes?
19. What do results with isotopes indicate?
20. What do the symptoms of potassium deficiency indicate about the direction of potassium translocation?
21. Is the translocation of carbohydrates in the direction expected from the laws of diffusion?
22. Is the rate what would be expected from the laws of diffusion?
23. What theory explains the translocation of solutes in the xylem?
24. What theory explains the translocation of solutes in the phloem?
25. What is the nature of the force according to the pressure-flow theory?
26. What would be the direction of flow according to this theory?
27. Would different substances move in different directions or the same direction?
28. What other theory is there?
29. In which theory are plasmodesmata required?
30. In which theory is diffusion involved?
31. What is the sugar that is translocated?
32. What is the evidence for metabolism?
33. What is the evidence against pressure flow?
34. What is the evidence against cytoplasmic streaming flow?

SPECIFIC REFERENCES

Anderssen, F. G. 1929. Some seasonal changes in the tracheal sap of pear and apricot trees. Plant Physiol. 4:459-476.

Barnes, R. L. 1963. Organic nitrogen compounds in tree xylem sap. Forest Sci. 9:98-102.

Barr, C. E., and T. C. Broyer. 1964. Effect of light on sodium influx, membrane potential and protoplasmic streaming in *Nitella*. Plant Physiol. 39:48-52.

Biddulph, O., and J. Markle. 1944. Translocation of radiophosphorus in the phloem of the cotton plant. Amer. J. Bot. 31:65-70.

Biddulph, O., and R. Cory. 1965. Translocation of C^{14} metabolites in the phloem of the bean plant. Plant Physiol. 40:119-129.

Bollard, E. G. 1957. Translocation of organic nitrogen in the xylem. Aust. J. Biol. Sci. 10:292-301.

Buvat, R. 1963. Infrastructure et différenciation des cellules criblée de *Cucurbita pepo*. Compt. Rend. Acad. Sci. Paris 256:5193-5195.

Canny, M. J. 1962. The translocation profile; sucrose and carbon dioxide. Ann. Bot. 26:181-196.

Canny, M. J., and O. M. Phillips. 1963. Quantitative aspects of a theory of translocation. Ann. Bot. 27:379-402.

Chen, S. L. 1951. Simultaneous movement of P^{32}

and C^{14} in opposite directions in phloem tissue. Amer. J. Bot. **38**:203-211.

Crafts, A. S., and H. B. Currier. 1963. On sieve tube function. Protoplasma **57**:188-202.

Eschrich. W. 1967. Bidirektionelle Translokation in Siebröhren. Planta **73**:37-49.

Hartt, C. E., H. P. Kortschak, A. J. Forbes, and G. O. Burr. 1963. Translocation of C^{14} in sugarcane. Plant Physiol. **38**:305-318.

Hatch, M. D., and K. T. Glasziou. 1964. Direct evidence for translocation of sucrose in sugarcane leaves and stems. Plant Physiol. **39**:180-184.

Joy, K. W., and A. J. Antcliff. 1966. Translocation of amino acids in sugar beet. Nature (London) **211**:210-211.

Klimovitskaya, Z. M. 1963. The role of the root system in the process of transferring manganese in plants. [Transl. title.] Ref. Zh. Biol. No. 8G36. Moscow.

Moose, C. A. 1938. Chemical and spectroscopic analysis of phloem exudate and parenchyma sap from several species of plants. Plant Physiol. **13**:365-380.

Murmanis, L., and R. F. Evert. 1966. Some aspects of sieve cell ultrastructure in *Pinus strobus*. Amer. J. Bot. **53**:1065-1078.

Palmquist, E. M. 1939. The path of fluorescein movement in the kidney bean, *Phaseolus vulgaris*. Amer. J. Bot. **26**:665-667.

Stout, P. R., and D. R. Hoagland. 1939. Upward and lateral movement of salt in certain plants as indicated by radioactive isotopes of potassium, sodium, and phosphorus absorbed by roots. Amer. J. Bot. **26**:320-324.

Thaine, R. 1964. Protoplast structure in sieve-tube elements. New Phytol. **63**:236-243.

Wanner, H. 1953. Die Zusammensetzung des Siebröhrensaftes; Kohlenhydrate. Ber. Schweiz. Bot. Ges. **63**:162-168.

Willenbrink, J. 1966. Zur lokalen Hemmung des Assimilattransports durch Blausäure. Z. Pflanzenphysiol. **55**:119-130.

Wooding, F. B. P. 1967. Fine structure and development of phloem sieve tube content. Protoplasma **64**:315-324.

GENERAL REFERENCES

Arisz, W. H. 1952. Transport of organic compounds. Ann. Rev. Plant Physiol. **3**:109-130.

Biddulph, O. 1959. Translocation of inorganic solutes, p. 553-603. *In* F. C. Steward. Plant physiology. Vol. II. Academic Press, Inc., New York.

Bollard, E. G. 1960. Transport in the xylem. Ann. Rev. Plant Physiol. **11**:141-166.

Crafts, A. S. 1951. Movement of assimilates, viruses, growth regulators, and chemical indicators in plants. Bot. Rev. **17**:203-284.

Crafts, A. S. 1961. Translocation in plants. Holt, Rinehart & Winston, Inc., New York.

Curtis, O. F. 1935. The translocation of solutes in plants. McGraw-Hill Book Co., New York.

Curtis, O. F., and D. G. Clark. 1950. An introduction to plant physiology. McGraw-Hill Book Co., New York.

Esau, K. 1966. Explorations of the food conducting system in plants. Amer. Sci. **54**:141-157.

Kursanov, A. L. 1961. The transport of organic substances in plants. Endeavour **20**:19-25.

Kursanov, A. L. 1963. Metabolism and the transport of organic substances in the phloem. Adv. Bot. Res. **1**:209-278.

Mason, T. G., and E. Phillis, 1937. The migration of solutes. Bot. Rev. **3**:47-71.

Münch, E. 1931. Die Stoffbewegungen in der Pflanze. Gustav Fischer Verlag, Stuttgart.

Swanson, C. A. 1959. Translocation of organic solutes, 481-551. *In* F. C. Steward. Plant physiology. Vol. II. Academic Press, Inc., New York.

Zeigler, H. 1963. Der Ferntransport organischer Stoffe in den Pflanzen. Naturwissenschaften **50**:177-186.

Zimmerman, M. H. 1960. Transport in the phloem. Ann. Rev. Plant Physiol. **11**:167-190.

chapter 11

Exchange of gases

PLANT PARTS INVOLVED

Two components of the transfer of materials have been considered: (1) absorption of water and solutes from the soil and (2) translocation of water and solutes within the plant. The two main remaining components are (3) absorption of gases and (4) release of gases. These two can be considered together as the exchange of gases. Since protoplasm is freely permeable to gases, any gas in the external atmosphere may be expected to diffuse into the plant. This is why plants have been injured or killed by leaks from the gas mains or by gaseous sulfur compounds from industrial plants. Similarly, a continuous exchange of carbon dioxide, oxygen, and water vapor takes place between all parts of the plant and its environment. Thus, the fine absorbing roots obtain oxygen from the soil and use it in respiration, which must proceed at a rapid rate since the respiration supported by it supplies the roots with all the energy for metabolism, growth, and active absorption of solutes. The carbon dioxide produced in respiration rapidly diffuses out of the cells into the surrounding soil. There are some plants that grow normally, although the root medium is deficient in oxygen, because the oxygen is transferred to the roots from the aerial part through the intercellular spaces. This is particularly true of some aquatic plants with intercellular spaces that make up as much as 70% of the volume of their tissues, as compared with much lower values (e.g., 20%) in most land plants.

The older roots and the stem are protected against excessive gas exchange by a more or less impermeable layer of suberin or cork. Nevertheless, there is also some exchange of carbon dioxide, oxygen, and water vapor between the stem and its environment. In woody plants this occurs primarily through openings in the bark known as *lenticels* since the corky covering of the stems is highly impermeable to these gases. As a result of this low permeability, the gas in the trunk may be of very different composition from that of the air (Table 11-1). The oxygen of the intercellular spaces becomes greatly depleted because of respiratory utilization, and the carbon dioxide content rises correspondingly. This is evidence that the stem is not a very efficient gas exchanger because the exchange lags far behind the production and utilization.

Table 11-1. Carbon dioxide and oxygen content of gas in trunks of paloverde (*Parkinsonia microphylla*)*

Date		Carbon dioxide (%)	Oxygen (%)
Jan.	15–16, 1931	3.4	15.3
	17–18	4.0	14.6
	19–20	3.8	14.9
Mar.	3–4	8.5	13.8
	5–6	8.8	13.7
	7–8	9.4	13.1
	9–10	10.7	12.9
	11–12	12.0	12.6
Apr.	20–21	7.3	14.5
	22–23	7.8	15.2
	23–24	9.1	14.3
	25–26	9.0	14.6

*Adapted from MacDougal, D. T., and E. B. Working. 1933. Carnegie Institute of Washington, Pub. No. 441. Washington, D. C.

The leaf is a much more efficient gas exchanger. It is the organ responsible for the major gas exchange of the plant; and since most leaves are covered by a layer of relatively impermeable cuticle, the main exchange is usually through the *stomata*. Each stoma consists of two guard cells that impede gas exchange when tightly pressed together (stomata closed) but permit free exchange when separated by a pore (stomata open).

STOMATAL MOVEMENT

In general, the stomata tend to show a diurnal periodicity, closing at night and opening during the day (Fig. 11-1). That this periodicity is related to light is easily shown since darkening the plant leads to closure; yet there are many exceptions (Loftfield, 1921). Some stomata open nor-

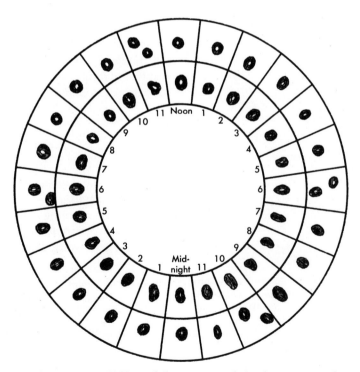

Fig. 11-1. Stomatal opening in alfalfa at different times of the day. Inner circle represents the lower leaf surface, and the outer circle represents the upper leaf surface. (Redrawn from Loftfield, J. V. G. 1921. Carnegie Institute of Washington, Pub. No. 314. Washington, D. C.)

mally at night, others close frequently at noon when the light intensity is at a maximum. Wilting may cause closure regardless of the light factor. There is also evidence of an internal rhythm in plants that favors opening at certain times of the day and closure at others (Meidner and Mansfield, 1965). Thus the effect of light on stomatal opening is apparently indirect through its control of some other factor. In order to explain the action of such indirect factors, the mechanism of stomatal movement must first be understood.

The walls of the two guard cells of a stoma are unevenly thickened. Increased turgor leads to opening by producing a bulge in the thinner outer walls and a consequent pulling apart of the thicker inner walls bordering the aperture. Decreased turgor leads to closure by the springlike recoil of the thicker inner walls (Fig. 11-2). Since turgor is caused by hydrostatic pressure, the stomatal movement is controlled

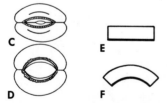

Fig. 11-2. Effect of changes in the turgor pressure of guard cells on stomatal opening. **A** and **B,** Cross section and half surface view of the stoma of *Amaryllis formosissima.* **C** and **D,** Surface views of closed and open stoma. **E** and **F,** Diagrammatic scheme of turgor curvature of a guard cell when the inner wall is thicker than the outer wall. (Redrawn from Benecke, W., and L. Jost. 1924. Pflanzenphysiologie. Gustav Fischer Verlag, Stuttgart.)

by water exchange, and any factor that alters the water content of the guard cells will affect their turgor and, therefore, the stomatal opening. The maximum possible turgor pressure of a cell is almost equal to the osmotic pressure of the cell sap. Any increase in the osmotic pressure resulting from an increase in solutes will therefore favor opening, any decrease will favor closure. However, it must be remembered that a change in solutes by itself affects only the potential turgor pressure, and the actual change in turgor pressure can occur only as a result of water movement into or out of the cell. Any real change in turgor pressure, on the other hand, must be accompanied by a change in degree of opening, provided that the cell is capable of a further stretch or shrinkage. But there is another factor: Stomatal opening is possible only if the guard cells can push back the epidermal cells adjacent to them. These other epidermal cells, as long as they are turgid, exert a back pressure on the guard cells; and if all the epidermal cells were to increase equally in osmotic potential and turgor pressure, no stomatal opening would result. Therefore, we can generalize that in all cases of stomatal opening

$$P_{tg} > P_{te}$$

where P_{tg} = turgor pressure of guard cells
P_{te} = turgor pressure of surrounding epidermal cells

Since stomatal opening normally occurs when the leaves are at maximum turgor (e.g., in early morning) and since the maximum possible turgor pressure is given by the osmotic pressure, it follows that:

$$P_{og} > P_{oe}$$

where P_{og} = osmotic pressure of guard cells
P_{oe} = osmotic pressure of surrounding epidermal cells

Since a cell's osmotic pressure is proportional to its cell sap concentration, this means that in a turgid leaf with open stomata, the cell sap concentration of the guard cells must be higher than that of the surrounding epidermal cells. This relationship is favored by the existence of chloroplasts in the guard cells and by their absence from the other epidermal cells. The guard cells can therefore synthesize organic solutes; the other nonphotosynthesizing epidermal cells must receive them from photosynthesizing cells.

Similarly, if all the epidermal cells showed corresponding decreases in turgor pressure, no closure could occur. In actual fact a simultaneous water loss from all the epidermal cells may result in an increased stomatal opening if the guard cells lose less than the other epidermal cells. For this reason, *incipient* wilting may actually cause increased stomatal opening, although more pronounced wilting will cause closure. This is because the guard cells are the last of the epidermal cells to reach zero turgor pressure. Finally, of course, excessive loss of water may cause stomatal opening because of drying and consequent collapse of the guard cells.

Thus, the opening and closing of stomata may be controlled solely by water movement into or out of the guard cells without an accompanying change in solutes. Stålfelt (1956) calls this *hydroactive* (opening or closure) as opposed to *photoactive* (opening or closure), in which these water movements are themselves controlled by light through its action on cell solute content. On the same basis, opening or closure in the dark, because of its control of cell solute content, may be called *scotoactive*. The mechanism of the photoactive and scotoactive movements is still not fully understood. It is necessary, therefore, to consider the possibilities. There is no question as to the nature of the force involved since the only one available is that caused by turgor pressure. Also, since there is no such thing as active water absorption (see Chapter 8), the turgor changes must be controlled by changes in solutes. There is ample direct evidence of this since the osmotic pressure of the guard cells is high (10 to 30 atm or higher) when the stomata are open and low (5 to 10 atm) when they are closed. Three lines of evidence prove that this photoactive opening is controlled

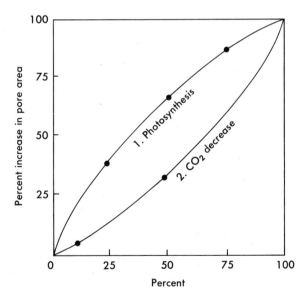

Fig. 11-3. Increase in stomatal aperture with (1) increase in photosynthesis as light is increased and (2) artificial decrease in CO_2 concentration of air at constant light of 50 footcandles. (From Scarth, G. W., and M. Shaw. 1951. Plant Physiol. **26**:207-225.)

112   *Biophysics*

by photosynthesis: (1) the relation of the two to light is similar (Fig. 11-3); (2) the relative effectiveness of different wavelengths of light is identical for stomatal opening and photosynthesis; and (3) no opening occurs in the absence of chlorophyll (e.g., in albino plants, only guard cells with chlorophyll are functional). The overall effect of light and dark may therefore be represented as in Fig. 11-4, A.

The major problem is the explanation of step 2. Photosynthesis may, of course, increase the solute concentration directly because of the synthesis of soluble carbohydrates. But the large increases in concentration (2 to 3 times) that occur in the short time (5 to 30 min) usually required for stomatal opening can hardly be caused by photosynthesis, particularly in view of the small amount of chlorophyll in the guard cells. In some stomata at least part of the increase in solutes is caused by a conversion of starch to sugar. Conversely, in the dark the loss of solutes in these stomata is at least partly caused by a reverse conversion of sugar to starch.

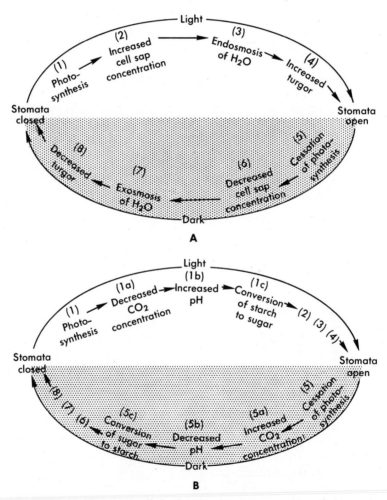

Fig. 11-4. A, Changes in the guard cells leading to photoactive opening (light) and scotoactive closing (dark) of the stomata. **B,** Classical (Scarth) hypothesis of the mechanism of stomatal opening and closing.

According to the classical theory of stomatal movement championed by Scarth (1932), the process is controlled by the steps shown in Fig. 11-4, *B*. Three facts are in favor of this scheme:

1. Artificial increases in pH induced by ammonia or ammonium hydroxide lead to stomatal opening.
2. Artificial atmospheres with low carbon dioxide concentrations (0.01% instead of the normal 0.03%) such as would occur because of photosynthesis also induce stomatal opening (Table 11-2 and Fig. 11-3). Direct measurements on leaves have shown that photosynthesizing leaves are capable of maintaining the carbon dioxide concentration in the intercellular spaces even below this value (Table 11-2). Maize is extreme in this respect, being able to maintain a minimal intercellular carbon dioxide concentration indistinguishable from zero (Meidner and Mansfield, 1965).
3. An enzyme (phosphorylase) capable of converting sugar to starch is present in the guard cells (Yin and Tung, 1948). In the presence of this enzyme at pH 5, the ratio of starch to sugar phosphate is 3.5 times that at pH 7 (Peat, 1951). Presumably, the mesophyll cells control starch to sugar conversion by a different enzyme system that does not show this pH response,

for example, a uridine diphosphate (UDP) or adenosine diphosphate (ADP) glucosyl transferase for starch synthesis and amylase for starch hydrolysis.

According to the original idea, the rise in pH was directly caused by the decreased concentration of carbon dioxide and therefore of carbonic acid in the cell sap. More recent evidence has shown that this part of the theory must be modified. It is true that low carbon dioxide concentrations lead to stomatal opening. But this occurs with very low carbon dioxide concentrations, for example, a decrease from 0.03 to 0.01%. A cell in equilibrium with air having 0.03% carbon dioxide will lose very little carbon dioxide when the content of the air drops to 0.01%. Furthermore, carbon dioxide forms the very weak carbonic acid, which has a very small effect on the pH of the buffered cell sap. Indirectly, however, the carbon dioxide concentration in the cell may have a very pronounced effect. The organic acids that are mainly responsible for the acidity of the cell are in metabolic equilibrium with carbon dioxide (see Chapter 13). Increasing the carbon dioxide content of the cell leads to an increase in organic acids caused by carboxylation. Removing carbon dioxide from the cell leads to a breakdown of the organic acids (decarboxylation). Consequently, when photosynthesis by the leaf cells lowers the carbon dioxide content of the intercellular spaces from 0.03% or higher to 0.01% by converting it into carbohydrate, this will shift the equilibrium toward breakdown of organic acids. An extreme example of this reversible change is provided by the succulents. They store organic acids at night (see Fig. 3-4) and convert them to carbohydrates during the day by photosynthetic assimilation of the carbon dioxide released by decarboxylation. The pH of all the cells of succulents, therefore, rises during daylight (by as much as 2 to 3 pH units) as predicted by the preceding hypothesis.

Table 11-2. Relation of stomatal opening to carbon dioxide concentration outside and inside the leaf*

External CO_2 concentration (ppm)	Internal CO_2 concentration (ppm)	Stomatal openings ($\% > 1\mu$)
8		100
100	40 (app)	70
200	100	26
400	240	12

*From Moss, D. N., and S. L. Rawlins. 1963. Nature (London) **197**:1320-1321.

Many objections have been raised against the Scarth theory:

1. The starch $\rightleftharpoons$ sugar changes do not occur in all stomata. In monocotyledons such as onion the guard cells do not form starch under any circumstances. But such plants contain large amounts of fructosans, and it is reasonable to assume that these function in the same way in the onion (and perhaps in other monocotyledons) as starch does in other plants (Ketellapper, 1963).

2. Although the relative effectiveness of different wavelengths of light is in general the same for photosynthesis as for stomatal opening, there are some quantitative differences; for example, blue seems to be relatively more effective for stomatal opening. It has therefore been suggested that the light effect may be partly caused by its effect on permeability. However, permeability can affect only the *speed* of opening and cannot explain the effect of light on the degree of opening. The differences in light response must therefore have some other explanation, as yet unknown.

3. In tobacco the opening is apparently related to the metabolism of glycolic acid (Zelitch, 1963; Waggoner and Zelitch, 1965). But this substance may be simply an intermediate in sugar synthesis, or it may produce its effect by controlling pH, as mentioned earlier.

It has also been suggested that active absorption of sugars by the guard cells may lead to opening, although no valid evidence has been produced. The known kinetics of the process, however, eliminate this as a possibility (Levitt, 1967). It has been suggested that active water absorption is involved; but since plants cannot absorb water actively (see Chapter 9), this suggestion must be discarded.

Therefore, although objections have been raised against the Scarth theory of photoactive stomatal opening, it is still the only useful theory available because it alone at present is able to explain all the known facts. Another kind of stomatal opening occurs at night and may therefore be called scotoactive opening. The scotoactive opening of succulent plants is apparently caused by the accumulation of organic acids synthesized from carbon dioxide in the dark (Nishida, 1963). They accumulate to a sufficient degree to produce osmotic opening of the stomata. In the daytime the carbon dioxide in the leaf is photosynthesized, leading to decarboxylation of the accumulated organic acids. When their concentration drops to a sufficient degree, the stomata close. However, for some time after this a sufficient concentration of organic acid must be maintained in the protoplasm by diffusion from the vacuole to produce a pH low enough to favor starch formation and stomatal closure. Only after all, or nearly all, the organic acid is metabolized can the pH rise to the level favoring starch to sugar hydrolysis and photoactive stomatal opening. Therefore, these succulents are able to continue photosynthesis during daylight, although their stomata remain closed for all or most of the daylight hours. This is, of course, an adaptive character that prevents excessive loss of water in dry climates.

However, although the Scarth cycle explains the main known facts, it tells nothing about the intermediate metabolic reactions that lead to the increase in solutes. The photosynthetic production of carbohydrates is a complex series of chemical reactions (see Chapter 15). It is not surprising, therefore, that several metabolic inhibitors greatly affect stomatal opening. Metabolic inhibitors such as alpha-hydroxysulfonates and sodium azide prevent opening. Others (phenylmercuric acetate, alkenylsuccinic acids) also prevent opening but apparently by a different mechanism. It has been suggested that a change in permeability of the guard cells may be involved (Zelitch, 1965), but as shown earlier, this effect cannot explain a change in degree of opening. It is more likely that these substances also exert their effects metabolically.

Most investigators consider only the role of stomatal opening in permitting photosynthesis by allowing carbon dioxide to diffuse into the leaf. They completely overlook the equally important role of permitting respiration by allowing oxygen to diffuse into the leaf. Even when night opening is recognized, no mention is made of the fact that oxygen deficiency in the dark can lead to wide opening. It is therefore necessary to explain the mechanism of stomatal opening resulting from oxygen deficiency in the dark as well as that resulting from carbon dioxide deficiency in the light. Unfortunately, little work has been done on this problem, and except for the case of succulents just mentioned, little or nothing is known of the mechanism of night opening, which may or may not be caused by oxygen deficiency.

DIFFUSION OF GASES THROUGH SMALL OPENINGS

The rapid rate of diffusion of gases into and out of the leaf is understandable on the basis of Fick's law of diffusion: The diffusion constant of gases is high (diffusion for water vapor is about 10,000 times diffusion for liquid water), the area across which diffusion is occurring is large, the concentration difference is large, and the distance is small. But the area is large only if we consider the whole leaf area. When, however, we remember that most of the gas movement occurs through the open stomata, the high rate of diffusion may be questioned because the area of the stomata is commonly only about 1% of the leaf area. Yet a leaf may lose almost as much water as a free water surface of the same total area. This means that the rate of water loss per unit stomatal area is almost 100 times that from a free water surface. It has also been shown that a photosynthesizing leaf absorbs carbon dioxide about as rapidly as potassium hydroxide solution of the same total surface. One must therefore conclude that either the laws of diffusion must be modified or else some other process speeds up the diffusion from the stomata. Brown and Escombe (1900) settled this problem by showing that the loss of water from a free water surface is almost unaltered when it is covered by a perforated sheet, even though the area across which diffusion occurs is reduced to a small fraction. This can be explained by the overlapping of the diffusion fronts (Fig. 11-5). The molecules passing through the pores at the perimeter can diffuse more freely in all directions because of the absence of other interfering molecules. They therefore spread out later-

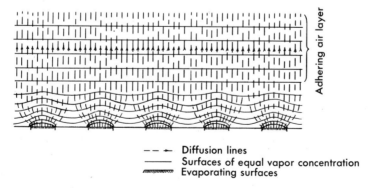

--- → **Diffusion lines**
——— **Surfaces of equal vapor concentration**
〰〰〰 **Evaporating surfaces**

Fig. 11-5. Overlapping of diffusion fronts above pores in a septum. (From Bange, G. G. J. 1953. Acta Bot. Neerl. 2:255-297.)

ally and soon meet with similar molecules from adjacent pores, forming a solid front a short distance above the pores. Expressed quantitatively, this means that the diffusion across small openings is proportional to the perimeter rather than the area of the opening.

The reason for this is that the area factor in Fick's law is only an approximation which really consists of two components: (1) the molecules that move in a direction normal to the surface (or aperture), and (2) those that can "spill over" the perimeter. Since the former varies with the area and the latter varies with the perimeter, Fick's law should be changed from

$$\frac{s}{t} \propto a$$

to

$$\frac{s}{t} \propto (a+p)$$

where 　a = area
　　　　p = perimeter
　　　　s = amount of substance diffusing (moles)
　　　　t = time (sec)

In the case of circular apertures, since $a = \pi r^2$ and $p = 2\pi r$:

$$a \propto r^2$$
$$p \propto r$$

If the proportions of the two components are about equal in the case of an aperture 1 mm in diameter, that is, half of the molecules diffuse normal to the surface and half "spill over" the perimeter, then increasing the aperture diameter to 100 mm would increase the area component 100^2 or 10^4 times and the perimeter component only 100 or 10^2 times; and the area component will now account for 99% of the diffusion. Conversely, if the aperture is decreased to 0.01 mm, the area component is decreased to $(0.01)^2$, or 10^{-4}, times, the perimeter component to 0.01, or 10^{-2}, times, and the perimeter component will now account for 99% of the diffusion. The conclusion, therefore, is that strictly speaking:

$$\frac{s}{t} \propto (a+p)$$

But for large apertures p is negligible:

$$\frac{s}{t} \propto a$$

For small apertures a is negligible:

$$\frac{s}{t} \propto p$$

However, it must be understood that as soon as overlapping occurs, the diffusion limit is reached, and any further increase in the perimeter of the apertures cannot lead to an increase in diffusion rate. Thus if the stomata are as close together as 10 to 20 times their diameters, maximum diffusion occurs. Doubling the diameter of their openings will have no effect on the diffusion rate. This fact must be taken into consideration when explaining the control of gas exchange by stomatal movement.

It must be emphasized that the preceding perimeter (or diameter) law (derived from Stefan's law) applies strictly only when the thickness of the membrane (or the length of the pore) is approximately zero, or when the ratio of the diameter to the pore length is constant (Lee and Gates, 1964). Stomata do not meet these requirements because the pore length is really the distance from the stomatal aperture to the evaporating cell surface at the bottom of the substomatal cavity. This distance is many times the widest stomatal aperture. However, if the vapor-pressure drop in this substomatal cavity is small enough, the stomata may obey the perimeter law reasonably well. Although no reliable measurements of this drop have yet been made, it seems reasonable to accept the assumption that it is small in a turgid plant under normal conditions. This follows from the fact that the boundary layer of moist air adjacent to the leaf surface (Fig. 11-5) is about 100 times the distance from the stomatal pore to the evaporating cell surface (see later). Therefore, the vapor-pressure drop within the substomatal cavity must be of the order of 1% of the total vapor-pressure drop between the leaf cells and the external atmosphere.

TRANSPIRATION

Since the quantitative aspects of oxygen and carbon dioxide exchange have been investigated primarily from the point of view of respiration and photosynthesis and will be considered under these headings, water vapor is the only gas that will be discussed at this point. The loss of water vapor from the plant has long attracted the attention of physiologists and has been called transpiration. This process can be defined as gaseous water diffusion driven by the net radiation absorbed by the leaf (Gates, 1966). Other parts of the plant be-

side the leaf can, of course, transpire, and transpiration can occur during short periods without absorption of radiation. However, the major transpiration is from the leaves, and over any considerable period of time the energy utilized is balanced by absorption of radiant energy. The diffusion may occur through the leaf cuticle or through the stomata. The former is called *cuticular transpiration* and the latter, *stomatal transpiration*. As a general rule, transpiration is higher when the stomata are open, lower when they are closed. In other words, the stomatal transpiration is usually higher

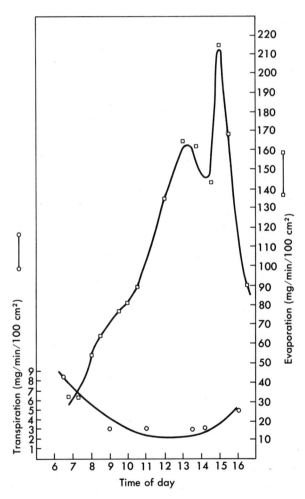

Fig. 11-6. Diurnal course of evaporation (Piche evaporimeter) and of transpiration from *Jatropha phyllacantha*. (From Ferri, M. G. 1955. Faculdad de Filosofia Ciencias e Letras, Universidade de São Paulo, São Paulo, Brazil. Bol. 195. Bot. 12.)

than the cuticular transpiration—commonly about 4 times as high on a sunny day. But there are all gradations from a maximum of several hundred to a minimum of nearly one. In the case of succulents, cuticular transpiration is essentially zero. A joint of a prickly pear can be kept in a dry room for a year without losing all of its water. Stomatal control of transpiration in such succulents is so effective that the transpirational curve at midday may be the reverse of the evaporation curve from a free water surface (Fig. 11-6). Pineapple shows this same reversal (Ekern, 1965). On the other hand, a shade plant (such as wood sorrel) may lose as much water through the epidermal cells as through the open stomata because the cuticle layer is almost nonexistent.

Many early workers believed that stomatal movement completely controlled transpiration. But this was soon found to be untrue. Transpiration rate may actually increase when stomata begin to close and may decrease when they open wider. This is sometimes because when the diameters of the stomatal openings are one tenth to one twentieth their distance apart, no further opening can affect the rate of diffusion through them. But it may also result from another variable factor that always controls the rate of diffusion—the concentration difference ($C_1 - C_2$ in Fick's law), or in the case of transpiration, the vapor-pressure difference. The larger the difference between the vapor pressure of the leaf and that of the surrounding air the greater is the transpiration rate. Both the relative humidity and the temperature may affect the vapor pressure of the air and the leaf. Consequently, they may both alter the transpiration rate.

Wind may also affect the transpiration rate. This is because of the boundary layer of moist air adjacent to the leaf surface. In still air this layer will be 1.57 cm thick above a circular leaf 10 cm in diameter (Lee and Gates, 1964). In turbulent air it is much

thinner. Since the distance x in Fick's law must include this boundary layer and since diffusion is inversely proportional to x, transpiration should be (and usually is) much higher in turbulent than in still air. However, the reverse effect may be obtained if the turbulence results in stomatal closure. Furthermore, even in turbulent air there remains a stationary boundary layer of moist air (Lee and Gates, 1964), and once this minimum layer is reached, any further increase in wind velocity is without effect. Wind will also affect the relationship between stomatal aperture and transpiration. In still air, stomatal transpiration may reach its maximum at about one fourth of full stomatal aperture because of the overlapping vapor fronts. In turbulent air, transpiration will continue to increase up to full stomatal aperture (Fig. 11-7).

There has been some controversy as to whether transpiration is a necessary evil or whether it is in some way essential for the

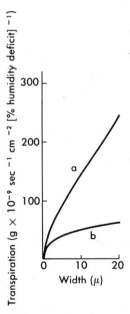

Fig. 11-7. Relation between transpiration and stomatal width in *Zebrina* in **a,** wind and **b,** still air. (From Bange, G. G. J. 1953. Acta Bot. Neerl. 2:255-297.)

welfare of the plant. From the former point of view, the function of the stomata is to permit exchange of carbon dioxide and oxygen in the processes of photosynthesis and respiration. At the same time, water is lost through the open stomata because there is no way of stopping the passage of water vapor while permitting the carbon dioxide and oxygen exchange. It may be objected that transpiration is useful to the plant by inducing the rise of water and indirectly its absorption from the soil. But it must be realized that although by far most of the water absorption is caused by transpirational pull, this part of the water is all wasted because transpiration can induce water absorption only by reducing cell turgor. The initial cell turgor, which is required for growth and other processes, results from root pressure. Transpiration must cause a loss of some of this water before it can lead to its replacement. The net gain in water caused by transpiration therefore can never be greater than zero. Thus, transpiration cannot be thought of as essential for water uptake.

From the other point of view, transpiration fulfills three main functions.

Transpiration helps maintain optimum turgidity

When a land plant is grown in a saturated atmosphere, it develops a softer, more watery type of growth. This presumably results from the larger turgor pressure, causing excessive cell enlargement. The larger cells will also have thinner walls because of the excessive stretch. Aquatic plants avoid this danger by having low osmotic pressures (3 to 5 atm in many cases of freshwater plants). Terrestrial plants avoid it because they are seldom at or near full turgor, and even though their cell sap concentrations may be high (osmotic pressures up to 200 atm in the case of some halophytes), transpirational loss keeps the turgor pressure well below the osmotic pressure.

Transpiration reduces leaf temperature

In full sunlight on a hot day, a leaf's temperature may be in danger of rising to the killing point. The evaporation of water is a cooling process resulting in a loss of 580 cal/g evaporated, and it may help to prevent such heat injury. It has been shown that the actual cooling effect of transpiration in most climates is not so great as some workers have thought. At the high temperatures most commonly found (30 to 40° C) this will not amount to more than 3 to 5° C (Curtis, 1938), but even this small quantity may save a plant from heat killing. It is particularly important since the surface temperature of a sunlit leaf may rise to as much as 20° C above the air temperature (Gates, 1963); and the lower surface may have a mean temperature only 1.5° C lower than that of the upper surface. The temperature of shaded leaves may be as much as 3 to 4° C below the air temperature. In exceptionally hot climates, however, where air temperatures may reach 50° C, some specially adapted plants are able to cool their leaves 10 to 15° C below the air temperature by having very high transpiration (and absorption) rates (Lange, 1959). In the case of these few exceptional plants, transpiration is unquestionably essential for survival since their killing points are below the maximum atmospheric temperatures.

Transpiration promotes translocation and uptake of solutes

The promotion of translocation and uptake of solutes by transpiration has been the subject of the most controversy. As already seen, the mineral elements absorbed from the soil are translocated mainly, if not solely, in the vessels of the xylem, rising in the transpiration stream. The question is sometimes raised concerning whether or not the speed of translocation of these solutes depends on the rate of transpiration. By an analogy (first proposed by van den Honert) with an endless belt delivering coal, we can see that the rate of delivery

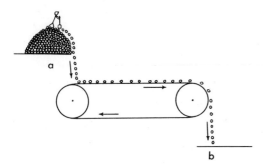

Fig. 11-8. Effect of transpiration on the rate of translocation of solutes in xylem: the analogy with an endless belt delivering coal. Rate of delivery to *b* (translocation) depends on the rate of drop from *a* (absorption), regardless of the rate of movement of the endless belt (transpirational stream).

normally depends not on the speed of rotation of the endless belt but on the rate of transfer to it (Fig. 11-8). If the substances enter the plant by active absorption, the amount reaching the leaves per unit time will depend not on the speed of the transpiration stream (or of the endless belt) but on the rate of absorption (or of deposit on the belt). On the other hand, if we consider individual branches of a plant coming off the main transpiration stream, the one with the most rapid tributary will tap off the largest amount of solutes from the main stream. Consequently, we might expect a more rapidly transpiring branch of a plant to receive a larger fraction of the absorbed nutrients. If, however, the transpiration (or endless belt) is so slowed down that the solutes accumulate in roots (or the coal piles up at the beginning of the belt), the absorption rate will be stopped or at least greatly retarded because of a backflow of accumulated solutes into the soil. Under these conditions, transpiration rate may be crucial to absorption by preventing this excessive accumulation.

Many investigators have actually found that increased transpiration rate does increase the uptake of substances from the root medium, but others have failed to find any such effect. The problem appears to

have been resolved as a result of the many investigations in recent years. The following generalizations now appear valid:

1. At low external ion concentrations (1 mM or less) the effect of transpiration is negligible; at high external concentrations it may be significant.
2. In the case of ions absorbed actively, especially those that are quickly metabolized, the effect of transpiration is slight even at high concentrations (80 mM).
3. In the case of some plants (e.g., maize, beans) the effect of transpiration seems to be slight, even at relatively high concentrations; in other plants it appears greater.

When significant effects of transpiration on ion uptake have been found, this has led to the conclusion that active absorption occurs only in excised roots and that passive absorption is characteristic of the whole plant as a result of transpiration. Actually, however, the effect of transpiration is probably mainly, if not solely, a result of its ability to reduce the amount of energy needed to absorb the solutes actively. It does this by quickly transporting the actively absorbed solutes away from the absorbing organs (the roots) to the leaves. This lowers the gradient against which the absorption would have to take place and therefore decreases the energy expended per ion absorbed. It has been suggested (Hylmö, 1953) that the increased tension produced by higher transpiration rates pulls the external solution through the intercellular spaces along the cell walls of the epidermal and cortical cells, in this way bypassing the more resistant semipermeable membranes of the protoplasm in all these cells. It would then be necessary for the solution to pass through the protoplasm of only one layer of cells, the endodermis, before being pulled into the xylem. Actually, however, the protoplasm would not be completely bypassed. On the contrary, the nutrient solution would be dragged through

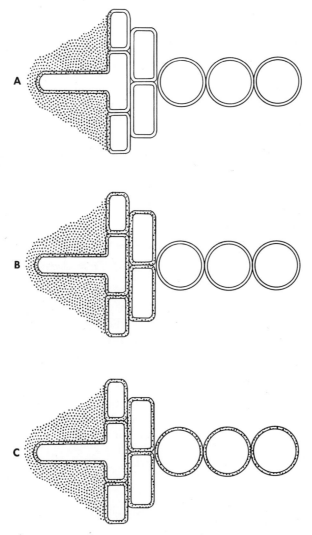

Fig. 11-9. Effect of transpiration on active absorption of solutes by roots. **A,** No transpiration. Only outer surface of epidermal cells come in contact with soil solutes. **B,** Slow transpiration. Outermost two layers in contact with soil solutes. **C,** Rapid transpiration. Several layers of cells in contact with soil solutes. Stippled cell walls contain soil solution, but the others do not.

the cell walls by the transpirational pull and would be brought into contact with the protoplasm of many more cells (Fig. 11-9). According to the "symplast" concept, introduced by Arisz, the protoplasm of all the living cells of a plant act as a single unit because of the plasmodesmal connections between cells. Therefore, once nutrient has been actively absorbed into the protoplasm of a single cell, it can move freely across the plasmodesmata from living cell to living cell until it is excreted into the vessels.

These interactions between transpiration and absorption point to an important principle: Numerous physiological processes occur in the plant simultaneously. Therefore, no matter how much we learn about each by studying it independently, it must be ultimately investigated when under the influence of the others in order to understand it fully. This principle recurs in plant physiology again and again.

The actual rate of transpiration may be measured in several ways. The water given off may be collected in drying towers (e.g., by a desiccant such as calcium chloride), and the increase in weight of the towers will be equal to the weight of water lost. This method has been replaced in recent years by the infrared absorption method, which permits a continuous record of the transpiration rate. The water content of the air is measured before and after it passes over the plant by the amount of infrared radiation absorbed. But these methods require enclosing the plants, which is difficult to do without altering the transpiration rate. Other methods are also available, but the simplest is to weigh a potted plant at intervals. The loss of weight is essentially equal to the loss of water from the plant since other factors (e.g., carbon assimilation) usually have a negligible effect on the weight compared to the effect of water loss. In the field, similar weight measurements have been made on shoots immediately after excision from the plant. The validity of this method has been the subject of much controversy.

Recent attempts to control transpiration have led to the development of antitranspirants—substances that reduce the rate of transpiration. These are of three main kinds: (1) apolar substances (e.g., cetyl alcohol, oxyethylene docosanol); (2) plastics (e.g., polyethylenes); and (3) substances that lead to stomatal closure (e.g., phenylmercuric acetate, Zelitch, 1963). All of these have led to some success, although their usefulness is still a matter of controversy. A new group of the third type—alkenylsuccinic acids (e.g., decenyl succinic acid)—have been found to increase the permeability of the root cells (Zelitch, 1964; Kuiper, 1964). However, the exact mechanism of their effect on stomata is unknown.

RESISTANCE TO FLOW OF WATER

Water enters the roots from the soil, flows up the plant, and evaporates from the shoot surface into the atmosphere. The question has long been asked: What portion of this continuous stream offers the major resistance to flow? In order to answer this question, an analogue of Ohm's law has been applied (van den Honert, 1948):

$$\frac{dm}{dt} = \frac{P}{R}$$

where $\frac{dm}{dt}$ = amount of electricity or water flowing per unit time

P = electric potential or water pressure drop

R = resistance to flow

In other words, at any point in the "stream," the amount of water flowing must be directly proportional to the pressure and inversely proportional to the resistance to flow. Since the stream is continuous, if the plant is in the steady state (loss equals absorption), the amount of water flowing past any cross-sectional area per unit time $\left(\frac{dm}{dt}\right)$ must be the same at all cross sections of the plant normal to the stream. Therefore, $\frac{P}{R}$ must be constant throughout the plant, that is:

$$\frac{P_1 - P_0}{R_r} = \frac{P_2 - P_1}{R_x} = \frac{P_3 - P_2}{R_1} = \frac{P_4 - P_3}{R_g}$$

where $P_1 - P_0$, $P_2 - P_1$, $P_3 - P_2$, $P_4 - P_3$, are the pressure drops across the various resistances

R_r = resistance across the root

R_x = resistance across the xylem

R_1 = resistance across the leaf

R_g = resistance across the gaseous phase above the leaf

In the case of water flowing through the plant, the various pressure drops were taken to represent the decreases in water potential. Since the greatest decrease in water potential occurs in the gaseous phase, it was concluded that the major resistance to flow must reside just above the leaf surface (where the largest drop in vapor pressure occurs). This conclusion is opposed to all physiological experience (Kramer, 1949), which indicates that the major resistance to the liquid flow is at the plasma membranes of those living cells across which the water must flow, that is, in the root and the leaf. The reason for the error is because the above analogue applies only to liquid flow within the plant, since it uses the hydro-static (osmotic) pressure gradient for bulk flow. The vapor flow within and above the plant is a diffusion flow and is, therefore, controlled by the vapor pressure gradient. Since the maximum possible vapor pressures (0.04 atm) are much smaller than the hydrostatic pressures (several atmospheres), the resistance to the vapor flow must also be much smaller than the resistance to the liquid flow.

Direct measurements have, in fact, revealed that the external, boundary layer resistance is generally less than 10% of the internal leaf resistance in corn (Impens et al., 1967). The clearest evidence is produced by stripping the epidermis from *Senecio* plants (Rawlins[*]). This causes rapid wilting (because of increased vapor flow), showing that the greatest resistance to flow is provided by the epidermal layer and not by the gaseous phase above it, which remains at the leaf surface whether or not the epidermis is present. It also proves that once the epidermal resistance is removed (e.g., by wide-open stomata), the major resistance to flow must occur in the liquid phase of water flow behind the surface of the evaporating leaf cells. Since the only resistance to flow in the vessels is the viscosity of the water, the major resistance must be at the plasma membranes in the living cells of the root and leaf, which offer a resistance to flow above that resulting from the viscosity of the water.

---

[*]Rawlins, S. L. (U. S. Salinity Laboratory, Riverside, Calif.): Personal communication.

QUESTIONS

1. What gases are exchanged between the plant and its environment?
2. What plant parts are involved in this exchange?
3. Does the plasma membrane interfere with this exchange?
4. What barriers do interfere with this exchange?
5. Do the cuticular or suberized layers completely stop exchange?
6. What evidence is there that exchange is impeded?
7. What plant organ is able to control the exchange?
8. How does the plant organ do this?
9. What force is responsible for the opening of the stomata?
10. What force is responsible for the closing?
11. What must occur in order for the opening force to change?
12. What controls the movement of water into or out of the guard cells?
13. In general, what is the relation of stomatal opening to light?
14. What are the exceptions to this relationship?
15. Is the degree of opening in any way affected by cells other than the guard cells?
16. What effect does wilting have on stomatal opening?
17. What chemical change in the guard cells may lead to stomatal opening?
18. What change in gas content may initiate this chemical change?
19. Trace the series of steps leading to stomatal opening when a leaf is exposed to light.
20. How does the rate of transpiration (per unit of leaf area) compare with the rate of evaporation from a free water surface?
21. What fraction of the leaf area is covered with stomata?
22. How does the rate of water loss per

unit stomatal area compare with that from a free water surface?

23. Does this mean that transpiration is too rapid to be a diffusion process?

24. How can this rate of water loss per unit of stomatal area be explained?

25. Suppose two beakers have diameters of 3 inches and 6 inches, respectively. If both are filled with water, how much weight would the 6-inch one lose during the time when the 3-inch one lost 2 g water?

26. Suppose a single stoma of 1μ opening loses 2 g water, how much would it lose in the same time when open to 2μ?

27. Does this mean that transpiration is always proportional to the stomatal aperture?

28. When is transpiration not proportional to the stomatal aperture?

29. What is meant by transpiration (as opposed to guttation)?

30. What is the difference between stomatal and cuticular transpiration?

31. Under what conditions could transpiration increase without the occurrence of a change in stomatal opening?

32. What effect does wind have on transpiration?

33. What is the primary role of stomatal opening in the physiology of the plant?

34. Can transpiration benefit the plant?

35. In what ways can transpiration benefit the plant?

SPECIFIC REFERENCES

Bange, G. G. J. 1953. On the quantitative explanation of stomatal transpiration. Acta Bot. Neerl. 2:255-297.

Benecke, W., and L. Jost. 1924. Pflanzenphysiologie. I. Gustav Fischer Verlag, Stuttgart.

Brown, H. T., and F. Escombe. 1900. Static diffusion of gases and liquids in relation to the assimilation of carbon and translocation in plants. Phil. Trans. Roy. Soc. [B] 193:223-291.

Curtis, O. F. 1938. Wallace and Clum, "Leaf temperatures": A critical analysis with additional data. Amer. J. Bot. 25:761-771.

Decker, J. P., and B. F. Wetzel. 1957. A method for measuring transpiration of intact plants under controlled light, humidity, and temperature. Forest Sci. 3:350-354.

Ekern, P. G. 1965. Evapotranspiration of pineapple in Hawaii. Plant Physiol. 40:736-739.

Ferri, M. G. 1955. Contribuição ao conhecimento da ecologia do cerrado e da caatinga. Faculdad de Filosofia Ciencias e Letras, Universidade de São Paulo, São Paulo, Brazil. Bol. 195. Bot. 12.

Gale, J., and R. M. Hagan. 1966. Plant antitranspirants. Ann. Rev. Plant Physiol. 17:269-282.

Gates, D. M. 1963. Leaf temperature and energy exchange. Arch. Meteorol. Geophys. Bioklim. 12:321-336.

Gates, D. M. 1966. Transpiration and energy exchanges. Quart. Rev. Biol. 41:353-364.

Heath, O. V. S., and F. L. Milthorpe. 1950. Studies in stomatal behavior. V. The role of carbon dioxide in the light response of stomata. J. Exp. Bot. 1:227-243.

Honert, T. H. van den. 1948. Water transport in plants as a catenary process. Discuss. Faraday Soc. 3:146-153.

Hylmö, B. 1953. Transpiration and ion absorption. Physiol. Plant. 6:333-405.

Impens, I. I., D. W. Stewart, L. H. Allen, Jr., and E. R. Lemon. 1967. Diffusive resistance at, and transpiration rates from leaves in situ within the vegetative canopy of a corn crop. Plant Physiol. 42:99-104.

Kuiper, P. J. C. 1964. Water transport across root cell membranes; effect of alkenyl-succinic acids. Science 143:690-691.

Lange, O. L. 1959. Untersuchungen über Wärmehaushalt und Hitzeresistenz mauretanischer Wüsten-und Savannenpflanzen. Flora 147:595-651.

Lee, R., and D. M. Gates. 1964. Diffusion resistance in leaves as related to their stomatal anatomy and microstructure. Amer. J. Bot. 51:963-975.

Loftfield, J. V. G. 1921. The behavior of stomata. Carnegie Institute of Washington, Pub. No. 314. Washington, D. C.

MacDougal, D. T., and E. B. Working. 1933. The pneumatic system of plants, especially trees. Carnegie Institute of Washington, Pub. No. 441. Washington, D. C.

Meidner, H., and T. A. Mansfield. 1965. Stomatal responses to illumination. Biol. Rev. 40:493-509.

Moss, D. N., and S. L. Rawlins. 1963. Concentration of carbon dioxide inside leaves. Nature (London) 197:1320-1321.

Nishida, K. 1963. Studies on stomatal movement of crassulacean plants in relation to the acid metabolism. Physiol. Plant. 16:281-298.

Peat. S. 1951. Biological transformations of starch. Adv. Enzymol. 11:339-375.

Scarth, G. W., and M. Shaw. 1951. Stomatal movement and photosynthesis in Pelargonium. Plant Physiol. **26**:207-225.

Slatyer, R. O., and J. F. Bierhuizen. 1964. The influence of several transpiration suppressants on transpiration, photosynthesis, and water use efficiency of cotton leaves. Aust. J. Biol. Sci. **17**: 131-146.

Stålfelt, M. G. 1966. The role of the epidermal cells in the stomatal movement. Physiol. Plant. **19**: 241-256.

Waggoner, P. E., and I. Zelitch. 1965. Transpiration and the stomata of leaves. Science **150**: 1413-1420.

Yin, H. C., and Y. T. Tung. 1948. Phosphorylase in guard cells. Science **108**:87-88.

Zelitch, I. 1964. Reduction of transpiration of leaves through stomatal closure induced by alkenylsuccinic acids. Science **143**:692-693.

GENERAL REFERENCES

Ketellapper, H. J. 1963. Stomatal physiology. Ann. Rev. Plant Physiol. **14**:249-270.

Kramer, P. J. 1949. Plant and soil water relationships. McGraw-Hill Book Co., New York.

Levitt, J. 1967. The mechanism of stomatal action. Planta **74**:101-118.

Scarth, G. W. 1932. Mechanism of the action of light and other factors on stomatal movement. Plant Physiol. **7**:481-504.

Stålfelt, M. G. 1956. Die stomatäre Transpiration und die Physiologie der Spaltöffnungen, p. 351-426. *In* W. Ruhland. Handbuch der Pflanzenphysiologie. Vol. 3. Springer-Verlag, Berlin.

Zelitch, I. (Ed.). 1963. Stomata and water relations in plants. Conn. Agric. Exp. Sta. Bull. No. 664.

Zelitch, I. 1965. Environmental and chemical control of stomatal movement in leaves. Biol. Rev. **40**:463-482.

part **III**

BIOCHEMISTRY

chapter **12**

Nutrition

A. Autotrophy as opposed to heterotrophy
B. Mineral nutrition
   1. Essentiality
   2. Roles of elements
   3. Toxic effects

AUTOTROPHY AS OPPOSED TO HETEROTROPHY

The nutrition of any organism includes all the substances that must be supplied from the outside in order for it to complete its life cycle normally. Those organisms that require both organic and inorganic substances are called *heterotrophic;* those that need be supplied only with the inorganic substances are called *autotrophic* ("self-feeding") since they make their own organic substances. Animals and nongreen plants are heterotrophic and so are the albino mutants of green plants, which can be grown to maturity if supplied with sugar, for example, by immersing the cut ends of their leaves in sugar solutions and allowing the sugar to be taken up in the transpiration stream. Of course, they must also be supplied with the same mineral nutrients as the normal green plant. When nongreen parts (e.g., roots, stem tips, embryos) are excised (severed) from the green parts of the plant, they can be made to grow only when supplied with at least a sugar and, in many cases, several additional organic substances (Table 12-1). Therefore, these parts of the green plant are heterotrophic. This fact has become more important to the plant physiologist in recent years because of the usefulness of such organ, tissue, or cell cultures as tools for investigating the growth and

development of the plant. With favorable material and suitable techniques, a single culture of maize endosperm has yielded more than 5000 g tissue in 5 months (Graebe and Novelli, 1966). In many cases it has not yet been found possible to grow such cultures in a completely defined medium, and organic nutrients of incompletely known chemical constitution must be added (e.g., coconut milk). The heterotrophic nutrition of such cultures, therefore, is not completely understood. In some cases the cultures may be maintained and subcultured indefinitely. In others (e.g., wheat roots) the cultures eventually die for unknown reasons.

The normal, green plant, on the other hand, is autotrophic; that is, it can synthesize all its essential organic substances, provided it is supplied with all the essential inorganic elements and grown under normal conditions (e.g., in the light). Nevertheless, the line between autotrophism and heterotrophism is not a sharp one. The difference may sometimes be quantitative rather than qualitative because a plant may be able to synthesize a substance but in insufficient quantity or too slowly to support normal growth and development. For this

129

reason, even the normally autotrophic higher green plant may become partially heterotrophic under unfavorable growing conditions. When, for instance, such a plant is grown autotrophically at too high a temperature (but below a directly lethal temperature), its growth may be impaired or even stopped completely. In some cases, at least, this is caused by its inability to synthesize certain organic compounds (e.g., vitamins or amino acids) in adequate quantities at high temperatures (Langridge, 1963), and it must be supplied with these substances in order to grow normally. Injuries resulting from these metabolic deficiences have been called "biochemical lesions."

Under normal growing conditions, however, all higher green plants are autotrophic, and their nutrition is therefore solely inorganic. In fact, it is commonly called mineral nutrition, and the elements absorbed by the roots (excepting carbon, hydrogen, and oxygen) are called mineral elements since they are obtained either directly or indirectly from the minerals in the soil. The mineral nutrition of plants therefore embraces all the elements essential to the plant's existence, with the exception of carbon, hydrogen and oxygen. Nitrogen, strictly speaking, is not a mineral element, but it is included with these since it is obtained by the plant from the soil solids.

MINERAL NUTRITION
Essentiality

The first and perhaps most fundamental problem of mineral nutrition is to determine which chemical elements are essential to the life and growth of the plant. As a start, one might analyze the plant to find out which are present (Table 12-2). When this is done, the number of elements found depends on the plant, the medium in which it was grown, and the completeness of the analysis. However, this tells us merely that the elements not found in the plant (assuming that the methods of analysis are

Table 12-1. White's nutrient medium for growing excised roots*

Substance	Milligrams per liter
$MgSO_4$	360.0
$Ca(NO_3)_2$	200.0
Na_2SO_4	200.0
KNO_3	80.0
KCl	65.0
$NaH_2PO_4 \cdot H_2O$	16.5
$Fe_2(SO_4)_3$	2.5
$MnSO_4$	4.5
$ZnSO_4$	1.5
H_3BO_3	1.5
KI	0.75
Sucrose	2000.0
Glycine	3.0
Nicotinic acid	0.5
Pyridoxine	0.1
Thiamine	0.1

*From White, P. R. 1943. A handbook of plant tissue culture. The Ronald Press Co., New York.

Table 12-2. Analysis of stems, leaves, cob, grain, and roots of corn*

Element	Percent of total dry weight
O	44.4
C	43.6
H	6.2
N	1.46
P	0.2
K	0.92
Ca	0.23
Mg	0.18
S	0.17
Fe	0.08
Si	1.17
Al	0.11
Cl	0.14
Mn	0.035
Undetermined	0.93

*From Miller, E. C. 1938. Plant physiology. McGraw-Hill Book Co., New York.

adequate) are not essential in appreciable quantities. The mere presence of an element in a plant does not prove its essentiality. Thus, if a large enough number of plants from different regions were analyzed, certain elements might sometimes be present or sometimes be absent from the plant. Actually, some sixty elements have been found in plants (Robinson and Edgington, 1945), and undoubtedly all the rest of the now 100-odd known elements could be made to occur in the plant by supplying them to the roots.

A more direct approach to the problem of essentiality is to grow plants in the complete absence of a given element. If the plant grows normally, the element is evidently not essential; if it fails to grow normally, the element is usually considered essential. In order to be certain that the element is truly essential, it is necessary to show not only that (1) a deficiency of the element makes it impossible for a plant to complete its vegetative or reproductive cycle, but also that (2) it cannot be replaced by any other element, and that (3) the effect is not simply the result of interaction with (e.g., detoxification of) other nonessential elements, organisms, etc. outside the plant. These three requirements have been called the *criteria of essentiality* (Arnon and Stout, 1939). However, it has recently been shown that some of the accepted essential elements can at least be partially replaced by others (e.g. magnesium by manganese, potassium by rubidium, etc., Hewitt, 1951).

The search for essential elements has required the development of a technique for growing the plant in the absence of impurities. This has meant the elimination of the natural root medium—the soil. In its place a pure sand, gravel, or best of all, simple solutions have been used. The latter so-called water or solution cultures offer the greatest difficulty since they fail to supply the normal support for the roots and the normal amounts of oxygen. This requires development of adequate artificial support and aeration. A successful water or solution culture therefore requires a certain amount of engineering. This system of growing plants in the absence of soil has been used to some extent on a practical scale and has been named *hydroponics* (Gericke, 1940).

Using only the first criterion of essentiality (i.e., no normal growth in the absence of the element), the following seven elements were long ago found to be essential: nitrogen, phosphorus, potassium, calcium, magnesium, sulfur, and iron. Thus, it was discovered that plants grow normally in four-salt solutions containing potassium nitrate, monopotassium phosphate, calcium chloride, and magnesium sulfate (or other salts supplying these six elements) with a small amount of ferric chloride (or other iron salt). Even three-salt solutions (monopotassium phosphate, calcium nitrate, magnesium sulfate) plus some iron were soon found adequate. Many formulas for such solutions have been adopted, some of them named after workers who developed them (e.g., Knop's, Cronin's, Tottingham's, Shive's). However, the early solutions were made with salts that were not pure. Small quantities of other salts were therefore invariably present. Since it was already shown that one element (iron) was needed only in a small quantity, it was logical to suspect that other elements might also be essential in trace amounts. It was soon found that the better the purification of the basic three or four salts the poorer was the growth. This was followed up by direct evidence that other elements are needed in small amounts. They are just as essential for the life and growth of the plants as the preceding seven elements, and therefore no growth is possible in their complete absence. However, in view of the small amounts needed, they are usually grouped (with iron) separately from the preceding six major, or macronutrient, elements and are called trace, minor, or micronutrient elements. By the late 1930's the following six had been conclu-

Table 12-3. Quantities of mineral elements used in complete nutrient solutions*

Major ion	Quantities (moles)	Trace element	Quantities (ppm)
NO_3^-	0.005–0.010	Fe	0.5
PO_4^-	0.00025–0.002	B	0.25 (or more)
$SO_4^=$	0.001–0.010	Mn	0.25
K^+	0.002–0.010	Zn	0.25
Ca^{++}	0.002–0.005	Cu	0.02
Mg^{++}	0.001–0.010	Mo	0.01

*Adapted from Robbins, W. R. 1946. Soil Sci. **62**:3-22.

sively proved to be trace essential elements: iron, manganese, copper, zinc, boron, and molybdenum. They are needed in the nutrient solution in amounts ranging from 0.5 ppm to 0.01 ppm (Table 12-3). Larger quantities may cause injury or death. Recent work indicates that chlorine (one of the most ubiquitous of elements) is a seventh trace essential element, although usually present in the plant in as large quantities as the major essential elements (Broyer et al., 1954). There are reasons for believing that cobalt is also essential, although the classical methods have failed to prove this because of the extremely small traces in which it seems to be needed. It is certainly essential for leguminous plants when these are not supplied with nitrogen compounds. But this is really a requirement for the symbiotic bacteria since no cobalt need be knowingly added when nitrogen compounds are supplied (Evans and Kliewer, 1964). Cobalt is also known to be essential for at least some algae. In the case of higher plants it is possible that enough cobalt is present as impurities to satisfy the needs of the plant, and therefore, the amount added knowingly has no effect. Indirect evidence favors this conclusion for the following reasons:

1. The amounts known to be present as impurities even in the most carefully purified media are of the order of the amount needed to supply the plant with the cobalt-containing vitamins (B_{12}).

2. The addition of cobalamin (vitamin B_{12}) under certain conditions furthers the synthesis of an amino acid (serine) in the primary root of *Zea mays* (Graser, 1965).

Direct evidence has also been produced in one case. Recent experiments that very carefully excluded contamination by traces of cobalt in the air have succeeded in demonstrating a significant decrease in growth when cobalt was not knowingly added (Wilson and Nicholas, 1967).

Some investigators have produced evidence for the essentiality of a few other elements (e.g., aluminum, silicon, selenium) at least for certain plants (Table 12-4), but other workers have failed to obtain positive results. This is partly explainable by the frequently observed stimulating effect of nonessential elements on the growth of plants, for instance by counteracting the toxicity of other elements present in the nutrient medium. Elements such as aluminum and silicon have been called *ballast elements* (Frey-Wyssling, 1945) because they are normally present in large amounts, although the plant can be grown perfectly normally without them (as far as can be determined). Some plants are *accumulators;* that is, they concentrate large quantities of certain elements (e.g., selenium, aluminum) in their tissues.

Complete absence of any one of the essential elements (major or trace) will completely stop growth, but under normal conditions the elements are never com-

Table 12-4. Essentiality of elements

Essential for all higher plants	Essential for all higher plants but not for all algae or fungi	Apparently essential for some species		
		Element	Species	Reference

1. Major (or macro)
 Cationic Anionic
   K   N
   Mg   P
   Ca   S
2. Trace (or micro)
 Cationic Anionic
   Fe   B
   Mn   Mo
   Cu   Cl*
   Co*
	Ca
	B
	Cl (?)
		V
		Al
		Si
		Na
		
		Se
		I
			Scenedesmus
			Ferns
			Diatoms
			Blue-green alga
			Atriplex
			Astragalus sp.
			Polysiphonia
				Arnon and Wessel (1953)
				Tauböck (1942)
				Lewin (1955)
				Allen and Arnon (1955)
				Brownell (1965)
				Trelease and Trelease (1938)
				Fries (1966)

*Cl and Co have not been tested on enough higher plants yet, but recent results indicate that they are essential (Broyer et al., 1954; Wilson and Nicholas, 1967).

pletely absent. If, however, they are present in less than optimum quantities, growth will occur but abnormalities known as *deficiency diseases* will develop. In animals this term is used for diseases caused by vitamin deficiencies. Since plants are autotrophs, they show disease symptoms only for deficiencies of mineral elements. The symptoms of such diseases are more or less specific for each element (Bear and Coleman, 1941), although it is sometimes difficult to distinguish the differences. Furthermore, different plants will show somewhat different deficiency symptoms.

The commonest type of deficiency symptom is *chlorosis,* a reduction in the amount of green pigment in the leaf. But chlorosis can occur as a result of any one of several deficiencies (nitrogen, magnesium, iron, manganese, etc.). However, the type of chlorosis is sometimes different for different elements. Thus it is a uniform loss of color in the case of nitrogen deficiency but more pronounced between the veins in the case of iron deficiency. The trace elements have given rise to a particularly large number of deficiency diseases, for example, celery crack, heart rot and dry rot of sugar beets, internal browning of cauliflower (B defi-

ciencies), little leaf or rosette of fruit trees (zinc deficiency), marsh spot of pea seeds (manganese deficiency), etc.

The deficiencies found in the plant may not correspond with those in the soil. Thus a lime-induced chlorosis may actually be an iron deficiency because the high pH has made the iron in the soil unavailable to the plant. Sometimes the ratios of the elements in the root medium may be more important than the absolute quantities (e.g., the iron-manganese and the calcium-boron ratios). A method that overcomes the unavailability of iron because of high pH is to apply chelated iron, that is, iron combined with an organic compound in such a way that it cannot ionize and therefore cannot be precipitated (Stewart and Leonard, 1952). One of the most commonly used of such substances is ethylenediaminetetraacetate (EDTA).

Some deficiency symptoms occur first on the oldest leaves (e.g., nitrogen, potassium, phosphorus, magnesium); others occur first on the youngest leaves (sulfur, calcium, iron, manganese, boron, copper, zinc). Elements of the first group are mobile and are transferred from the older inactive leaves to the younger growing leaves. The elements of the second group are immobile and re-main in the old leaves, causing the newly developing leaves to show the deficiency although the older ones do not (Müller, 1949). However, the preceding separation into clear-cut groups of mobile and immobile elements is not to be taken too rigidly since some elements are partially mobile (e.g., sulfur) and others are completely mobile (e.g., potassium).

The following quantitative evaluation of mobility in the phloem of *Yucca flaccida* has been calculated from the ratio of $\dfrac{\text{concentration in exudate}}{\text{concentration in leaves}}$ (Tammes and van Die, 1966):

1.	K	0.89		7.	Na	0.12
2.	P	0.81		8.	Cu	0.09
3.	N	0.61		9.	Mg	0.03
4.	B	0.29		10.	Ca	0.01
5.	Zn	0.27		11.	Mn	0.01
6.	Fe	0.12				

But mobility to regions of exudation may be different from mobility to growing regions of the plant. This seems to be particularly true of magnesium, which is largely mobile to growing regions and immobile in the preceding test. Iron is one of the most immobile ions in the plant according to Burleson and Cunningham (1963). Manganese, on the other hand, may be immobile in old leaves but is mobile in young leaves (Klimovitskaya and Vizir, 1963).

Even some of the most immobile elements can be translocated, although only in extremely small amounts (e.g., calcium, Biddulph et al., 1959). It is perhaps because of this poor translocation of calcium that many seeds and pollen grains are deficient in this element and cannot germinate normally unless supplied with it. Similarly, manganese, one of the "immobile" elements, apparently can be redistributed, although to a very small degree compared to potassium and nitrogen (Vose, 1963). The differences in mobility, therefore, are quantitative rather than qualitative, and if

a grouping is desirable, the following three groups would perhaps serve as a rough approximation: nearly completely mobile—K, Cl; largely mobile—N, P, Mg, S; largely immobile—Ca, Fe, Mn, B, Cu, Zn, Mo.

Roles of elements

Difficult as it sometimes has been to prove the essentiality of a nutrient element, it is often much more difficult to find out why it is essential, that is, its role in the physiology of the plant. If the element is a constituent of a substance whose function in the plant is known, this, of course, explains why that element is essential. However, it must always be remembered that a single element may play many different roles in the physiology of the plant. Nitrogen, for instance, is one of the elements in proteins and nucleic acids; life is therefore impossible without nitrogen. But it is also a constituent of phospholipids, some vitamins, and chlorophyll, all of which play definite roles in the physiology of the plant. Nitrogen also occurs in many substances with unknown roles in the plant (e.g., plant bases). Phosphorus, like nitrogen, is a necessary constituent of many vitally important substances: nucleoproteins, phospholipids, enzyme components, etc. Sulfur occurs in proteins and some vitamins as well as in other substances such as glutathione. The importance of sulfolipids has recently been discovered (Davies et al., 1965). Magnesium is a constituent of chlorophyll, but this cannot be its only function since it is also essential for nongreen plants, primarily because it is an activator of many enzymes. Calcium occurs in the pectate that cements the walls of adjacent cells together, but unlike magnesium, it is not essential for some nongreen lower plants. However, its role is obviously much more fundamental than as a mere cementer of cells. There is reason to believe that protoplasm cannot exist in the absence of certain cations. In most organisms calcium is undoubtedly one of these ions and potassium is another. Since protoplasm probably functions only at a pH of about 6.5 to 7, the protoplasmic proteins with their relatively low isoelectric points (mostly 4.5 to 5) must be in the acidic or anionic form. They must therefore form salts with the cations. Apparently, K^+ and Ca^{++} and perhaps Mg^{++} are the cations primarily involved, and they must be present in a certain ratio to maintain the normal, native, and active state of the proteins. Unlike the other major essential elements, potassium is not a constituent of any known essential organic substance but occurs almost solely in the ionic form. The K^+ ion is, however, essential as an activator in protein synthesis (Evans and Sorger, 1966). The fact that it is needed in so much larger quantity than ions such as Ca^{++} and Mg^{++} may be related to the fact that a balanced solution for living cells consists of about 10 parts of potassium to 1 part of calcium or magnesium. It is also needed in unusually large quantities as an enzyme activator (0.01 to 0.1 normal, Evans and Sorger, 1966) because so small a percentage remains attached to the enzyme on account of the high dissociation of its salts. According to earlier concepts, a "balanced ion atmosphere" was needed to maintain the properties of protoplasm. This view was based primarily on the success of balanced solutions in keeping cells alive and on colloidal principles. However, modern evidence points to the organelles of protoplasm as requiring protection by cations; for example, K^+ protects ribosomes (Ts'o, 1962). This may result from neutralization of the negative charge of the proteins, enabling them to attach to RNA. Mg^{++} stabilizes the configuration of the ribosomal RNA (Bayley, 1966; Sager and Hamilton, 1967).

The major cationic elements are therefore needed in much larger quantities than the trace cationic elements, probably because of their role in neutralizing the acid proteins and other organic acids produced continuously in the metabolism of the cell (see Chapter 14). In the case of magnesium a second role is as a constituent of chlorophyll.

The role of the trace elements is largely understood: Iron, copper, and zinc are con-

Table 12-5. Organic substances required by living plant cells and tissues and the essential mineral elements contained by them

Element	Cell wall Pectins	Protoplasm Proteins (enzymes)	Nucleic acids	Lipids*	Chlorophyll	Enzyme cofactors
1. Major						
N		X	X	X	X	X
P			X	X		X
S		X		X		X
Ca	X	X (protein salt?)				X
Mg	(X)	X (protein salt?)			X	X
K		X (protein salt?)				X
2. Trace						
Fe, Cu, Mn, Zn, Mo,						X
B						Inhibitor?

*The listed elements are needed only for specific lipids.

stituents of many enzymes (e.g., peroxidase, catalase, and cytochromes in the case of iron; tyrosinase and ascorbic acid oxidase in the case of copper; carbonic anhydrase in the case of zinc), and manganese is an activator of several others. It is now known that molybdenum is a cofactor of the enzymes of nitrate assimilation. However, in spite of the great amount of work on boron, its functions are completely unknown. According to one theory, boron forms a complex with sugars that penetrates into living cells more readily than do free sugars and is therefore more readily translocated to the growing cells (Gauch and Dugger, 1953); but this theory is opposed by most experimental evidence, and many other explanations have been proposed. According to Lee and Aronoff (1967), borate inhibits an enzyme system, which in the absence of boron, leads to excess formation of phenolic acids and to necrosis and eventual death of the plant.

The main substances required for the life of the plant and the essential nutrient elements they contain are summarized in Table 12-5.

Toxic effects

All the trace essential elements are toxic when supplied to the plant in quantities considerably above the optimum. Borates have actually been used as weed killers, and copper has long been known to be a cause of the toxicity of standard distilled water to living cells. Some plants are much more sensitive to such effects than others. Algae are readily killed by low concentrations of copper (e.g., 1 ppm). Higher plants suffer injury in acid soils frequently because of excess absorption of manganese. Even the major essential elements may be injurious in high concentrations, sometimes because of the osmotic effect leading to dehydration and perhaps even plasmolysis of the cells. The toxicity of single salt solutions (e.g., potassium chloride) has already been discussed (see Chapter 7) and is over-

come by antagonism (e.g., by calcium chloride). Even nonessential elements may counteract (or antagonize) such toxicity. This is one reason why some nonessential elements can sometimes stimulate the growth of plants. Thus aluminum stimulates growth when copper is present in toxic quantities. On the other hand, some nonessential (or possibly essential) elements are highly toxic (e.g., mercury, selenium, iodine, cadmium, cobalt, nickel, thallium).

It must therefore be concluded that all the mineral elements, whether essential or not, are capable of being toxic to plants if supplied in sufficient concentration and in the absence of a sufficient quantity of an antagonizing element. For optimum nutrition a mineral nutrient solution must therefore consist of both an optimum concentration of each salt and an optimum balance of the ions composing the salts.

QUESTIONS

1. What is an autotroph?
2. What is a heterotroph?
3. To which groups do higher plants belong?
4. Are there any exceptions?
5. Are any parts of the higher plant heterotrophic?
6. What must be supplied to excised roots beside inorganic substances in order for them to grow?
7. What is usually meant by the nutrition of higher plants?
8. How many elements are found in a normal plant?
9. How many salts must the higher plant be supplied with in order to provide all the major essential elements?
10. How many essential "mineral" elements are known?
11. Is there any difference in essentiality between major and trace elements?
12. What are the criteria of essentiality?
13. What elements are likely to be added to the essential list?

14. Will these be major or trace essential elements?
15. What is a deficiency disease?
16. What are some of the symptoms of a deficiency disease?
17. Why, in general, are elements essential?
18. Give an example of why elements are essential.
19. Why are the trace elements essential?
20. Why are trace elements essential in smaller quantities than the major elements?
21. Are there any elements whose essentiality is not understood?
22. Are any essential elements toxic?
23. Which are toxic in lower concentrations, the major or the trace essential elements?
24. Are any nonessential elements toxic?
25. Can nonessential elements counteract toxicity of essential elements?
26. Of what importance is toxicity in a search for new essential elements?

SPECIFIC REFERENCES

Allen, M. B., and D. I. Arnon. 1955. Studies on nitrogen-fixing blue-green algae. I. The sodium requirement of *Anabaena cylindrica*. Physiol. Plant. **8**:653-660.

Arnon, D. I., and P. R. Stout. 1939. The essentiality of certain elements in minute quantity for plants with special reference to copper. Plant Physiol. **14**:371-375.

Arnon, D. I., and G. Wessel. 1953. Vanadium as an essential element for green plants. Nature (London) **172**:1039-1040.

Bayley, S. T. 1966. Composition of ribosomes of an extremely halophilic bacterium. J. Molec. Biol. **15**:420-427.

Biddulph, O., R. Cory, and S. B. Biddulph. 1959. Translocation of calcium in the bean plant. Plant Physiol. **34**:512-519.

Brownell, P. F. 1965. Sodium as an essential micronutrient element for a higher plant (*Atriplex vesicaria*). Plant Physiol. **40**:460-468.

Broyer, T. C., A. B. Carlton, C. M. Johnson, and P. R. Stout. 1954. Chlorine—a micronutrient element for higher plants. Plant Physiol. **24**:526-532.

Burleson, C. A., and H. G. Cunningham. 1963. Iron status and needs of the southern region. Plant Food Rev. **9**:9-13.

Davies, W. H., E. I. Mercer, and T. W. Goodwin.

1965. The occurrence and intracellular distribution of the plant sulfolipid in maize, runner beans, plant tissue cultures and *Euglena gracilis*. Phytochemistry **4**:741-749.

Evans, H. S., and M. Kliewer. 1964. Vitamin B_{12} compounds in relation to the requirements of cobalt for higher plants and nitrogen-fixing organisms. Ann. N. Y. Acad. Sci. **112**:735-755.

Fries, L. 1966. Influence of iodine and bromine on growth of some red algae in axenic culture. Physiol. Plant. **19**:800-808.

Gauch, H. G., and W. M. Dugger, Jr. 1953. The role of boron in the translocation of sucrose. Plant Physiol. **28**:457-466.

Graebe, J. E., and G. D. Novelli. 1966. A practical method for large scale plant tissue culture. Exp. Cell Res. **41**:509-520.

Graser, H. 1965. Förderung der Serin-synthese durch Vitamin B_{12} (Cobalamin). Naturwissenschaften **52**:108-109.

Klimovitskaya, Z. M., and L. K. Vizir. 1963. The content and distribution of manganese in plants. [Transl. title.] Ref. Zh. Biol. No. 16872. Moscow.

Langridge, J. 1963. Biochemical aspects of temperature response. Ann. Rev. Plant Physiol. **14**: 441-462.

Lee, S., and S. Aronoff. 1967. Boron in plants; a biochemical role. Science **158**:798-799.

Lewin, J. C. 1955. Silicon metabolism in diatoms. II. Sources of silicon for growth of *Navicula pelliculosa*. Plant Physiol. **30**:129-134.

Müller, D. 1949. Die physiologische Grundlage für die Mangelsymptome der Pflanzen. Physiol. Plant. **2**:11-23.

Robbins, W. R. 1946. Growing plants in sand cultures for experimental work. Soil Sci. **62**:3-22.

Robinson, W. O., and G. Edgington. 1945. Minor elements in plants and some accumulator plants. Soil Sci. **60**:15-28.

Sager, R., and M. G. Hamilton. 1967. Cytoplasmic and chloroplast ribosomes of Chlamydomonas; ultracentrifugal characterization. Science **157**: 709-711.

Stewart, I., and C. D. Leonard. 1952. Iron chlorosis —its possible causes and control. Citrus Mag. **14**(10):22-25.

Tammes, P. M. L., and J. van Die. 1966. Studies on phloem exudation from *Yucca flaccida* Haw. IV. Translocation of macro- and micronutrients by the phloem sap stream. Proc. Kon. Nederl. Akad. Wet. [Biol. Med.] **69**:655-659.

Taüböck, K. 1942. Über die Lebensnotwendigkeit des Aluminiums für Pteridophyten. Bot. Arch. **43**:291-304.

Trelease, S. F., and H. M. Trelease. 1938. Selenium as a stimulating and possibly essential element for indicator plants. Amer. J. Bot. **25**:372-380.

Ts'o, P. O. P. 1962. The ribosomes—ribonucleoprotein particles. Ann. Rev. Plant Physiol. **13**: 45-80.

Vose, P. B. 1963. The translocation and redistribution of manganese in Avena. J. Exp. Bot. **14**:448-457.

White, P. R. 1943. A handbook of plant tissue culture. The Ronald Press Co., New York.

Wilson, S. B., and D. J. D. Nicholas. 1967. A cobalt requirement for non-nodulated legumes and for wheat. Phytochemistry **6**:1057-1066.

GENERAL REFERENCES

Bear, F. E., and R. Coleman (Ed.). 1941. Hunger signs in crops. American Society Agronomy and National Fertilizer Association, Washington, D. C.

Evans, H. S., and G. J. Sorger. 1966. Role of mineral elements with emphasis on the univalent cations. Ann. Rev. Plant Physiol. **17**:47-76.

Frey-Wyssling, A. 1945. Ernährung und Stoffwechsel der Pflanzen. Buchergilde Gutenberg, Zurich.

Gericke, W. F. 1940. Soilless gardening. Prentice-Hall, Inc., Englewood Cliffs, N. J.

Hewitt, E. J. 1951. The role of the mineral elements in plant nutrition. Ann. Rev. Plant Physiol. **2**:25-52.

Hewitt, E. J. 1952. Sand and water culture methods used in the study of plant nutrition. Tech. Commun. Commonw. Bur. of Hort. Plantn. Crops No. 22, East Malling.

Hoagland, D. R. 1944. Inorganic plant nutrition. Chronica Botanica Co., Waltham, Mass.

Miller, E. C. 1938. Plant physiology. McGraw-Hill Book Co., New York.

Stiles, W. 1946. Trace elements in plants and animals. Cambridge University Press, London.

Wallace, T. 1953. The diagnosis of mineral deficiencies in plants by visual symptoms. Chemical Publishing Co., Inc., New York.

chapter 13

Metabolism

A. Anabolism and catabolism
B. Changes in chemical structure
  1. Hydrolysis-condensation reactions
  2. Group transfer reactions
  3. Group removal or addition reactions
  4. Electron transfer or oxidation-reduction reactions
  5. Isomerization
  6. Synthesizing reactions
C. Energy changes
D. Enzymes

ANABOLISM AND CATABOLISM

The nutrition of the plant should really include all the raw materials that must be supplied to support its normal growth and development. Beside the mineral elements already discussed, the autotrophic plant requires an adequate supply of carbon dioxide, oxygen, and water. However, the plant must first convert all these raw materials into complex organic substances before they can support its growth and development. Conversely, many of these complex organic substances must be subsequently broken down to simpler substances in order to support certain phases of the plant's growth and development. All these chemical changes that take place in the plant are referred to as its *metabolism*. The metabolism of the plant may be conveniently divided in two parts (Fig. 13-1). All the chemical changes leading to syntheses (building of more complex from less complex substances) comprise its *anabolism*. All those leading to breakdown of more complex to less complex substances are included in its *catabolism*. The many intermediate substances produced in the process are the intermediate metabolites. The many chemical reactions in which these are involved constitute the intermediate metabolism of the plant. The tremendous number of chemical changes that constantly must

be taking place becomes evident from the vast number of chemical substances found in the plant. All except the few that have not yet changed on entering from the environment have arisen from synthesis or breakdown in the plant—primarily in the protoplasm. Many of the substances are so complex that their synthesis may have required dozens of chemical reactions. Furthermore, although many of the substances in the plant are stable (e.g., substances in the cell wall, such as cellulose), others, and in particular the fundamental structural and functional units of the protoplasm (some of the nucleic acids and proteins), *are continuously being broken down and resynthesized*. This is proved by feeding the plant with radioactive carbon or heavy nitrogen because the substances formerly present in the protoplasm soon become labeled by these isotopes and, therefore, must have been broken down and resynthesized. By these methods, protein turnover

Metabolism

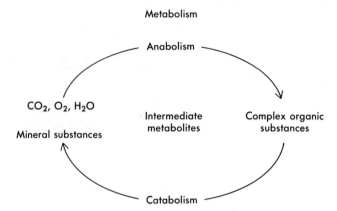

Fig. 13-1. The interdependence of anabolism and catabolism in the overall metabolism of the plant.

rates of 0.5 to 1.5% per hour have been found in rapidly growing leaves (Hellebust and Bidwell, 1963).

An actively growing plant enlarges its body by rapidly building new cells and tissues. It must therefore rapidly synthesize from the raw materials that it absorbs from its environment the building blocks needed to produce these new cells and tissues. Hence, an actively growing plant must metabolize actively. But what if it is not producing new cells and tissues and therefore is not growing? Does metabolism then cease? According to the laws of thermodynamics, free energy tends toward a minimum. The complex chemical substances that make up the body of even a non-growing plant are at a higher free energy level than the raw materials in its environment from which they were synthesized. Therefore, catabolic (breakdown) reactions must take place, leading to a decrease in free energy and ultimately to death of the plant. Even in the dead (aseptic) plant, the breakdown reactions can be expected to continue until equilibrium is reached with its environment. At the normal temperatures of a living plant these changes would lead to relatively rapid death. Consequently, in order for the nongrowing plant to survive, it

must prevent or at least slow down this approach to equilibrium. Therefore, even in a nongrowing plant, anabolic changes must be constantly taking place, counteracting the destructive catabolic changes. When the two are occurring at the same rates, the plant is in the steady state and is neither growing nor decreasing in mass. When anabolic changes exceed the catabolic, it is either storing complex reserves or growing. When catabolic changes exceed the anabolic, it is losing mass; that is, it is either using up its reserves more rapidly than it is growing or it is dying. An exact steady state is, of course, difficult to attain. It is approached most closely in dormant (nongrowing) plants. But even then, catabolism may slightly exceed anabolism (e.g., in seeds, tubers, etc.), and death will eventually occur; or anabolism may slightly exceed catabolism (e.g., some evergreens in winter on bright sunny days), and the reserves will increase.

It therefore must be concluded that whether or not a plant is growing, as long as it is alive it must be metabolizing; and this must include both catabolism and anabolism. Even when it dies, chemical changes will continue; but these will involve only breakdown reactions, and even they will not be in the ordered, controlled manner characteristic of the living plant. The term *metabolism* should therefore be restricted to the ordered series of chemical changes under the control of the living cell. It is usually assumed that no life is possible without metabolism, and no metabolism is possible without life.

Both of these assumptions may be questioned, however. It is now possible to follow with cell extracts and, therefore, in nonliving material, many of the individual metabolic reactions that normally occur in living cells. This means that portions of the plant's metabolism are possible in the absence of life. Conversely, some biologists have asked the philosophical question: If metabolism comes to a complete stop, is the organism dead? A possible test case is provided by dormant, air-dry seeds, which metabolize extremely slowly even at normal temperatures. It is now known that such seeds can be cooled to the lowest temperatures so far attained (a small fraction of a degree above absolute zero) without injury. When rewarmed and supplied with water, they germinate and grow normally. Presumably, if absolute zero were attained, they would also survive; yet all chemical reactions within them would cease and, by definition, so would metabolism. It is a philosophical question whether the seeds would be dead or dormant at absolute zero and whether they would be revived from death or merely activated from the dormant state when rewarmed and moistened. From the practical point of view, however, the preservation of life at such low temperatures is of prime importance. Recent results indicate that seeds kept in the frozen condition (in the permafrost) for 10,000 years are still alive and are now growing normally (see Chapter 22).

In spite of the preceding apparent exception, it is certainly true that under normal conditions there is no life without metabolism; and, therefore, life cannot be understood without a thorough knowledge of metabolism. Two aspects of metabolism are intimately interrelated but are more conveniently discussed separately: the changes in chemical structure and the energy changes.

CHANGES IN CHEMICAL STRUCTURE

The vast number of chemical substances found in plants might lead to the expectation of a correspondingly large number of kinds of chemical reactions. Fortunately, there are only six main types of reactions.

Hydrolysis-condensation reactions

$$R^1OR^2 + H_2O \underset{\text{Condensation}}{\overset{\text{Hydrolysis}}{\rightleftharpoons}} R^1OH + R^2OH$$

where R^1 and R^2 = different chemical radicals (e.g., a methyl group, a benzene ring, etc.). Hydrolysis is a breakdown reaction in which water participates and a large molecule is broken down into two or more smaller ones. Condensation is the reverse synthesis of a large molecule from two smaller ones, with the release of water. At equilibrium ($\rightleftharpoons$), reactants and products exist in proportions characteristic of each specific reaction. Because of the high concentration of water in normally metabolizing cells, the hydrolytic products on the right will be at much higher concentration than the condensed reactants on the left *when the reaction is at equilibrium.* Therefore, the hydrolysis tends to go to completion. But this may not hold true in the plant since the metabolic process of which this reaction is a component may not be at equilibrium. Many different kinds of substances may be hydrolyzed, for example, esters, glycosides, peptides, etc. Consequently, hydrolyses are responsible for the cleavage of many kinds of macromolecules. When the macromolecules are insoluble food reserves (e.g., fats, carbohydrates, proteins) and the hydrolysis converts them into soluble substances, the process is called *digestion.* Many carbohydrates are broken down in this way, for example:

$$Starch + H_2O \rightleftharpoons Maltose$$

It was formerly thought that the reverse reaction (condensation) was responsible for the syntheses of macromolecules; but it is now known that these syntheses are generally caused by the group transfer, addition, or synthesizing reactions described next. There are, however, some cases where reversal of hydrolysis is the normal mode of synthesis (e.g., ATP synthesis, see Chapter 14).

Group transfer reactions

$$R^1OR^2 + R^3OH \rightleftharpoons R^1OH + R^3OR^2$$

This reaction, in its simplest form, is basically similar to a hydrolysis, but instead of water, a more complex chemical hydroxide is involved. Many different kinds of chemical groups may be transferred: C-groups (methyl, hexose, etc.), N-groups (amino, etc.) P-groups, S-groups. One of the most important of these is the transfer of P-groups, a vast number of which occur in the plant; for example:

$$ATP + ribose \rightleftharpoons ADP + ribose\ phosphate$$
where ATP = adenosine triphosphate
ADP = adenosine diphosphate

Group removal or addition reactions

$$R^1R^2H \rightleftharpoons R^1H + R^2$$

One of the most important of these involves the formation and cleavage of C-C bonds:

$$RCOOH \underset{\text{Carboxylation}}{\overset{\text{Decarboxylation}}{\rightleftharpoons}} RH + CO_2$$

These reactions are basic to many metabolic processes. The breakdown of carbohydrates in respiration is largely caused by a series of decarboxylations. The synthesis of carbohydrates in photosynthesis is largely caused by a series of carboxylations. Besides carbon dioxide, other simple molecules (e.g., water, ammonia) may also be removed from or added to the more complex organic molecules in the plant.

Electron transfer or oxidation-reduction reactions

$$A + B \rightleftharpoons A^+ + B^-$$

These reactions are among the commonest in all metabolism. A series of electron transfers occurs in each of the most important metabolic processes—respiration, photosynthesis, N-metabolism, S-metabolism. This sequence of electron transfers between adjacent molecules has led to the term *electron transport,* indicating successive transfers between a series of molecules rather than a single transfer from one molecule to a second neighboring one.

The substance (A) that loses the electron is the electron donor and is said to be oxidized (or is the reducing agent or reductant). The substance (B) that gains the electron is the electron acceptor and is said to be reduced (or is the oxidizing agent or oxidant). In order for such a transfer to occur between atoms, the electron acceptor must be more electronegative than the electron donor (see Table 4-3).

Ions are not necessarily formed as a result of the electron transfer because the electron may be accompanied by a proton in the form of a hydrogen atom, resulting in hydrogenation of the electron acceptor and dehydrogenation of the electron donor:

$$AH_2 + B \rightleftharpoons A + BH_2$$

The substance AH_2 is oxidized to A, whereas B is reduced to BH_2. This kind of oxidation-reduction reaction depends on the lower electronegativity (see Table 4-3) of the hydrogen atom than that of the other atoms (carbon, nitrogen, oxygen, sulfur) with which it is normally combined in plant substances. The removal of a hydrogen atom is therefore equivalent to removal of the electron that it had donated to the

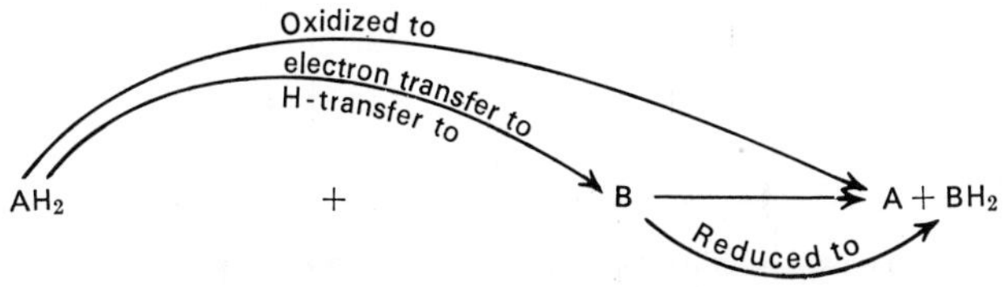

1. Reducing substance (AH_2) (reductant) becomes oxidized due to	Oxidizing substance (B) (oxidant) becomes reduced due to
2. Electron loss from an electron donor or	Electron gain by an electron acceptor or
3. H-loss from an H-donor	H-gain by an H-acceptor

Fig. 13-2. Three equivalent terminologies for electron transfer.

carbon (or other) atom to which it was bonded, and the addition of a hydrogen atom to a carbon (or nitrogen, oxygen, sulfur) atom is equivalent to a transfer of an electron to the carbon or other atom. The apparent hydrogen transfer may actually be a true electron transfer since in some cases only the electron may be passed from substance to substance, the proton remaining temporarily in solution as an H^+:

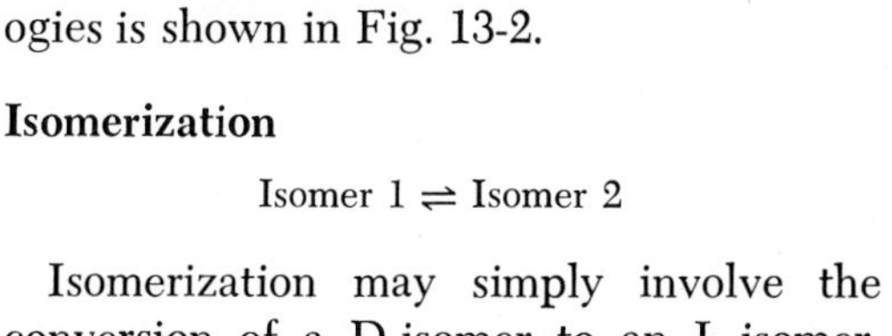

The equivalence of the different terminologies is shown in Fig. 13-2.

Isomerization

$$\text{Isomer 1} \rightleftharpoons \text{Isomer 2}$$

Isomerization may simply involve the conversion of a D-isomer to an L-isomer, or it may result from the shift of a chemical group, for example, in the conversion from a 2-phospho-compound to a 3-phospho-compound or the conversion of one pentose to another.

Synthesizing reactions

Synthesis of complex substances may involve transfer and addition reactions, but they are also commonly caused by the following synthesizing reaction:

$$A + B + NPPP \rightleftharpoons AB + NP + PP$$

where N = nucleoside (a purine or pyrimidine combined with a pentose)
P = phosphate
Nucleoside phosphate is called a nucleotide

Thus (1) the making and breaking of the electron pairs of covalent bonds and (2) the electron transfer between atoms are the fundamental characteristics of metabolism and therefore of life. The six elements (hydrogen, carbon, nitrogen, oxygen, sulfur, phosphorus) have mobile electrons and form multiple bonds more easily than all the other elements in the periodic table (Pullman and Pullman, 1963). It is therefore not surprising that the elements hydrogen, carbon, nitrogen and oxygen make up 90% of living systems and that sulfur and phosphorus are next in quantity. Since double bonds are caused by mobile electrons, this also explains the common occurrence of conjugated substances (i.e., substances with alternating single and double bonds) in living cells. Conjugated substances are typically colored.

ENERGY CHANGES

As stated earlier, no life has been found without metabolism. In the case of a growing plant this follows logically from energy considerations. There are four kinds of work associated with growth: (1) mechanical work such as the growth upward of the plant body against gravity and the creation of new surfaces, (2) "osmotic" work, or the absorption of substances against a diffusion gradient followed by osmosis; (3) electrical work, when the substances are in the ionic form and are absorbed against an electrical potential gradient; and (4) chemical work involved in the synthesis of the many substances in its cells. Probably 99% of the plant's energy requirements are for the last of these—chemical work. According to the first law of thermodynamics, all these four kinds of work must involve a transfer of energy from some other source. The only source directly available to the plant is the energy released in its metabolism. According to the second law of thermodynamics, free energy (energy available to do work) tends toward a minimum. Yet the free energy of the living plant actually increases during growth because of its increased content of many substances at a high level of free energy. This is especially true of protoplasm, the vital properties of which depend on its content of many substances in a highly unstable state. The second law dictates that these substances must be continuously changing to the more stable, lower free energy states. This would lead to death of the plant if unopposed. But the plant is a thermodynamically open system; it can absorb both mass and free energy from its environment. Therefore, the continuously formed substances of low free energy

are just as continuously reconverted to the high energy state by the expenditure of metabolic energy (the only source of extra free energy constantly available to the plant). This metabolic energy, in turn, is derived by the green plant from the radiant energy of sunlight, which is absorbed daily and stored as metabolizable, chemical energy. A specific example is provided by the proteins of the living protoplasm. In the native state (in which they function normally) they exist as large molecules folded in a highly ordered manner (see Chapter 16). From the second law of thermodynamics we would expect them to change to the unfolded (less ordered) denatured state. These denatured proteins must then be renatured, commonly by hydrolysis to amino acids (see Chapter 16), followed by recombination to form the native proteins. A steady state of breakdown and resynthesis is thus maintained *at the expense of metabolic energy.*

This cycle of changes is illustrated in Fig. 13-3.

In all the preceding six classes of chemical reactions, energy changes accompany the changes in chemical composition. The reversibility of the reaction depends on these energy changes. Hydrolyses usually involve a large release of energy and, therefore, go nearly to completion. This is really why they are not readily reversible. Group removal reactions (e.g., decarboxylations) usually involve a small release of energy and are, therefore, readily reversible. Thus, these reactions can go in either direction and may be controlled by the end product. A decrease in carbon dioxide content of the air may be expected to lead to decarboxylation and an increase, to lead to carboxylation. These energy changes are readily explained because changes in chemical structure involve electron displacements and therefore the performance of work. Thus, an addition of energy to an atom may be expected to result in an electron displacement followed by a change in struc-

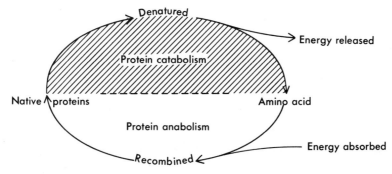

Fig. 13-3. Protein metabolism in living cells. Native proteins unfold to the denatured form followed by hydrolysis to amino acids accompanied by a loss of free energy. This is counteracted by a continuous synthesis of native protein at the expense of amino acids and free energy.

ture. When at its lowest free energy level, the molecule is said to be in the "ground state," and all the electrons are distributed in pairs with antiparallel spins. If an electron is displaced from its orbit ("excited"), an "empty" molecular orbit results and the free energy of the molecule increases since it can do work by allowing an electron to fall into this empty orbit. It can therefore undergo chemical reaction.

In any one chemical reaction there is normally both a displacement of electron orbits *from* one atomic nucleus and a displacement *toward* another atomic nucleus. If the former involves more work than the latter, the net result will be a gain in free energy by the products formed as compared with the reactants, and the reaction is *endergonic, or energy is absorbed*. Endergonic reactions cannot take place unless free energy is supplied from some external source. Conversely, if the displacement toward the second nucleus involves more work than displacement from the first nucleus, there will be a net loss of free energy by the products formed, and the reaction is *exergonic, or energy is released*. Exergonic reactions are *spontaneous*. This means that they can occur without a *net* supply of free energy. The syntheses of more complex from less complex molecules in the plant are endergonic; the breakdown reactions are exergonic. We can now define anabolism

and catabolism energetically. That part of the metabolism which results in net syntheses and therefore consists mainly of endergonic reactions is called *anabolism;* the part consisting of net breakdown (exergonic) reactions is called *catabolism*. However, the two are intimately related because the endergonic reactions cannot take place unless free energy is supplied from some other source. But it must be remembered that many reactions are taking place in the plant at any one time. Therefore, it is at least conceivable that the endergonic reactions may take place at the expense of the free energy released by the exergonic reactions. When this happens, endergonic reactions are called *driven reactions;* and since the two kinds are energetically linked, they are also called *coupled reactions*. Thus the fixation of nitrogen by microorganisms

$$N_2 + 3H_2O \rightleftharpoons 2NH_3 + 3/2O_2$$

is an endergonic reaction that cannot go of itself because there is a free energy *increase* of 162 kcals (Wohl and James, 1942). If, however, the oxygen produced can be used to oxidize glucose, the free energy *released* by this reaction is 172 kcals. The total (coupled) reaction is

$$N_2 + 3/2H_2O + \frac{1}{4}C_6H_{12}O_6 \rightleftharpoons 2NH_3 + 3/2CO_2$$

where $C_6H_{12}O_6$ = glucose. This combination of reactions is accompanied by a net free energy decrease of 10 kcal (172 − 162

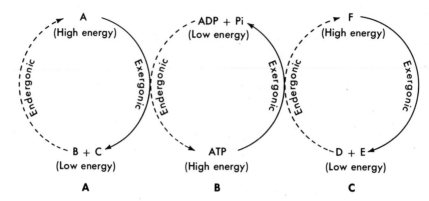

Fig. 13-4. Schematic representation of high energy phosphate link, **B**, between the driving, **A**, and the driven, **C**, reaction. The exergonic reaction in **A** drives the endergonic reaction in **B**. The exergonic reaction in **B** drives the endergonic reaction in **C**.

kcal) and is therefore thermodynamically spontaneous.

Thus, in general, the anabolism of plants is possible only if catabolism is simultaneously taking place. This must always be remembered, although the two are studied separately for purposes of classification. It is, of course, true that the free energy needed to drive some endergonic reactions may be obtained from sources other than exergonic reactions. In the plant this is true of the photochemical reactions in the photosynthetic process, the free energy for which is obtained from sunlight.

Plant metabolism, and therefore plant life, is thus driven by the electronic energy of exergonic reactions

Linked or coupled reactions $\begin{cases} 1) \ A \rightarrow B + C + \text{energy (exergonic)} \\ 2) \ \text{Energy} + D + E \rightarrow F \ (\text{endergonic or driven}) \end{cases}$

where A, D, E are reactants and B, C, F are products. But the energy is not transferred directly. When an exergonic reaction occurs by itself, the electronic energy is converted to heat and is therefore lost. In the plant much of the energy of exergonic reactions can be trapped in the unstable, or "high energy," bonds of high energy phosphates, for instance, by linking two inorganic orthophosphates, molecules of low free energy in an anhydride link of high

free energy with the formation of pyrophosphate (P-O-P). The major form of this pyrophosphate is in adenosine triphosphate (ATP). The high energy pyrophosphate can then combine with other substances in such a way as to induce them to undergo endergonic reactions at the expense of the high energy pyrophosphate, which is then converted back to low energy inorganic phosphate. The preceding reactions would therefore have to be linked to another pair of reactions

$$\text{ADP} + \text{P}_i \rightleftharpoons \text{ATP}$$

where P_i = inorganic phosphate. The energetic relationship between these reactions is indicated in Fig. 13-4.

The ATP contains the exergonic energy that would otherwise be given off as heat and transfers it to D + E, driving the reaction to product F and regenerating the low free energy inorganic phosphate. The cycle can be repeated over and over, and the high energy phosphates can receive the energy from many different substances and can transfer it to many others. By analogy with a dynamo (Fig. 13-5), the energy released in metabolism can be thought of as being "brushed off" as high energy phosphate groups and conducted along the adenylic acid (AMP) or adenosine diphosphate (ADP) "wire" to drive the "motors" of the

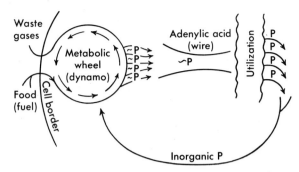

Fig. 13-5. Analogy between utilization of metabolic energy by means of high energy phosphates ($\sim P$) and conversion of mechanical energy of dynamo into electrical energy. (From Lipmann, F. 1941. Adv. Enzymol. 1:99-162.)

endergonic reactions. The waste, low energy inorganic phosphate must then be reconverted to high energy phosphate by the metabolic wheel ("the dynamo"). This explains why the syntheses of macromolecules are not produced in the plant by a simple reversal of hydrolysis, that is, by condensation. The condensation reactions are endergonic, and since they are not linked with reactions involving high energy phosphates, they are not receiving the free energy needed to drive them.

The transfer of energy by means of high energy phosphates is required for many energy-consuming processes in the plant—the movement of substances against a gradient (active absorption), the streaming of cytoplasm, the creation of new surfaces, etc. But because of their instability, the high energy phosphates cannot be stored in large enough quantities for future needs of the plant. Therefore, the excess high energy phosphate is used to synthesize the more stable carbohydrates, fats, etc., which can retain the chemical energy until it is needed, at which time it must be reconverted to high energy phosphates before being used. The transfer of metabolic (chemical) energy over long distances is by means of these stable storage products, for example sucrose (see Chapter 10).

But what of these stable substances? How

can they retain large quantities of chemical energy and how can they liberate this energy when it is needed? It must be realized that exergonic reactions are *spontaneous* only from the thermodynamic point of view; that is, they can take place without a *net* supply of energy. This does not mean that the reactions will take place unaided but simply that they are downhill reactions as far as the net energy change is concerned. However, the potentially reactive molecules are made up of both positively and negatively charged components and, therefore, may either attract or repel each other (Linford, 1966). When the molecules of two substances capable of reacting approach each other, the first effect is an attraction, which results in a closer approach (Fig. 13-6, from *c* to *b*). Even if the reaction between the two is exergonic and therefore spontaneous, the forces of repulsion now prevent a closer approach and the reaction does not occur. In order for them to approach close enough (Fig. 13-6, from *b* to *a*) to react chemically, work must be done on them by some outside force. The energy just sufficient to do this work is called the *activation energy* (Fig. 13-7, ΔE_1^*). When the energy level of the molecules is raised by this amount, the reaction takes place and the products are formed with a net *release* of energy—the *energy of*

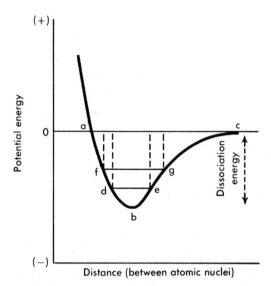

Fig. 13-6. Attraction and repulsion between molecules (or other chemical particles) as they approach each other. On approaching from c to b there is a decrease in energy; that is, they attract each other. From b to a there is an increase in energy; therefore, no further approach is possible without adding energy to the system, that is, doing work on the molecules. (From Linford, J. H. 1966. An introduction to energetics with applications to biology. Butterworth & Co. [Publishers], Ltd., London.)

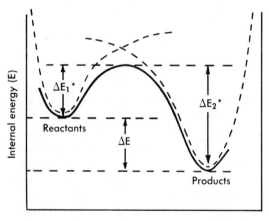

Fig. 13-7. Energy changes in a spontaneous reaction. The internal energy of the molecules must be increased by ΔE_1^* (the activation energy) before the reaction can begin. (From Linford, J. H. 1966. An introduction to energetics with applications to biology. Butterworth & Co. [Publishers], Ltd., London.)

reaction (Fig. 13-7, ΔE). The total energy released is the sum (Fig. 13-7, ΔE_2^*) of the activation energy and the energy of reaction. The reverse (endergonic) reaction will occur only if this total sum of the two amounts of energy is supplied to the products. The activation energy, therefore, constitutes the barrier to the progress of spontaneous (exergonic) reactions; and the more stable the reactants the larger is the energy barrier. The activation energy may be supplied by heating the reactants.

ENZYMES

Stable substances can be made to undergo exergonic reactions without heating in the presence of a highly reactive substance acting in extremely small quantities and known as a *catalyst*. If the reaction involves two substances, the catalyst may speed it up by simply adsorbing these substances, bringing them close enough together to reduce the energy barrier. Thus, when a stove is turned on, the gas does not burn until the activation energy is supplied by a match. Instead of a match, platinum black may be used. It acts as a catalyst by adsorbing the gases from the stove and the air, in this way starting the reaction without a supply of heat because it lowers the activation energy.

The catalyst may form an intermediate compound with the substance, converting it into a substance that is unstable. The energy barrier is in this way reduced to a small enough value for the energy of the molecules to overcome; and they can now form products spontaneously by dropping to a lower energy level (Fig. 13-7). When two substances react, the steps will then be as follows:

$$A + catalyst \rightleftharpoons cat\ A$$
$$cat\ A + B \rightleftharpoons C + D + catalyst + energy$$
$$\text{Net reaction: } A + B \xrightarrow{\text{catalyst}} C + D + energy$$

The net result is that the catalyst has permitted the reaction to take place by removing or lowering the energy barrier (energy of activation). It is active in small quantities because it is regenerated after each reaction and able to combine with another molecule.

In the presence of such a catalyst, the exergonic reaction proceeds more or less rapidly with the release of free energy. The chemical substance or substances resulting from the first reaction may in turn be capable of reacting exergonically with another substance in the presence of a second catalyst, which is different from the first. Thus, a whole chain of reactions may occur in a definite order, each link in the chain controlled by a specific catalyst. As a result, a complex substance may be broken down, step by step, to simpler and simpler substances, and at each step a packet of energy will be released, which may be used to drive an endergonic reaction

$$A \underset{}{\overset{cat_a}{\rightleftharpoons}} 2B + energy$$
$$B \underset{}{\overset{cat_b}{\rightleftharpoons}} C + D + energy$$
$$C \underset{}{\overset{cat_c}{\rightleftharpoons}} E + F + energy$$
$$D \underset{}{\overset{cat_d}{\rightleftharpoons}} G + H + energy$$

Net result: $A \underset{}{\overset{cat_{a-d}}{\rightarrow}} E + F + G + H + 4$ energy packets

where cat_a, cat_b, etc. are each specific catalysts for a specific reacting substance.

By this gradual, orderly release of energy at the right time and in the right place, the plant is able to perform all its energy-consuming processes: synthesis of substances, active absorption of substances, cell growth, maintenance of gradients, protoplasmic streaming, etc. But none of these energy-consuming processes would be possible in the absence of the catalysts necessary for the energy-releasing reactions; and since there are many such reactions, there must also be many catalysts because they are highly specific. The organic catalysts that control the metabolism of living organisms are known as *enzymes*. They can induce reaction velocities as much as 10 times as rapidly as those produced by the best of the known nonenzymatic catalysts. The mechanism by means of which such high velocities are brought about is still unknown (Koshland, 1963*a*). Because of their high activity, their presence can be easily revealed. In many cases it is necessary only to show that substance A will not break down of itself but will when mixed with a plant juice. This indicates that the juice contains an enzyme that can catalyze the breakdown of A. For more rigorous proof this preliminary test is followed by other more exacting ones. Some reactions may conceivably occur in the plant without catalysis by enzymes; but new enzymes are still being discovered, and some may later be found to control reactions now believed to be nonenzymatic.

In recent years new enzymes have been discovered at the rate of thirty to forty a year. About 900 are known and more than 100 of these have been sufficiently purified to be crystallized. These have all proved to be globular proteins with molecular weights from about 10,000 to millions (Koshland, 1963*a*).

Some of these are enzymatically active without any nonprotein component, but most enzymes consist of two components: a protein and a nonprotein, or cofactor (Fig. 13-8). The nonprotein component may be firmly attached to the protein, in which case it is called a *prosthetic group;* or it may be readily removed (e.g., by dialysis), in which case it is called a *coenzyme.* If the nonprotein component is a metal ion, it is called an *activator* since the protein is inactive in the absence of the metal ion. The hydrolases (see later) generally do not have coenzymes, although they frequently require metal activators. The chemical substance(s) that reacts in the presence of the enzyme to form some other substance(s) (the product(s)) is known as the *substrate.*

The terminology of enzymes was arbitrary at first. Later, the tendency grew to name an enzyme after either the substrate

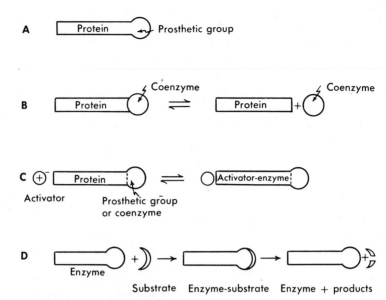

Fig. 13-8. Diagrammatic representation of enzyme systems. **A,** Protein plus nonremovable component. **B,** Protein plus removable component. **C,** Protein plus one or two removable components. **D,** Enzyme-substrate reaction.

Table 13-1. Some plant enzymes*

EC no. (Enzyme Commission)	Group	Systematic name	Trivial name	Active groups and cofactors	Reaction
	Hydrolases				
3.1.1.3		Glycerol ester hydrolase	Lipase	Ca^{++}	A triglyceride + H_2O = diglyceride + fatty acid
	Transferases				
2.3.1.13		UDP glucose: D-fructose 2-glycosyltransferase	UDP glucose-fructose glucosyltransferase, sucrose-UDP glycosyl-transferase		UDP glucose + D-fructose = UDP + sucrose
	Oxidoreductases				
1.10.3.1		o-Diphenol : O_2 oxidoreductase	Catechol oxidase	Cu	2 o-diphenol + O_2 = 2 o-quinone + $2H_2O$
	Lyases				
4.1.1.1		2-Oxo-acid carboxy-lyase	Pyruvate decarboxylase (formerly known as carboxylase)	T.M^{++}	A 2-oxo-acid = aldehyde + CO_2

*From Dixon, M., and E. C. Webb. 1964. Enzymes. Academic Press, Inc., New York.
T. = thiamine pyrophosphate; M^{++} = divalent cation.

or the kind of reaction, adding the suffix -ase. Now all the known enzymes have been given "systematic names" by the 1961 Commission of the International Union of Biochemistry (Dixon and Webb, 1964), consisting of the substrate followed by the kind of reaction plus the ending -ase (Table 13-1). Since many of these names are unwieldy, the old "trivial names" may still be used. The commission established six main classes of enzymes based on the main kinds of reactions they control. Ninety per cent of these enzymes control the first four main types of reactions mentioned earlier:

1. Hydrolytic reactions are controlled by hydrolases (3)
2. Group transfer reactions, by transferases (2)
3. Oxidation-reduction reactions by oxidoreductases (1)
4. Group removal and addition reactions by lyases (4)

5. Isomerization reactions by isomerases (5)
6. Synthesizing reactions by ligases (previously called synthetases) (6)

The numeral in parenthesis after each name is the group number agreed on by the Committee on Enzyme Nomenclature.

In contrast to the proteins of enzymes that are numbered in the many hundreds, there are only twenty-four to thirty-six cofactors (Table 13-2). Most of the coenzymes are nucleotides (purine or pyrimidine—sugarphosphates, see Chapter 16) or related substances, and each may be a cofactor for a large number of different enzymes. The amounts present may therefore be considerable, for example, as high as 1 mg NADPH (nicotinamide-adenine dinucleotide phosphate) per gram fresh weight in maize seedlings (Fritz et al., 1963). Other cofactors, however, are present in much lower quantities. In the case

Table 13-2. Some examples of enzyme cofactors

Kind	Name	Abbreviation	Kind of enzyme associated with cofactor
1. Coenzyme	Nicotinamide-adenine dinucleotide (coenzyme I)	NAD^+ (DPN)	O-R
	Nicotinamide-adenine dinucleotide phosphate (coenzyme II)	$NADP^+$ (TPN)	O-R
	Ascorbic acid	AA	O-R
	Gluthathione	GSH	O-R
	Coenzyme A	CoA	Transferase
2. Prosthetic groups	Flavin-adenine mononucleotide	FMN	O-R
	Flavin-adenine dinucleotide	FAD	O-R
	Thiamine pyrophosphate	TPP	Lyase
	Pyridoxine phosphate	PP	Transferase, hydrolase
3. Activators		K^+	Hydrolase, transferase
		Mg^{++}	Transferase, hydrolase, lyase, ligase
		Ca^{++}	Hydrolase
		Zn^{++}	Hydrolase
		Cu^{++}	O-R
		Mn^{++}	O-R, transferase hydrolase, lyase
		Fe^{++}	O-R, hydrolase
		Co^{++}	O-R
		Mo^{++}	O-R

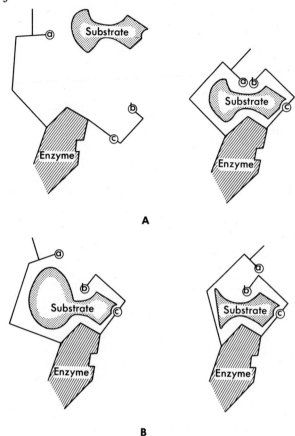

Fig. 13-9. Specificity of enzyme for substrate. **A,** Good fit between the substrate and reactive groups of the enzyme; reaction proceeds. **B,** Poor fit between substrate and reactive groups of the enzyme; no reaction. *a, b,* and *c,* represent either cofactors or reactive groups of the protein molecule ("active sites"). (From Koshland, D. E., Jr. 1963. Sympos. Quant. Biol. **28**:473-482.)

of heterotrophic plants or plant parts (as well as other organisms) some of these co-enzymes must be supplied in the form of vitamins in the nutrient medium. Because of the combination of the protein with the substrate, the latter is activated; but only parts of the enzyme, known as "active sites," react with the substrate. The actual mechanism is still not completely understood (Koshland, 1963*a*), but the active sites normally include the cofactors and, in many cases, one or more sulfhydryl (SH) groups on the protein (Barron, 1951). It is the precise arrangement of these active sites that is responsible for the *specificity* of the

enzyme, that is, its ability to catalyze the reaction of a single substance or up to a few dozen related substances (Fig. 13-9). Thus several enzymes may contain the same prosthetic group (Table 13-2), but because of the difference in the protein component, each will act specifically on a different substance. For instance, tyrosinase and ascorbic acid oxidase both have copper as a prosthetic group, yet the former acts on tyrosine (and several similar substances) but not on ascorbic acid; the latter acts on ascorbic acid but not on tyrosine. The specificity may be narrow in some cases but may extend to a number of closely related

Table 13-3. Relative distribution of enzymes and prosthetic groups*

Enzyme group	Number of enzymes known	Pyri-doxal	Flavin	Metal	Haem	Thiamine pyrophos-phate	Biotin	Coba-lamine
I. Oxidore-ductases	222	1	34	24	9	2	0	0
II. Trans-ferases	238	18	0	0	0	1	1	0
III. Hydrolases	213	1	0	4	0	0	0	0
IV. Lyases	117	29	1	1	0	4	0	1
V. Isomerases	47	4	0	0	0	0	0	1
VI. Ligases	47	0	0	0	0	0	4	0
Total	884	53	35	29	9	7	5	2

*From Smith, M. H. 1967. Nature (London) 213:627-628.

substances in others. The following example illustrates some of the preceding terminology:

$$\text{Malate} + \text{Malate dehydrogenase} + \text{NADP}^+ + \text{Mn}^{++} =$$
(substrate)   (enzyme)   (coenzyme)   (activator)

$$\text{Pyruvate} + \text{CO}_2 + \text{NADPH} + \text{Malate dehydrogenase} + \text{Mn}^{++}$$
(products)   (reduced coenzyme)   (enzyme)   (activator)

where NADP^+ = nicotinamide-adenine dinucleotide phosphate
NADPH = reduced nicotinamide-adenine dinucleotide phosphate
Mn = manganese

The very rapid action of enzymes may be illustrated by catalase, which controls the following reaction:

$$2H_2O_2 \xrightleftharpoons{\text{Catalase}} 2H_2O + O_2$$

where H_2O_2 = hydrogen peroxide

One molecule of catalase can decompose 1000 molecules of hydrogen peroxide per second! The concentration of the enzyme affects the speed of the reaction but not the position of equilibrium, and the final quantity of product is unchanged, except when the enzyme concentration is very high.

Because of the protein portion, enzymes are *heat labile;* that is, they are destroyed by 60° to 100° C. The coenzyme component is heat stable, being unaffected by boiling. The enzyme is frequently inactive at tem-peratures near freezing or in an acid pH in the case of some (e.g., trypsin). Low temperature inactivation is reversible, and so is pH inactivation in some cases if the pH change is not too extreme. Poisons may inactivate many enzymes when present in very small quantities; for example, 0.0001 M cyanide inactivates many respiratory enzymes (containing heavy metals). These substances are called *inhibitors*. Other inhibitors are sulfides, azides, fluorides, iodoacetates, etc. Some inhibitors are somewhat specific; that is, they inhibit only one or a few enzymes. Frequently, they inhibit by combining with the prosthetic group or with the sulfhydryl group of the protein, thus preventing it from combining with the substrate. Some of the newer inhibitors actually prevent synthesis of the enzyme protein (see Chapter 16).

An important group of inhibitors are the structural analogues—substances so similar chemically to the substrate that the enzyme cannot distinguish between them. In order to be an inhibitor, the structural analogue must not be made to react by the enzyme but must remain attached to it. If the normal reaction is represented as follows:

$$1. \ E + S \rightleftharpoons ES; \quad ES \rightarrow E + P$$

Inhibition may occur in the following two ways:

$$2. \ E + I \rightleftharpoons EI$$
$$3. \ E + I \rightarrow EI$$

where   E = enzyme
         S = substrate
         P = products
         I = inhibitor

In 2 the inhibition is reversible (i.e., it can be overcome by increasing the amount of substrate); in 3 it is irreversible regardless of substrate concentration.

The structural analogue is also called an antimetabolite if it causes signs of deficiency of the subtrate (or essential metabolite, Hochster and Quastel, 1963).

One of the most amazing facts about enzymes is the tremendous number of different kinds that must coexist in the extremely small space occupied by the protoplasm of a single cell. The number is so large that it has even been suggested that all the cytoplasmic proteins are enzymes. Each must be able to control its own particular reaction without interference from—indeed, with the aid of—other enzymatically controlled reactions. This orderly arrangement would not be possible if all the enzymes were mixed haphazardly in solution. It is not surprising, then, that so many of them have been found to be attached to particles in the cytoplasm, for example, the mitochondria. If each one occupies its own place on such a particle, an orderly series of reactions can take place there. Such groups have been called multienzyme systems.

On the other hand, the enzyme content of a cell is not fixed but varies with the conditions. Many enzymes are *adaptive;* that is, they accumulate in the presence of substrate and disappear in its absence. Thus, the enzyme nitrate reductase increases in quantity if nitrate is supplied to the plant and decreases in quantity as the nitrate is metabolized and unreplaced. Other enzymes are *constitutive*—they occur regardless of the presence or absence of substrate.

Enzymes have frequently been reported outside the protoplasm of the higher plant cell and even outside the plant organ; but it is difficult to prove that this occurs in the case of the normal, uninjured cell. The frequently reported occurrence of cytochrome oxidase on the walls of plant cells is believed to be an artifact, although in its stead, peroxidase has been identified on the walls of lignifying cells (Lipetz, 1965).

Plant metabolism includes such a vast number of chemical reactions that it is neither possible nor desirable to consider each reaction individually. But it is possible to deal with them in groups since they fit into metabolic processes. An understanding of metabolism requires the study of these individual subsidiary processes, into which it can be divided. The most important of these can be classified as shown on the opposite page.

A complete understanding of plant metabolism would also include S-metabolism and P-metabolism, as well as that of all the other essential elements. Although some of these metabolic branches are distinct from each other, others are intimately related. The preceding classification must therefore be considered as one of convenience rather than representing completely separate categories in the plant.

QUESTIONS

1. What is meant by metabolism?
2. What is meant by anabolism and catabolism?

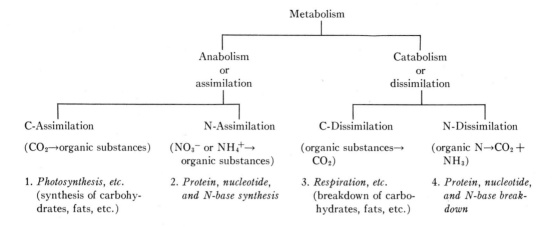

3. How many different kinds of reactions occur in the plant?
4. Name one kind of reaction.
5. Give an example of one kind of reaction.
~6. Name a second reaction and give an example, etc.
7. Which kinds of reactions are not readily reversible? Why?
8. What is an oxidation?
9. What is a reduction?
10. Is there any necessary relation between oxidation and reduction?
11. What chemical element is commonly transferred in an oxidation?
12. What are exergonic and endergonic reactions?
13. How are they related to anabolism and catabolism?
14. How are they related to each other?
15. What are spontaneous reactions?
16. Do they take place unaided?
17. What is activation energy?
18. What role do catalysts play in reactions?
19. What are the catalysts in the plant?
20. What is the chemical nature of pure enzymes?
21. What other component does it usually require for activity?
22. What are these components called?
23. Which part is primarily responsible for the chemical activity?

24. What is enzyme specificity?
25. Which component is responsible for it?
26. What is the effect of high temperature on enzymes?
27. What is the effect of low temperature on enzymes?
28. How do these effects differ?
29. Are there any other factors that affect enzyme activity?
30. Which component is heat labile?
31. Can two different enzymes have the same cofactor?
32. What is the substrate?
33. How are enzymes proved to be present in the plant?
34. How are enzymes named?

SPECIFIC REFERENCES

Barron, E. S. 1951. Thiol groups of biological importance. Adv. Enzymol. **11**:201-266.

Bernfeld, P. 1951. Enzymes of starch degradation and synthesis. Adv. Enzymol. **12**:379-428.

Fritz, G. J., E. R. Stout, and D. E. Leister. 1963. Estimation of nicotinamide nucleotide coenzymes in etiolated maize seedlings. Plant Physiol. **38**: 642-648.

Hellebust, J. A., and R. G. S. Bidwell. 1963. Protein turnover in wheat and snapdragon leaves. Canad. J. Bot. **41**:969-983.

Koshland, D. E., Jr. 1963a. Correlation of structure and function in enzyme action. Science **142**: 1533-1541.

Koshland, D. E., Jr. 1963b. The role of flexibility in enzyme action. Sympos. Quant. Biol. **28**:473-482.

Lipetz, J. 1965. Peroxidase interference with cytochrome oxidase localization in plant cells. J. Histochem. Cytochem. 13:300-301.

Lipmann, F. 1941. Metabolic generation and utilization of phosphate bond energy. Adv. Enzymol. 1:99-162.

Smith, M. H. 1967. Relative distribution of enzymes and prosthetic groups. Nature (London) 213:627-628.

Theorell, H. 1956. Nature and mode of action of oxidation enzymes. Science 124:467-472.

Wohl, K., and W. O. James. 1942. The energy changes associated with plant respiration. New Phytol. 41:230-256.

GENERAL REFERENCES

Dixon, M., and E. C. Webb. 1964. Enzymes. Academic Press, Inc., New York.

Enzyme nomenclature; recommendations (1964) of the International Union of Biochemists. 1965. Elsevier Publishing Co., Amsterdam.

Florkin, M., and E. M. Stotz. 1962. Comprehensive biochemistry. Elsevier Publishing Co., Amsterdam, New York, 15 vol. (+12 in preparation).

Green, D. E. 1940. Mechanisms of biological oxidations. Cambridge University Press, New York.

Hochster, R. M., and J. H. Quastel, 1963. Metabolic inhibitors. I. Academic Press, Inc., New York.

Lehninger, A. L. 1965. Bioenergetics. W. A. Benjamin, Inc. New York.

Linford, J. H. 1966. An introduction to energetics with applications to biology. Butterworth & Co. (Publishers), Ltd., London.

Northrop, J. H., M. Kunitz, and R. M. Herriott. 1948. Crystalline enzymes. Columbia University Press, New York.

Pullman, B., and A. Pullman. 1963. Quantum biochemistry. Interscience Publishers, New York.

Schwimmer, S., and A. B. Pardee. 1953. Principles and procedures in the isolation of enzyme. Adv. Enzymol. 14:375-409.

Sumner, J. B., and K. Myrback. 1950-1952. The enzymes. Academic Press, Inc., New York. 4 vol.

Walter, C. 1965. Steady-state applications in enzyme kinetics. The Ronald Press Co., New York.

chapter 14

Respiration

OVERALL PROCESS

The most important catabolic, energy-releasing process in the plant is the *oxidative breakdown of organic substances*, which is called respiration. Sugars are most commonly the starting point, and carbon dioxide and water are usually the end substances formed, although there are many exceptions. Respiration consists of a large number of chemical reactions, all of which are enzymatically controlled. For some purposes, however, it is possible to ignore the enzymes and even the individual chemical reactions and to consider the overall process. Thus, if sugar is respired, respiration can be represented by:

$$C_6H_{12}O_6 + 6O_2 \longrightarrow 6CO_2 + 6H_2O + 678 \text{ kcal}$$
where $C_6H_{12}O_6$ = glucose

This means that 180 g (1 mole) glucose (or some other hexose) is oxidized by 192 g (6 moles) of oxygen, releasing 264 g (6 moles) of carbon dioxide, 108 g (6 moles) of water, and 678 kcal of heat. Thus each gram of glucose respired releases 3.74 kcal of heat. It must be recognized that this equation represents only a *balance sheet*, showing the quantitative relations between the raw materials respired and the products formed. That it does *not* represent the true chemical changes is obvious from the now known fact that all the oxygen absorbed in the molecular form is reduced to water, al-

though the preceding general equation does not include enough water to account for this relation. The value of 678 kcal means that 113 kcal are released per mole of oxygen absorbed. This, however, is a theoretical value, and measured values may be significantly different, for example, 126, 141, and 97 kcal for germination of seeds of *Prosopis juliflora*, *Kochia scoparia*, and *Oxyria digyna*, respectively (Lacher et al., 1966). The difference from the theoretical value undoubtedly results from secondary reactions such as hydration.

If this equation is correct as a balance sheet, it should be possible to measure respiration by determining any one of the preceding five quantities. In practice it is usually most convenient to determine either the carbon dioxide released or the oxygen absorbed. If the correctness of the preceding equation is in doubt, it is necessary to measure both quantities. The *molar* ratio of carbon dioxide evolved per mole oxygen absorbed is then called the *respiratory quotient* (R.Q.):

$$\text{R.Q.} = \frac{\text{Moles } CO_2 \text{ evolved}}{\text{Moles } O_2 \text{ absorbed}}$$

If this ratio is 1, the preceding equation would appear to be correct; that is, glucose or some other carbohydrate is apparently being respired to carbon dioxide and water.

If substances other than glucose are respired, a different respiratory quotient usually results. Thus a fat would give:

$$2C_{51}H_{98}O_6 + 145O_2 \longrightarrow 102CO_2 + 98H_2O$$

where $C_{51}H_{98}O_6 =$ glycerol tripalmitate

The respiratory quotient is therefore 0.7. Similarly it can readily be calculated for organic acids, for example, 4 for oxalic acid ($C_2H_2O_4$) if it were respired. The value for proteins varies between 0.5 and 1.0, depending on the particular protein and the substances produced. Actually, the quotient can vary from zero (no carbon dioxide evolved) to infinity (no oxygen absorbed). Some values that have been found experimentally are listed in Table 14-1.

Table 14-1. Respiratory quotients (CO_2/O_2) in different plants*

	Respiratory quotient
Leaves rich in carbohydrates	1
Darkened shoots of *Opuntia* (prickly pear)	0.03
Germinating starchy seeds	1
Germinating linseed (high fat)	0.64
Germinating buckwheat (high protein)	0.50
Germinating peas	1.5–2.4

*Adapted from Thomas, M., et al. 1960. Plant physiology. J. & A. Churchill, Ltd., London.

A respiratory quotient of less than 1 may result from one of several causes:

1. A substrate of lower oxygen content than carbohydrate is being used.
2. A carbohydrate is incompletely oxidized, for example, with the formation of organic acids of higher oxygen content than that of the carbohydrates. This is characteristic of succulents such as *Opuntia* (Table 14-1).
3. Oxygen uptake occurs in processes other than respiration.
4. Carbon dioxide is absorbed in the process of dark carbon dioxide assimilation (see Chapter 15). This is also characteristic of succulents.

A respiratory quotient greater than 1 may mean that some high oxygen substances, such as organic acids, are being respired or that respiration is partially incomplete and therefore does not utilize molecular oxygen (see later).

In spite of these many exceptions, the respiratory quotient of higher plants is commonly 1. This means that the equation given on p. 157 is probably correct for most higher plants under normal conditions, and carbohydrates are the main substrate for respiration. Many other lines of evidence confirm this conclusion. Even when fats are the main reserve material, they are first converted to sugars, which are then respired. Consequently, an understanding of respiration requires some knowledge of the carbohydrates.

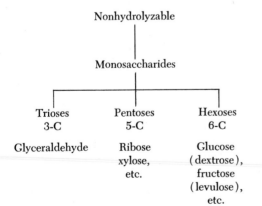

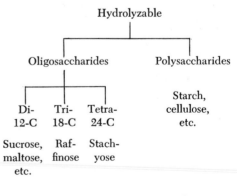

Many different carbohydrates occur in the plant—some soluble (sugars), others insoluble. Each has an empirical formula consisting of carbon, hydrogen, and oxygen, with the hydrogen and oxygen in the proportions in which they occur in water. But the properties of carbohydrates depend on their structural formulas. They consist of polyhydroxy (OH) aldehydes (substances with a CHO group) or ketones (substances with a CO group). In the case of the more complex carbohydrates, which are not themselves aldehydes or ketones, they give rise to polyhydroxy aldehydes and ketones (and to nothing else) when hydrolyzed.

The main carbohydrates in plants may be classified as shown at the bottom of the opposite page.

Complete hydrolysis of starch, cellulose, and maltose yields only the monosaccharide glucose. Hydrolysis of sucrose yields both glucose and fructose. Because of their hydroxyl groups, carbohydrates form esters with acids. The most important of these in metabolism are the phosphate esters which are formed as follows:

$$RCHOH + ATP \rightleftharpoons RCHOH_2PO_3 + ADP$$

where ATP = adenosine triphosphate
ADP = adenosine diphosphate
$RCHOH_2PO_3$ = sugar phosphate

This formation of a phosphate ester of the sugar is the first step in respiration. It is called *phosphorylation*.

GLYCOLYSIS

Many organisms are able to break down (or lyse) carbohydrates to carbon dioxide and some other substance (e.g., an alcohol or organic acid) without utilizing molecular oxygen. This process has been called by several names: *anaerobic respiration* because it may occur in the absence of air; *fermentation* because it is so often controlled by microorganisms or their enzymes (previously known as ferments); or *glycolysis* (glyco- means sugar) because it is a

breakdown of sugars. It is now known that all higher plants are capable of respiring anaerobically for varying lengths of time; but most die within 1 to 3 days in the complete absence of molecular oxygen. Since anaerobic respiration may occur even in the presence of oxygen, it must be defined as respiration without the *participation of molecular oxygen*.

When higher plants respire in the absence of oxygen, their respiration is usually partly or wholly alcoholic fermentation:

$$C_6H_{12}O_6 \longrightarrow 2C_2H_5OH + 2CO_2 + 28kcal$$
where C_2H_5OH = ethyl alcohol

Much less energy is released in this way than in complete aerobic oxidation to carbon dioxide and water. Substances other than ethyl alcohol may accumulate, for example, glycerol, lactic acid, acetaldehyde, etc. In the case of some plants (e.g., potato tubers), acids accumulate but no alcohol. If only alcohol is produced, the ratio of alcohol to carbon dioxide is 1.04. The actual ratios found vary from this value all the way down to zero (Table 14-2).

Since so many plants accumulate ethyl alcohol under anaerobic conditions, early workers suggested that the normal aerobic respiration involves, first, the anaerobic breakdown to alcohol and carbon dioxide, and, second, a further breakdown of the alcohol to carbon dioxide and water. That the argument does not hold for some plants

Table 14-2. Ratios of carbon dioxide to alcohol produced in plants respiring anaerobically*

	Ratio
Carrot roots	100:102
Oranges	100:70
Apples	100:42
Potato tubers	100:7
Alcoholic fermentation (theoretical)	100:104

*Adapted from Kostytchev, S. R. 1926. Lehrbuch der Pflanzenphysiologie. I. Springer-Verlag, Berlin.

(e.g., potato) follows from the near absence of alcohol under anaerobic conditions. Furthermore, it was soon shown that higher plants do not oxidize alcohol very readily.

It is now known that alcohol is not produced when the higher plant is respiring aerobically, although the first stage of aerobic respiration is the same as in alcoholic fermentation, up to a point preceding alcohol formation. Alcohol is thus an end product that occurs on a branch line instead of on the main line of reactions. This side branch operates only when respiration is anaerobic. The actual sequence in alcoholic fermentation consists of some twelve chemical reactions, including the following steps:

$$(1) \qquad\qquad (2) \qquad\qquad (3)$$
Hexose→ Hexosephosphates→ Triosephosphates→
$$(4) \qquad\qquad\qquad (5)$$
Phosphoglyceric acid→ Phosphopyruvic acid→
$$\quad (6) \qquad\qquad (7)$$
Pyruvic acid→ Acetaldehyde→ Ethyl alcohol

Each step in the reaction is controlled by an enzyme. The variety of kinds of enzymes involved is indicated by the number of each step or group of steps, each number corresponding to a specific enzyme group as follows:

(1) Transferase (transfer of phosphate from ATP to hexose)
(2) Hydrolase (lysis or splitting of C_6 to $2C_3$)
(3) Oxidoreductase (oxidation of aldehyde to carboxyl)
(4) Lyase (hydro-lyase; removal of molecule of water)
(5) Transferase (phosphotransferase: transfer of phosphate to ADP)
(6) Lyase (carboxy-lyase: removal of carbon dioxide)
(7) Oxidoreductase (reduction of aldehyde to alcohol)

The preceding series of reactions is called the Embden-Meyerhof-Parnas Pathway (EMP).

KREBS CYCLE

When the respiration is aerobic, glycolysis goes only as far as pyruvic acid. Acetaldehyde and alcohol are on the

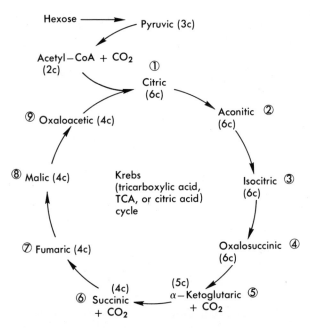

Fig. 14-1. Simplified version of Krebs cycle. The number of C-atoms in each organic acid is indicated in parentheses. The water molecules removed and the enzymes involved are not shown.

anaerobic branch line. Pyruvic acid is therefore the pivot. When aerobic respiration occurs, this substance is converted to another organic acid, which in its turn is changed to a third one, etc. (Fig. 14-1). Each molecule of pyruvic acid is thus broken down to carbon dioxide and water by passing through a series of some twelve chemical reactions, converting it successively into 6-, 5-, and 4-carbon acids. This cyclic series is known as the Krebs (or tricarboxylic acid, TCA, or citric acid) cycle.

Again, each step in the Krebs cycle is controlled by an enzyme; but not as many kinds of enzymes are required for the Krebs cycle proper as for glycolysis. Since the pyruvic acid is essentially broken down to carbon dioxide and water, the main enzymes are oxidoreductases and lyases (both carboxy-lyases and hydrolyases).

The first part of respiration sometimes goes through a pathway that differs from glycolysis by being aerobic and has been called the pentose phosphate pathway (PPP), or hexose monophosphate pathway (or shunt-HMP or HMS) since it includes these substances. The whole process of respiration may therefore be represented as shown in Fig. 14-2.

The oxidative breakdown of carbohydrate can therefore be described more correctly as follows:

Anaerobic phase:

1. $C_6H_{12}O_6 \longrightarrow 2CH_3 \cdot CO \cdot COOH + 4H$

Aerobic phase:

2. $2CH_3 \cdot CO \cdot COOH + 5O_2 \longrightarrow 6CO_2 + 4H_2O$
3. $\qquad\qquad 4H + O_2 \longrightarrow 2H_2O$

Total: $C_6H_{12}O_6 + 6O_2 \longrightarrow 6CO_2 + 6H_2O$
where $CH_3 \cdot CO \cdot COOH$ = pyruvic acid

ELECTRON TRANSPORT

The net result of the Krebs cycle is thus the breakdown of pyruvic acid to carbon dioxide and water. The carbon dioxide release is a group removal reaction, in which carboxylic acids are decarboxylated in the presence of carboxy-lyases. The formation of water is caused by oxidation-reduction reactions involving the removal of hydrogen atoms from the organic acids (or dehydrogenations) in the presence of oxidoreductase enzymes. This occurs, for instance, in the case of succinate:

$$\text{Succinate} \underset{}{\overset{\text{O-Rase}}{\rightleftharpoons}} \text{Fumarate} + 2H$$

These hydrogen atoms are ultimately combined with oxygen to form water. The dehydrogenation is also a removal of electrons since each hydrogen atom possesses an electron. It is therefore called an electron transport. Since neither the hydrogen atom nor its electron can exist free, each hydrogen or electron donor must be linked to a hydrogen or electron acceptor in order to become oxidized by the electron loss. Such a link cannot occur directly between the electron-donating organic acids and the electron-accepting oxygen, because the oxidation-reduction potentials of the two systems are too far apart; it is only when the potentials

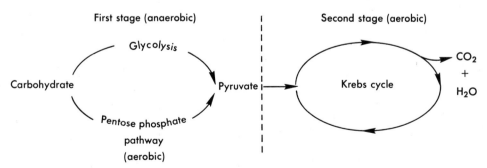

Fig. 14-2. The two main stages in the respiratory breakdown of carbohydrate to $CO_2 + H_2O$.

of two systems overlap, that one can oxidize or reduce the other. Therefore, a series of oxidation-reduction systems must mediate the electron transport from the organic acids to molecular oxygen. Each of these systems accepts electrons from another system at a higher reduction potential (lower oxidation potential) and transports them to a third system at a lower reduction potential (higher oxidation potential). In this way the electrons are transported step by step from a region of higher free energy to one of lower free energy, and the net result is a transfer of hydrogen atoms from the organic acids to molecular oxygen, forming water as follows:

$$\text{Organic acid} \xrightarrow{\quad H \quad} \text{System 1} \xrightarrow{\quad H \quad} \text{System 2} \xrightarrow{\quad H \quad} O_2$$

Each of these steps is controlled by a specific oxidoreductase enzyme, and in each case a cofactor is successively reduced by the system above it and oxidized by the system below it. There are many such systems, for example:

$$NADP^+ + 2H \rightleftharpoons NADPH$$
$$Cytochrome_{ox} + 2H \rightleftharpoons Cytochrome_{red}$$
$$Quinone + 2H \rightleftharpoons Hydroquinone, \text{ etc.}$$

where NADP$^+$ = nicotinamide-adenine dinucleotide phosphate
NADPH = reduced nicotinamide-adenine dinucleotide phosphate

OXIDATIVE PHOSPHORYLATION

There are, therefore, many side reactions associated with the Krebs cycle proper, without which the cycle would soon come to a stop. Most of these are electron transfer reactions. There are also some electron transfers, although fewer in number, associated with glycolysis. In both cases each individual electron-transfer reaction releases a small amount of free energy; and because of the greater number associated with the Krebs cycle, much more total free energy is released in the aerobic than in the anaerobic phase. However, the free energy of any chemical reaction is given off as heat, unless it is coupled to an energy-ab-

sorbing reaction. Therefore, the plant must have some mechanism of trapping this free energy before it becomes converted to unavailable heat energy and of using it to drive its energy-requiring processes. The trapping reaction is the conversion of low energy (e.g., inorganic) phosphates to high energy organic phosphates. This is the process illustrated earlier by analogy with a dynamo (the metabolic wheel), from which energy is brushed off and transported to the region of utilization by way of a substance known as adenylic acid (see Fig. 13-2). When the phosphate gives up its energy, low energy inorganic phosphate is regenerated, which may be again converted into the high energy form by the metabolic wheel. The phosphates take part in the Krebs cycle as well as in glycolysis, although this is not shown in Fig. 14-1. The importance of the high energy phosphates is dramatically illustrated by the use of certain substances (e.g., 2,4-dinitrophenol) that act as *uncouplers*. They prevent the conversion of respiratory energy to high energy phosphates by inhibiting the enzymes controlling these reactions. But they do not prevent the respiratory breakdown of the carbohydrates, and in fact the rate of this breakdown may actually increase, although the energy released is wasted, and therefore, growth and other energy utilizing processes are inhibited.

The minimum overall conversion of respiratory energy to high phosphate energy may be represented as follows (Lehninger, 1964):

$$C_6H_{12}O_6 + 6O_2 + 38ADP + 38H_3PO_4 \rightarrow 6CO_2 + 38ATP + 44H_2O + 382 \text{ kcal}$$

where H$_3$PO$_4$ = phosphoric acid (actually in the form of a phosphate)
ADP = adenosine diphosphate
ATP = adenosine triphosphate

The release of heat energy is much less than the 674 kcal given in the first overall equation. This means that nearly 300 kcal of energy that would have been given off as heat is trapped in the ATP and can be used to drive endergonic processes. The efficiency of energy conversion according to the preceding equation is therefore about 44%, but the exact relation in plants is unknown. It has even been suggested that the plant may be 90% efficient in the conversion of respiratory energy to high energy phosphate (Beevers, 1961). Recent measurements have led to values between the preceding two figures. Lacher and associates (1966) made use of the fact that dinitrophenol (DNP) is an uncoupler of oxidative phosphorylation (see earlier). It therefore prevents the phosphates from trapping any of the energy released by respiratory electron-transport reactions; and all the free energy is released as heat. By comparing the energy values in normal and uncoupled germinating seeds, they showed that 62% of the respiratory energy is trapped by oxidative phosphorylation.

Because of its high free energy, ATP is unstable and must be utilized shortly after it is formed. Therefore, the plant cannot store its energy in this form and must depend on the more stable carbohydrates as its main store of energy, which can be readily converted to the available (ATP) form as needed by the process of respiration.

It must be emphasized, therefore, that the main function of respiration is the production of high energy phosphates (mainly ATP), which may then be used to drive the endergonic processes of the plant. For some time, scientists have sought the mechanism by means of which the energy released by the exergonic (electron transport or oxidative) reactions of respiration is transferred to the phosphate, yielding ATP. In the case of glycolysis the mechanism is simple since the phosphates are directly involved in the electron transport. The high energy phosphates (e.g., glycerophosphate) are therefore intermediate products. In the case of the Krebs cycle, it is the free organic acids (or their salts) that transfer the electrons to a series of cofactors (see p. 162). There is no direct con-

nection between the electron transport chain and the phosphates. Therefore, there must either be some unknown phosphate intermediates, or another mechanism must be involved. One such mechanism has recently been proposed by Mitchell (1967). He calls it the "chemiosmotic hypothesis." It can be simply described as follows:

1. The electron transport leads to an accumulation of the oppositely charged protons (H^+) within the mitochondrial membrane, which is impermeable to it.
2. This leads to a higher $[H^+]$ inside than outside the mitochondrion and a higher $[OH^-]$ outside than inside it.
3. This activity gradient "pulls" the H^+ and OH^- out of the ADP and inorganic phosphate, leading to a condensation reaction (i.e., removal of water) and the formation of ATP.

This hypothesis has the added advantage that it can also explain the mechanism by means of which respiratory energy is utilized to produce ion transport (i.e., active absorption), which would essentially be a reversal of this process, the released H^+, for instance, exchanging for a K^+ ion from the outside.

RESPIRATORY ENZYMES

Since glycolysis and the Krebs cycle proper each involves some ten to twelve reactions, and this does not include the many side reactions of electron transport and phosphorylation, there must be several dozen enzymes that participate in the respiration of a single sugar molecule to carbon dioxide and water. The detailed treatment of all these enzymes is the province of a more advanced course in biochemistry.

Plant physiologists, however, have long been preoccupied with the problem of the "terminal enzymes" (Beevers, 1961). These are the enzymes that control the final electron (or hydrogen) transfer to molecular oxygen. There are four main groups of terminal enzymes:

1. When oxygen is the hydrogen acceptor, the enzymes are called oxidases.
2. If the oxygen molecule is incorporated into the substrate, the enzymes are oxygenases.
3. If hydrogen peroxide is the acceptor, they are peroxidases.
4. A final exceptional group, catalases, breaks down any hydrogen peroxide formed.

The second group is not important in plants. The other reactions may be represented as follows:

$$R(OH)_2 + O_2 \xrightarrow{\text{Oxidase}} R(O)_2 + H_2O_2$$

$$AH_2 + H_2O_2 \xrightarrow{\text{Peroxidase}} A + 2H_2O$$

$$H_2O_2 \xrightarrow{\text{Catalase}} H_2O + \frac{1}{2}O_2$$

Several oxidases are known, each of which controls the oxidation of a specific substrate:

$$\text{Ascorbic acid} \xrightarrow{\substack{\text{Ascorbic acid} \\ \text{oxidase}}} \text{Dehydroascorbic acid}$$

$$\text{Phenol} \xrightarrow{\text{Phenolase}} \text{Quinone}$$

$$\text{Reduced cytochrome} \xrightarrow{\substack{\text{Cytochrome} \\ \text{oxidase}}} \text{Oxidized cytochrome}$$

All of the oxidases are heavy metal enzymes because of the presence of iron or copper as cofactors. They are therefore inhibited by cyanides since these combine with heavy metals. This inhibition becomes progressively lower in aging organs. On the basis of the most recent evidence, cytochrome oxidase is the one involved in the terminal oxidation reaction, leading to the formation of high energy phosphates. The exact role of the other oxidases in plant metabolism is still not well understood.

All the enzymes controlling glycolysis and the pentose phosphate pathway appear to occur free in the cytoplasm. On the other hand, those responsible for the Krebs cycle

reactions have been found to occur only in the mitochondria. When these organelles are isolated from the plant, the oxidation by them of the intermediates in the Krebs cycle requires the addition of oxygen, ADP, and inorganic phosphate. This is because the previously described phosphorylation is linked to the electron transport of the oxidation reactions.

The mitochondrion is a complex structure in spite of its small size (Fig. 14-3). Some of the enzymes are loosely associated with it, others are tightly enclosed in it, and a third group are structurally bound to it (Klingenberg and Pfaff, 1966). It is presumably because of its specialized structure and the specific location of its many enzymes that the mitochondrion is capable of producing all the steps of the Krebs cycle and associated reactions in the correct order and at the necessary rates.

On the basis of the overall equation, the respiration of higher plants in its strictest sense consists only of the energy-releasing breakdown of sugars to carbon dioxide and water. This process, therefore, cannot include all of catabolism. There must frequently also be a hydrolytic breakdown (or *digestion*) of more complex, insoluble carbohydrates to sugars, of lipids to fatty acids, glycerol, etc., and of proteins to amino acids. The products of digestion may then be respired, or they may be resynthesized to other complex substances, or they may be translocated to other parts of the plant. Just as in the case of respiration, digestive processes are controlled by enzymes, most of which are hydrolases. In some cases it is impossible to distinguish between digestion and respiration, for example, when starch is phosphorylated to hexose phosphate, which is then broken down in glycolysis. Similary, although respiration in its narrower sense has been described previously as beginning with carbohydrates, in its broader sense this is not always the case, as shown earlier by the variation in respiratory quotients. In many seeds, for instance,

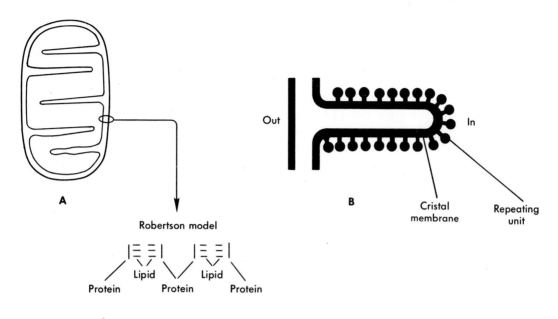

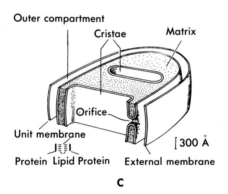

Fig. 14-3. Mitochondrion structure. **A** and **B**, Two-dimensional structure. **C**, Three-dimensional structure. (**A** and **B** from Lee, C. P., and L. Ernster and **C** from V. P. Whittaker. 1966. *In* J. M. Tager, et al. Regulation of metabolic processes in mitochondria. Elsevier Publishing Co., Amsterdam.)

the reserves are in the form of fats or proteins, and during germination these are respired. The reactions involved are less well understood than in the case of the more common respiratory breakdown of carbohydrates. Proteins, however, if respired to completion, may give rise to ammonia as well as to carbon dioxide, and this may lead to injury.

RESPIRATION RATES

Plant respiration releases large quantities of carbon dioxide into the atmosphere. The forest has been estimated to release about 3400 g of carbon dioxide per square meter per year and, therefore, to respire some 2100 g carbohydrate per square meter per year (Woodwell and Dykeman, 1966). Yet at certain times of the year the rate drops to a very low value, and at other times it rises to a correspondingly high value because the rate of respiration is affected by many factors, both internal and external. From the equation it is obvious that an ample supply of hexoses or other respirable material is essential for a rapid rate of carbon dioxide evolution. The same is true of the oxygen supply to the respiring cells, at least if respiration is aerobic. Yet it has been shown that, in many cases, oxygen actually decreases the rate of breakdown of sugars as compared with anaerobic respiration. This is called the *Pasteur effect.* An accumulation of the end products also reduces the rate. This fact is made use of in the storage of fruit. Air containing 10% carbon dioxide retards respiratory breakdown and therefore reduces sugar loss and prolongs the life of the fruit; but the oxygen content of the air must be maintained as high as normal (at the expense of the nitrogen) to prevent anaerobic respiration, which leads to injury.

The rate of respiration is, of course, greatly affected by temperature. A rise of 10° C usually increases the rate 2 to 3 times. This is the main basis for the low-temperature preservation of plant parts. Dropping the temperature from 30° to 0° C may reduce the rate of respiration to 1/27 (i.e. $[1/3]^3$), in this way preserving the reserves 27 times as long. Injury may greatly increase the respiration rate.

A striking variation in respiration rate may occur with age. Thus a developing fruit may show a significant drop during the early stages, followed by a rise, then another drop (Fig. 14-4). The first drop is perhaps more apparent than real since the cell enlargement results in about as much of a decrease in percent protoplasm (which is the only part of the cell capable of respiring) as the decrease in respiration per gram of total dry matter (Kidd and West, 1945). The subsequent rise and fall, however, are real changes since they occur when the fruit is already full size. The rise that accompanies ripening is called the *climacteric* and is associated with the production of ethylene. Other volatile products responsible for the flavor and aroma also reach a maximum at this time (e.g., methyl, ethyl, and amyl esters of formic, acetic, caproic, and caprylic acids). Artificial ripening may be induced by treating the unripe fruit with ethylene gas. Unripe bananas when treated for 5 days with air containing 1 part ethylene in 10,000 at 20° to 22° C, may ripen 20 days sooner than the untreated fruit. Even when artificially produced in this way, the ripening is always accompanied by an increased rate of respiration. This climacteric period is followed by the final period of *senescence,* during which the respiration process seems to be thrown out of gear. Not only does the total rate drop rapidly but ethyl alcohol and acetaldehyde steadily accumulate, the latter giving the poor taste. This seems to indicate that reactions involving molecular oxygen are hindered and that anaerobic respiration becomes more pronounced. Other metabolic changes also occur; for example, insoluble pectins that

Rate of CO₂ production

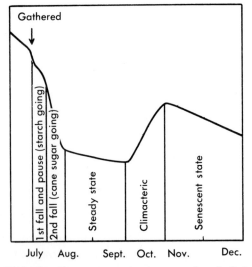

Fig. 14-4. Changes in respiration rate of apple fruit during growth and development. (From Kidd, F. R. 1935. Nature [London] **135:**327-330.)

normally cement cells together are converted to soluble pectin by the enzyme protopectinase. As a result, the cells separate and the flesh becomes mealy. In the case of fruit such as tomato and grapes, even the cell walls dissolve and the free protoplasts are released. These later break down. This whole self-digestive process is known as *autolysis.*

The effect of light on respiration rate has been a subject of much controversy for some time. It was once suggested that green cells cease respiring when they begin photosynthesizing in the light. Experiments with radioisotopes soon disproved this concept. They showed that respiration does continue in the light even in green cells, although the rate may increase, decrease, or remain unchanged, depending on the conditions. But such conclusions are based solely on measurements of evolution of labeled carbon dioxide or of absorption of labeled oxygen. Recent more detailed investigations, including the intermediates of respiration, have indicated that respiration is also changed qualitatively in the light; that is, the intermediate chemical steps may be different. The term *photorespiration* has therefore been coined for respiration in the light (Tregunna et al., 1966).

QUESTIONS

1. If a higher plant in the dark gives off 12 moles carbon dioxide in a certain time, how much oxygen would you expect it to have absorbed?
2. What substrate would have been used up?
3. How much substrate would have been used up?
4. How much heat would have been liberated?
5. What respiratory quotient is assumed?
6. What does this mean?
7. Between what limits may the respiratory quotient vary?
8. What is the respiratory quotient if fats are respired?
9. What is the respiratory quotient if proteins are respired?
10. What is the respiratory quotient if organic acids are respired?
11. What does a respiratory quotient of zero mean?
12. What does a respiratory quotient of infinity mean?
13. What is the major substrate for the respiration of higher plants?
14. What is a carbohydrate?
15. What kinds of carbohydrates are there?
16. Which carbohydrates are respired?
17. Which carbohydrates are not respired?
18. What is the first chemical step in respiration?
19. What is anaerobic respiration?
20. What is alcoholic fermentation?
21. What is glycolysis?
22. Is more, less, or the same amount of energy released in glycolysis than that in aerobic respiration?
23. What are the usual end products in higher plants?
24. How is respiration measured?
25. What relation is there between glycolysis and aerobic respiration?
26. How many chemical steps are there in glycolysis?
27. What mineral element is essential for glycolysis?
28. What are some of the intermediate substances formed?
29. What is the "pivot" substance?
30. What is the name for the 2nd or aerobic half of (aerobic) respiration?
31. Is there any other name for it?
32. What kind of chemical substances are involved?
33. Approximately how many chemical reactions are included?
34. What is the net result of the Krebs cycle?
35. What is the pentose phosphate pathway?
36. What role does electron transport play in respiration?
37. What is oxidative phosphorylation?

38. What role does it play in respiration?
39. How many enzymes are probably involved in respiration?
40. Where are the Krebs cycle enzymes located?
41. What effect does temperature have on respiration rate?
42. What is meant by the climacteric?
43. By senescence?
44. By autolysis?

SPECIFIC REFERENCES

Kidd, F. R. 1935. Respiration of fruits. Nature (London) **135**:327-330.

Kidd, F. R., and C. West. 1945. Respiratory activity and duration of life of apples gathered at different stages of development and subsequently maintained at a constant temperature. Plant Physiol. **20**:467-504.

Klingenberg, M., and E. Pfaff. 1966. Structural and functional compartmentation in mitochondria, p. 180-201. *In* J. M. Tager, S. Papa, E. Quagliariello, and E. C. Slater. Regulation of metabolic processes in mitochondria. Elsevier Publishing Co., Amsterdam.

Lacher, J. R., A. Amador, and K. Snow. 1966. Effect of dinitrophenol on the heats of respiration of germinating seeds of *Prosopis juliflora, Kochia scoparia, and Oxyria digyna*. Plant Physiol. **41**:1435-1438.

Lee, C. P., and L. Ernster. 1966. The energy-linked nicotinamide nucleotide transhydrogenase reaction, p. 218-234. *In* J. M. Tager, S. Papa, E. Quagliariello, and E. C. Slater. Regulation of metabolic processes in mitochondria. Elsevier Publishing Co., Amsterdam.

Mitchell, P. 1967. Proton-translocation phosphorylation in mitochondria, chloroplasts, and bacteria; natural fuel cells and solar cells. Fed. Proc. **26**:1370-1379.

Tregunna, E. B., G. Krotkov, and C. D. Nelson. 1966. Effect of oxygen on the rate of photorespiration in detached tobacco leaves. Physiol. Plant. **19**:723-733.

Whittaker, V. P. 1966. The ultrastructure of mitochondria, p. 1-27. *In* J. M. Tager, S. Papa, E. Quagliariello, and E. C. Slater. Regulation of metabolic processes in mitochondria. Elsevier Publishing Co., Amsterdam.

Woodwell, G. M., and W. R. Dykeman, 1966. Respiration of a forest measured by carbon dioxide accumulation during temperature inversions. Science **154**:1031-1034.

GENERAL REFERENCES

Beevers, H. 1961. Respiratory metabolism in plants. Harper & Row, Publishers, New York.

James, W. O. 1946. The respiration of plants. Ann. Rev. Biochem. **15**:417-434.

Kostytchev, S. R. 1926. Lehrbuch der Pflanzenphysiologie. I. Chemische Physiologie. Springer-Verlag, Berlin.

Lehninger, A. L. 1964. The mitochondrion; molecular basis of structure and function. W. A. Benjamin, Inc., New York.

Thomas, M., S. L. Ranson, and J. A. Richardson. 1960. Plant physiology. J. & A. Churchill, Ltd., London.

chapter **15**

Photosynthesis

ENERGY RELATIONS

Photosynthesis is a unique process in two distinct ways. Unlike respiration, which occurs in essentially all living cells of all living organisms, it is unique to the green cells of plants (including some bacteria). It is also unique in being the single, energy-trapping process on which all life (except that of a few chemosynthetic bacteria) depends, since it is the only biological process that converts large quantities of light energy into chemical energy. In the case of nonliving systems most of the photochemical reactions known to chemists are, in fact, exothermic and are therefore valueless for storing energy (Daniels, 1967). In these cases the light simply catalyzes the reaction. Similarly, the many other photophenomena that occur in all organisms do not convert light energy into chemical energy. Most of them are primarily processes of a cybernetic character—the photoproducts control chain reactions that affect or determine the response of an organism to the environment (Thomas, 1965).

The central role of photosynthesis in the energy cycle of life is illustrated in Fig. 15-1. Photosynthesis can therefore be defined as the *light-energy-trapping process on which all life depends.* Ever since its importance to all life was first realized, it has attracted the attention of many of the world's top scientists. The story of photosynthetic research is one of the most fascinating in the whole realm of biology. In an elementary text it is impossible to do justice to the romance of the history of research into the problem, to the many controversies that have raged and that still exist, or to the many details of the process that are now known.

Historically, photosynthetic research can be divided into two eras: B.I. (before isotopes) and A.I. (after isotopes). The first era extends to the mid-1930's. The information available in this era was so meager that it could be adequately covered in an elementary text. During the second era, however, and especially in the past decade, research has been so intense and varied that even active investigators of photosynthesis have difficulty in adequately explaining all of what is known about the subject. Furthermore, interpretations vary with the research school. In order to explain their

169

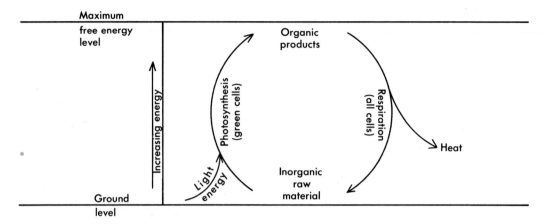

Fig. 15-1. Role of photosynthesis in the energy cycle of life.

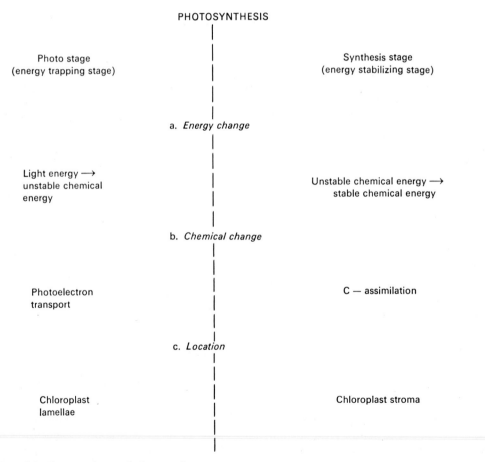

Fig. 15-2. The two phases of photosynthesis.

interpretations, investigators of photosynthesis are prone to propose "schemes." There is a dangerous tendency for textbook writers to forget that they *are* just schemes (i.e., hypotheses) and to accept them as scientific laws. We must, therefore, emphasize at this point that we are adopting those schemes that seem best supported by the evidence at this writing. They must not be accepted as the final word on the subject.

The chemical energy that the plant produces photosynthetically must be stable enough to store for future use. There are two main stages in the process: (1) the energy-trapping stage and (2) the energy-stabilizing stage. These two stages differ both with respect to the nature of their chemical changes and to their location in the cell (Fig. 15-2).

Historically, only the second stage was investigated during the first era. It was therefore common to find a chapter entitled Carbon Assimilation but none labelled Photosynthesis, especially in the European texts. For this historical reason and because of the relation of this second stage to respiration, which is covered in the previous chapter, it will be discussed first.

C-ASSIMILATION
"Dark" C-assimilation

The synthesis stage of photosynthesis is the anabolic counterpart of respiration because it is the synthesis of organic substances from carbon dioxide. It is therefore called photosynthetic C (carbon)-assimilation. This process is possible only in green plants; but all organisms—colorless as well as green plants and even animals—are able to assimilate carbon nonphotosynthetically in the dark. This is called "dark" C-assimilation since it can occur in the continuous absence of light, whereas photosynthetic C-assimilation can occur in the dark only immediately after a light period.

In both cases, carbon dioxide is the source of carbon, but the end products are normally different in higher plants. Photosynthetic C-assimilation results in the formation of carbohydrates, dark C-assimilation in organic acids. Whether or not it occurs as a component of photosynthesis, the synthesis of complex substances from carbon dioxide involves endergonic reactions that must be driven by a continuous supply of free energy. In the case of dark C-assimilation, the only energy available

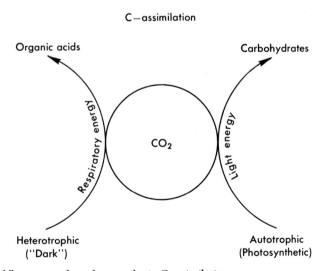

Fig. 15-3. "Dark" as opposed to photosynthetic C-assimilation.

is respiratory energy, and this drains the reserves of the organism. Dark C-assimilation is therefore a wasteful process. Photosynthetic C-assimilation, on the other hand, increases the reserves by utilizing light energy. The two processes are also called heterotrophic and autotrophic C-assimilation, respectively (Fig. 15-3). Dark, or heterotrophic, C-assimilation does, however, play a specific role in the metabolism of the plant in spite of its wastefulness. Succulents, for instance, which grow under conditions of water deficiency, retain their high water content by keeping their stomata closed during the daytime. This reduces transpiration to a minimum. However, it also prevents the uptake from the atmosphere of the carbon dioxide needed for photosynthesis. At night, when there is little danger of water loss, they open their stomata and accumulate carbon dioxide by dark C-assimilation. But the carbon dioxide is accumulated by carboxylation to the carboxyl radical (—COOH) and storage of the acids in the vacuole. Since carboxylases are present, there is an equilibrium reaction:

$$RCOOH \rightleftharpoons RH + CO_2$$

In the light the small amount of carbon dioxide is photosynthesized to sugars. This destroys the equilibrium, and carbon dioxide is formed from RCOOH. Photosynthesis can in this way continue in the light without an external supply of carbon dioxide (and therefore with the stomata closed) until all the RCOOH accumulated at night is used up in the synthesis of carbohydrates. Since the photosynthetic gain in energy exceeds the respiratory loss of the dark C-assimilation, the net result is a gain:

This is, of course, only making the best of a poor situation, and these succulents, as a rule, cannot assimilate as much carbon per 24 hr period as can a normally photosynthesizing plant when supplied with adequate water. Pineapple seems to be an exception since it is as high yielding a plant as any nonsucculent.

It is becoming apparent that dark C-assimilation may play an important role even in nonsucculents. *Avena* coleoptiles, for instance, show greater growth in the dark in the presence of 0.5% to 10% carbon dioxide than in its absence (Harrison, 1965). Excised tomato roots growing in liquid culture show 50% more growth when aerated with air than when aerated with carbon dioxide-free air (Splittstoesser, 1966). It is now believed that dark C-assimilation occurs in all plants (Walker, 1966). Furthermore, it may occur in the light as well as in the dark, although in the light it accounts for only 3% to 12% of the total assimilated.

Photosynthetic C-assimilation

In many respects photosynthetic carbon assimilation is the direct opposite of respiration and can actually be represented crudely by the same equation turned around:

$$6CO_2 + 6H_2O \xrightarrow[\text{Chlorophyll}]{\text{Light}} C_6H_{12}O_6 + 6O_2$$

$$\text{where } C_6H_{12}O_6 = \text{glucose}$$

This equation is also frequently written in the simplified form:

$$CO_2 + H_2O \xrightarrow[\text{Chlorophyll}]{\text{Light}} (CH_2O) + O_2$$

Night (N)		Day (D)
$\longleftarrow$		$\longrightarrow$
(dark)		(light)

Vacuole $\longleftarrow$ $\dfrac{\text{Respiratory}}{\text{energy}}$ $\boxed{RCOOH \rightleftharpoons RH + CO_2}$ $+ H_2O$ $\xrightarrow[\text{energy}]{\text{Light}}$ (CH_2O)

Small energy loss   Large energy gain

24 hr sum: small energy loss (N) + large energy gain (D) = moderate energy gain
where (CH_2O) = carbohydrate

Instead of carbohydrate being broken down into carbon dioxide and water (as happens in respiration), the latter are built up to carbohydrate; instead of oxygen being absorbed, it is released; instead of energy being released, it is absorbed in the form of light energy and converted into the chemical energy of the carbohydrate molecule; instead of oxidation of a carbon-containing compound and reduction of oxygen to water, carbon dioxide is reduced to hexose, and water is oxidized to oxygen.

Just as in the case of respiration, the rate of photosynthesis can be theoretically measured by determining any one of the quantities in the preceding equation; but the most easily determined are the carbon dioxide absorbed and oxygen released. The photosynthetic quotient is O_2/CO_2, and as in the case of the converse respiratory quotient (CO_2/O_2), is usually equal to 1. However, the photosynthetic quotient has not proved to be as valuable a ratio as the respiratory quotient partly because photosynthesis, unlike respiration, normally goes all the way or not at all (as far as gross measurements of oxygen and carbon dioxide are concerned) and partly because carbohydrates are nearly always the end point.

Just as in the case of respiration, the preceding equation for photosynthesis is simply a statement of the relative quantities of raw materials and final products. It reveals nothing about the individual chemical reactions that follow each other, step by step, in an orderly and integrated fashion. During the two decades from the 1930's to the 1950's, intensive investigations largely revealed what these chemical reactions are. This great progress in our understanding of the photosynthetic intermediates largely results from the use of isotopes combined with the newly developed chromatographic techniques. The important contributions of these "tagged" elements to our understanding of permeability and translocation have already been mentioned. In the case of photosynthesis, they have been responsible for opening up a whole new field of investigations.

The normal isotope of carbon is C^{12}, the heavy carbon (the stable isotope) is C^{13}, and the two radioactive isotopes are C^{11} and C^{14}. Only radioactive isotopes have been used in studies of photosynthesis. The first used (in the 1930's) was C^{11}, the only one available at this early date since only the cyclotron was producing isotopes. By use of this isotope, Rubin and associates were able to develop the techniques and concepts that served as the bases for the later, more successful investigations using C^{14}. Their lack of success resulted solely from the extremely short half-life of C^{11}— 22 min. This resulted in a useful life of about 3 hr, during which the whole investigation had to be completed. On the other hand, C^{14} has a half-life of over 5000 years, and therefore, continues to label substances for an unlimited time, as far as experiments are concerned. This isotope did not become available until the first atomic pile was built in the mid-1940's. The general procedure is as follows.

The plant is supplied with $C^{14}O_2$ in the light for a few seconds, or minutes, then quickly killed and analyzed. Those substances that contain C^{14} must have been synthesized by the plant from the $C^{14}O_2$. If the plant is allowed to use the $C^{14}O_2$ for different lengths of time, the substance formed in the shortest period is obviously the first in the chain of reactions. In this way it was proved that a phosphoric ester of a 3-carbon compound, *phosphoglycerate* (PGA), is the first detectable substance formed in photosynthesis, appearing within 2 to 5 sec. The carbon dioxide actually is fixed in the first reaction before being converted to phosphoglycerate. This substance must then be reduced to the triose, and two molecules of triose are finally combined to form hexose (Fig. 15-4). But again, as in respiration, many intermediate chemical reactions are involved.

Enough information is now available to construct a hypothetical series of reactions

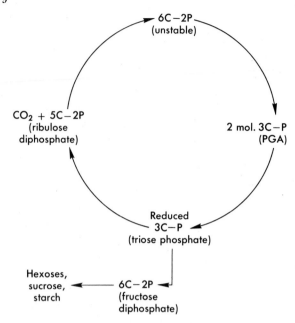

Fig. 15-4. Simplified photosynthetic C-assimilation cycle.

known as the Benson-Calvin (or simply, Calvin), or pentose phosphate, cycle (Fig. 15-5). As the latter name implies, it is to a large degree a reversal of the respiratory pentose phosphate path. However, other reactions are also important, and the cycle is constantly being modified and improved as new information accumulates (Stiller, 1962). Some of this evidence indicates that the Benson-Calvin cycle is not a fixed one applicable to all plants. Unfortunately, it has been worked out under nonphysiological conditions, that is, with much higher carbon dioxide concentrations than are normally available. At physiological carbon dioxide concentrations a glyoxylate cycle has also been found to occur in leaves (Zelitch, 1965). More recent work, however, reinforces the Benson-Calvin cycle and leads to the conclusion that the role of glycolic acid in photosynthesis is of limited importance (Plamondon and Bassham, 1966). It is apparently formed as an end product of photosynthesis from sugar phosphate, which is first formed from 3-phosphoglycerate (Hess and Tolbert, 1967). In sugar cane the first stable compounds are

malic and aspartic acid, which are converted to sucrose via 3-phosphoglycerate and hexosephosphate (Kortschak et al., 1965). The evidence therefore indicates that, in this plant, the carboxylation reaction gives rise only to 4C-dicarboxylic acids (Hatch and Slack, 1966). This new photosynthetic pathway has been found in several species of *Gramineae* but not in others and in only one species of *Cyperaceae,* among sixteen species from other families (Hatch et al., 1967). It has also been found in a dicot species (Osmond, 1967).

Just as in the case of respiration, photosynthesis includes not only changes in chemical composition but energy changes as well. The endergonic changes of the photostage are light induced and cannot be studied by use of C^{14}, as was done in the case of C-assimilation, which per se does not include a photochemical process and does not require light.

PHOTOELECTRON TRANSPORT

As in the case of any electron transport, the photoelectron transport of photosynthesis results from the presence of a series

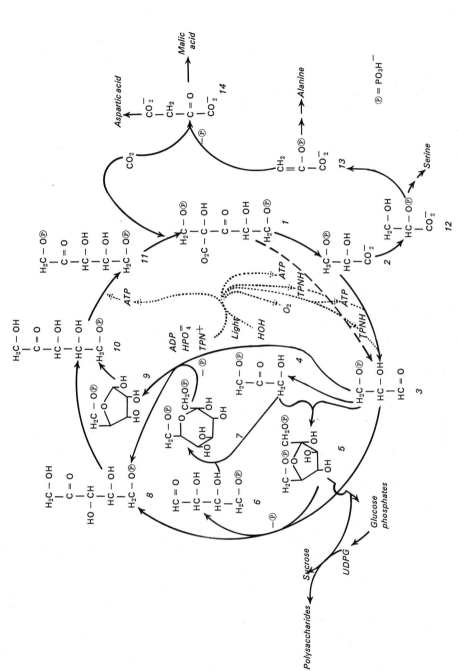

Fig. 15-5. Benson-Calvin, or pentose phosphate, cycle. (From Calvin, M., and J. A. Bassham. 1962. The photosynthesis of carbon compounds. W. A. Benjamin, Inc., New York.)

of substances capable of acting alternately as electron donors and acceptors (see p. 183). At one end of the chain of donors, oxygen is liberated.

Oxygen evolution

Although chlorophyll-containing bacteria can photosynthesize anaerobically without releasing oxygen, all the green, higher plants evolve oxygen during photosynthesis. From the general, overall equation given on p. 172, it is not clear whether this oxygen comes from the carbon dioxide, the water, or both. This question was answered by the first experiment on photosynthesis (by Rubin's group) in which isotopes were used. When heavy oxygen (O^{18}) was supplied to the plant as part of the water molecule, the oxygen evolved was found to be enriched in O^{18}. When it was supplied as part of the carbon dioxide molecule, the oxygen evolved was not enriched in O^{18}. Although this evidence is not as conclusive as it at first appeared (Brown and Frenkel, 1953), it agrees with other evidence indicating that the source of the oxygen evolved in photosynthesis is the water and not the carbon dioxide. Consequently, the equation as given on p. 172, is not correct and must be modified as follows:

$$6CO_2 + 12H_2O^* + \text{Light energy} \longrightarrow$$
$$C_6H_{12}O_6 + 6H_2O + 6O^*_2$$

The asterisk indicates the same atoms of oxygen. To balance the equation, twice as many water molecules must be involved as in the first overall equation. Even this equation is not necessarily correct. It simply shows the *minimum* number of water molecules. As far as the preceding evidence is concerned, there may be 18 or 24 or some other multiple on the left side of the equation and 12 or 18, etc., respectively, on the right side of the equation (Rabinowitch, 1945).

The oxygen-evolving reaction is therefore as follows:

$$4H_2O \xrightarrow[\text{Chlorophyll}]{\text{Light}} 4H + O_2 + 2H_2O$$

Since this implies the splitting of water by light into hydrogen and hydroxyl radicals, it was originally called the "photolysis" of water. However, it must not be assumed that this reaction is truly photochemical since the light is not absorbed by the water. More recently, therefore, the process has been thought of as a photo-oxidation of hydroxyl ions accompanied by a photoreduction of H^+, that is a photoelectron transport. It is still believed by some, however, that there is no unequivocal direct evidence for the participation of water in the primary photochemistry of photosynthesis (Clayton, 1965).

Unlike photosynthetic C-assimilation, the oxygen evolution can occur only in the light. It can, in fact, occur in the absence of any appreciable C-assimilation, in the so-called *Hill reaction*. This is the name given to the *photosynthetic release of molecular oxygen by free, illuminated chloroplasts* accompanied by reduction of an oxidizing substance (such as ferric salt, quinones, etc.). The free chloroplasts are obtained by grinding or "homogenizing" green leaves in buffer solutions and centrifuging the homogenate successively at two speeds, the first at a very slow speed to remove sediment, the second at a slightly more rapid speed to precipitate the chloroplasts. The latter are then resuspended in solutions of various kinds and exposed to light under conditions permitting measurement of oxygen evolution.

In this way it has been found possible to obtain nearly as rapid evolution of oxygen from isolated chloroplasts as from the normal leaf, although no carbon assimilation occurs. But it must be emphasized that the free chloroplast obtained in this manner is not a complete system in itself even for oxygen evolution. It must be supplied with an oxidizing agent which it can reduce in the light. That the evolution of oxygen is controlled by enzymes has been shown by the use of inhibitors such as urethane, which also inhibit photosynthesis in the normal leaf. By supplying the necessary enzymes and substrates not present in the free chloroplasts, it is possible to obtain carbon dioxide assimilation with free chloroplasts in the light (Vishniac and Ochoa, 1952).

The Hill reaction can be represented as follows:

$$A + H_2O \xrightarrow[\text{Chloroplasts}]{\text{Light}} AH_2 + \tfrac{1}{2}O_2$$

where A = oxidizing substance

The Hill reaction, of course, also proves that the oxygen is evolved from water and not from the carbon dioxide, since no (or very little) carbon dioxide is absorbed in the process.

NADP$^+$ reduction

The Hill reaction results in the oxidation of water to molecular oxygen. Although many substances can serve artificially as the electron acceptor or oxidant (becoming reduced in the process) the natural substance is NADP$^+$:

$$NADP^+ + H_2O \xrightarrow[\text{Chloroplasts}]{\text{Light}} NADPH + \tfrac{1}{2}O_2 + H^+$$

where NADP$^+$ = nictinamide–adenine dinucleotide phosphate
  NADPH = reduced nicotinamide–adenine dinucleotide phosphate

One result of the photostage of photosynthesis is, therefore, the formation of reduced NADP, which can then bring about the reduction of carbon dioxide to carbohydrate. But several other substances are also involved as intermediates in the electron transport from water to NADP (e.g., quinones, cytochromes, ferredoxin).

Photophosphorylation—the energy trap

It is now clear that, although the C-assimilation reactions of the Calvin cycle can all occur in the dark, they must all be preceded in the plant by photo-oxidation-reduction (or photoelectron transport) reactions, known as "light reactions." These culminate in the synthesis of high energy phosphate, which can then be used in the dark to drive the reactions of the Calvin cycle. Since this production of high energy phosphate differs from its respiratory formation by a dependence on light energy, it is called *photo*phosphorylation and may be represented as follows:

$$ADP + P_i \xrightarrow[\text{Chloroplasts}]{\text{Light}} ATP$$

where ADP = adenosine diphosphate
  ATP = adenosine triphosphate
  P$_i$ = inorganic phosphate

Two kinds of photophosphorylation have been identified in experiments with free chloroplasts. Cyclic photophosphorylation occurs independently of the "photolysis" of water and is therefore not accompanied by oxygen evolution. Noncyclic photophosphorylation is linked to both the formation of reduced pyridine nucleotide and the photolysis of water as follows:

$$NADP^+ + ADP + P_i + H_2O \xrightarrow[\text{Chlorophyll}]{\text{Light}}$$
$$NADPH + ATP + \tfrac{1}{2}O_2 + H^+$$

There has been some controversy as to whether or not the cyclic process occurs in the normal photosynthetic process (van Niel, 1962). According to Ramirez and associates (1968), it may serve as the source of ATP for biosynthetic processes occurring in chloroplasts that are not on the main photosynthetic path of carbohydrate synthesis but that branch off from this path, for example, protein, DNA, and RNA synthesis.

The hydrogen from the reduced pyridine nucleotide (NADPH) may then initiate the reduction of carbon dioxide, and the high energy phosphate (ATP) may supply the energy to drive the endergonic reactions of the Calvin cycle (see later). The two substances formed (NADPH and ATP) cannot be appreciably stored and are present only in catalytic amounts (Arnon, 1961).

PHOTOCHEMICAL REACTIONS

The visible radiations known as "light" occur from the violet at about 380 nm (nanometers, formerly called millimicrons [mμ]) to the far red at about 770 nm. Photosynthesis is light controlled because it can occur only when exposed to radiation within this range of wavelengths, except in photosynthetic bacteria which can photosynthesize in the short infrared up to about 900 nm. Infrared radiation has no effect on the photosynthesis of all other plants, other than the indirect temperature effect. Ultraviolet irradiation inhibits photosynthesis because of the destruction of essential substances such as plastoquinone (Mantai and Bishop, 1967). The true "light reactions" are, of course, the photochemical reactions. These are the direct result of the absorption of a photon of light by a molecule. The added energy from this photon lifts an electron from the "ground" (i.e., low energy) state to the "excited" (i.e., high energy) state, leaving behind a positive "hole." While in this excited state, the molecule can undergo a photochemical reaction. The molecules in the plant directly involved in the photochemical reactions are the photosynthetic pigments that absorb the light. They then transfer the light energy, via electron transfers, and set in motion all the above mentioned reactions—pyridine nucleotide reduction, photophosphorylation, photolysis of water, and C-assimilation. The light absorption leads to activation of chlorophyll in an extremely rapid process, probably occurring in about 10^{-15} sec (Kamen, 1963):

$$Chl + h\nu \longrightarrow Chl^*$$

where Chl = a chlorophyll molecule
$\quad$ hν = a photon of light
$\quad$ Chl* = an excited chlorophyll molecule

The excitation of the chlorophyll molecule may result in an electron transfer, ultimately to NADP:

$$2Chl^* + NADP^+ + H^+ \longrightarrow 2\ Chl^+ + NADPH$$

The chlorophyll molecule may then be returned to the unexcited state by electron transport, ultimately from water (or its OH^-):

$$Chl^+ + OH^- \longrightarrow Chl + \tfrac{1}{2}H_2O + \tfrac{1}{4}O_2$$

In this way the chlorophyll molecule would be regenerated in the low energy or nonexcited state after absorbing a quantum of light energy and using it to reduce a molecule of $NADP^+$ and to oxidize a molecule of water (Fig. 15-6). *The primary photochemical act is, therefore, the separation of oxidizing and reducing entities by light via the chlorophyll molecule.*

It has long been suspected that there is more than one photochemical reaction in photosynthesis. Direct evidence for this has now been produced by the "Emerson enhancement (or second Emerson) effect." This name refers to the greater photosynthetic yield in the simultaneous (or quickly alternating) presence of light of two wavelengths than the summation of the separate yields in each wavelength. The exact way in which these two photochemical reactions

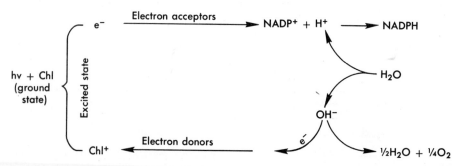

Fig. 15-6. Primary photochemical act. Arrows represent the direction of electron transport. (Redrawn from Clayton, R. K. 1965. Molecular physics in photosynthesis. Blaisdell Publishing Co., Division of Ginn & Co., Waltham, Mass.)

fit into the scheme is still being investigated. The most widely accepted current hypothesis explains the two primary photoacts as depending on two photosystems (Fig. 15-7). Photosystem I produces a strong reductant (leading to NADP reduction) and a weak oxidant. Photosystem II produces a weak reductant and a strong oxidant (leading to oxygen evolution). The two operate in series, the weak oxidant from photosystem I oxidizing the weak reductant from photosystem II by way of an electron transport system involving intermediate carriers (Schwartz, 1967). System I is most efficiently sensitized by red light of longer wavelengths (above 690 nm but not above 720 nm), system II by red light of shorter wavelength (below 690 nm).

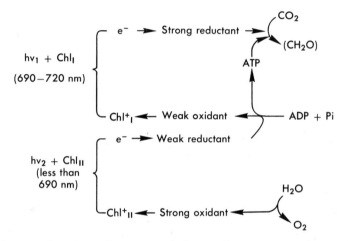

Fig. 15-7. The two photochemical reactions of photosynthesis. Arrows represent the direction of electron transport. (Redrawn from Clayton, R. K. 1965. Molecular physics in photosynthesis. Blaisdell Publishing Co., Division of Ginn & Co., Waltham, Mass.)

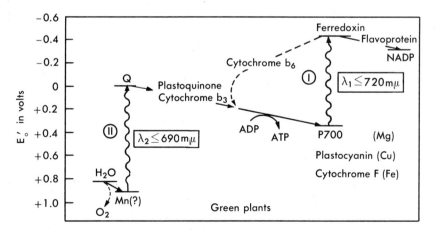

Fig. 15-8. Photosynthetic electron transfer scheme resulting in a rise in reduction potential from that of water to that of NADPH. (From Olson, J. M. 1967. Brookhaven Lect. Ser., Pub. No. 62, Brookhaven National Laboratory, Upton, N. Y.)

The two photosystems are presumably different chlorophyll-protein complexes. Photosystem I is supposed to be the more primitive one since it occurs in photosynthetic bacteria, whereas photosystem II does not (Olson, 1967). Two chlorophyll-protein complexes have, in fact, been isolated from the chloroplasts of several plants, and they are believed to represent systems I and II (Thornber et al., 1967). The net effect of this photoelectron transport system is to raise the oxidation-reduction potential from a maximum oxidized level in water (+800 mv) to a maximum reduced

level in NADPH (actually in ferredoxin, −400 mv). This involves an increase in energy because of the absorption of the two photons of light energy by the two chlorophyll systems (Fig. 15-8).

The further oxidation-reduction reactions leading to the "photolysis" of water, photophosphorylation, and NADP reduction are complete within about 10^{-4} sec from the instant of light absorption. It is only after these initial light-controlled reactions that the dark chemical reactions of C-assimilation (the Calvin cycle) can occur.

A simplified version of the relations be-

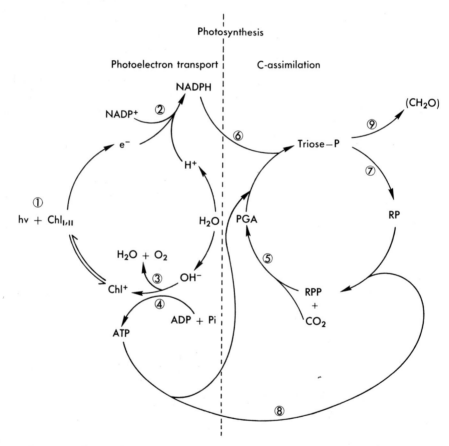

Fig. 15-9. Interrelations between photo stage (photoelectron transport) and synthesis stage (C-assimilation) of photosynthesis. The numbers correspond to the following reactions: **1,** primary photochemical reactions; **2,** NADP reduction; **3,** photolysis of water (or photo oxidation of hydroxyl ions); **4,** photophosphorylation; **5,** carboxylation; **6,** reduction (of PGA) with dephosphorylation of ATP; **7,** regeneration (of ribulose phosphate); **8,** phosphorylation (of ribulose phosphate to ribulose diphosphate); **9,** condensation of triose phosphate to hexose phosphate and release of carbohydrate.

tween the two stages of photosynthesis is shown in Fig. 15-9. If this representation of the photosynthetic process is correct, 3 molecules ATP and 2 molecules NADPH are required for every 1 molecule carbon dioxide assimilated.

The photolysis of water, the reduction of pyridine nucleotide, and photophosphorylation are represented as light reactions as opposed to the reactions of C-assimilation, which can all occur in the dark immediately after exposure to light. However, even these so-called light reactions have dark components (e.g., in the case of photophosphorylation, Hind and Jagendorf, 1963). Thus, they are light reactions only in the sense that they are more closely dependent on light in time and therefore have not been found to occur after the light exposure. For this reason they take place in the quantasomes, together with the primary light reactions, whereas the C-assimilation occurs in the stroma of the chloroplasts (Smith and French, 1963) and is, therefore, separated from the photochemical reactions in space as well as in time. It has been suggested that quantasomes are the "photosynthetic unit," a name given by Emerson and Arnold for the packet of chlorophyll molecules, all of which are capable of transferring the light energy they trap to a single reacting carbon dioxide molecule.

This photosynthetic unit is supposed to contain about 2500 chlorophyll molecules, consisting of four small units of pigment system I and one large unit of pigment system II (Gibbs, 1967). A heavy particle has been isolated that is active in oxygen evolution and enriched in chlorophyll b and manganese and is believed to be pigment system II. A lighter particle has been separated that is active in NADP reduction, is enriched in chlorophyll a, iron, and copper, and is believed to be pigment system I.

The exact function of the quantasome is, however, still in question. Its size does not agree with the suggested size of the photosynthetic unit. It is 100 Å square with a molecular weight of 2×10^6 and is a complex of about 50% protein and 17% chloroplast pigments, having only 230 chlorophyll a molecules (Vernon and Seeley, 1966). It is membrane bound and possesses adenosine triphosphatase activity. It has therefore also been called a photophosphorylase (Howell and Moudrianakis, 1967). It is, however, composed of smaller units that are thought to be enzyme complexes (Mühlethaler, 1966). The membrane of the chloroplast is layered on both sides with quantasomes and is, therefore, thought to be the actual site of photosynthesis.

According to another concept (Junge et al., 1968) there are three photosynthetic functional units:

1. The functional unit for the photoelectron transport system. This contains 10^3 chlorophyll molecules. The light absorbed by this unit is conducted to the two photochemically active chlorophyll molecules that generate the electron flow.

2. The functional unit for the proton transporting system. This includes the photoelectron transport system that generates the proton transport but is 100 times larger (10^5 chlorophyll molecules). It corresponds to one *thylakoid*, that is, one disc-shaped, closed vesicle surrounded by a membrane.

3. The functional unit of photophosphorylation. This also consists of one thylakoid, because the proton transport promotes ATP formation (according to the chemiosmotic theory, see Chapter 14).

One thylakoid, therefore, contains many photoelectron transport units, presumably in the plane of its membrane.

PIGMENTS

There are two main kinds of chlorophylls in higher plants (although each may occur in more than one form)—one bluish green (chlorophyll a), the other yellowish green (chlorophyll b). The empirical formulas are

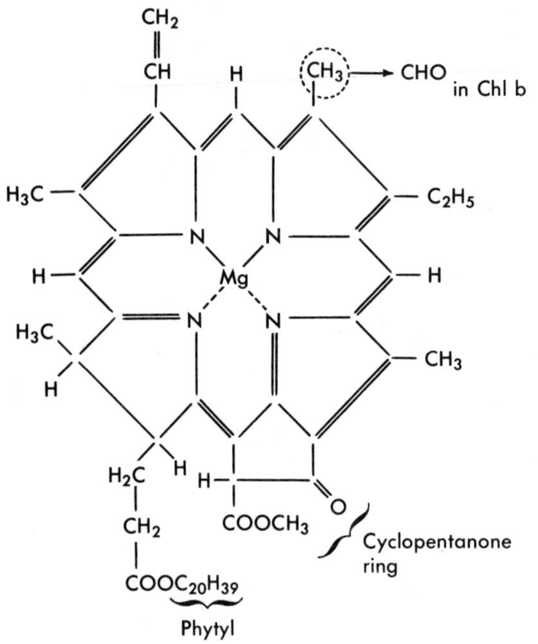

$C_{55}H_{72}O_5N_4M_g$ for chlorophyll *a*, the same with a CHO group in place of a CH_3 group for chlorophyll *b*. The chlorophyll molecule cannot be synthesized by the cell in the absence of iron. The exact structure of chlorophyll is now known (Fig. 15-10) and has been confirmed by its total synthesis by Woodward (Smith and French, 1963). The portion of the molecule that contains magnesium is chemically related to the heme of the blood. Both chlorophylls occur only in the chloroplasts (in higher plants), usually in the ratio of 3*a*:1*b*. With them are two yellow carotinoid pigments, carotin and xanthophyll, in about the same molar quantity as chlorophyll *b*. Carotin is a polyene hydrocarbon ($C_{40}H_{52}$) and xanthophyll is an oxygen derivative of it ($C_{40}H_{52}O_2$). The four are known as the *chloroplast pigments*. The carotinoids are just as efficient harvesters of light as the chlorophylls, and they apparently play two roles in photosynthesis (Clayton, 1965). But chlorophyll is irreplaceable as the initiator of the chemistry of photosynthesis.

Fig. 15-10. Chlorophyll molecule. (Redrawn from Clatyon, R. K. 1965. Molecular physics in photosynthesis. Blaisdell Publishing Co., Division of Ginn & Co., Waltham, Mass.)

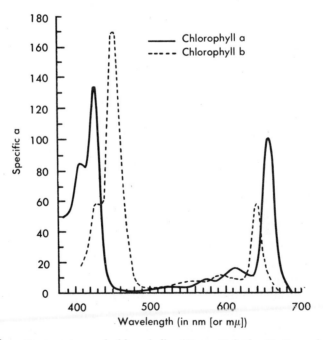

Fig. 15-11. Absorption spectrum of chlorophylls. (From Zscheile, F. P., and C. L. Comar. 1941. Bot. Gaz. **102**:463-481.)

That chlorophyll is the energy-absorbing substance can be shown in several ways. The *absorption spectrum* (the curve showing the percent absorption of light at each wavelength) of chlorophyll (Fig. 15-11) is similar to the *action spectrum* of photosynthesis (the curve showing the relative rates of the reaction when supplied with the same amount of radiant energy at each wavelength). In other words chlorophyll absorbs most intensely in the red and blue at the wavelengths that permit the most rapid photosynthesis; it absorbs least in the green, and this light (when at the same intensity as the other wavelengths) permits only the slowest rate of photosynthesis. It is interesting that all wavelengths of light do permit photosynthesis, although in view of the preceding absorption characteristics of chlorophyll, red is the most efficient. Chlorophyll possesses the property of *fluorescence*—it is able to absorb light of one wavelength and to emit light of a longer wavelength. Whatever wavelength it absorbs, it always fluoresces in the red (with a maximum at 684 nm). For this reason chlorophyll appears green by transmitted light, red by reflected light. This is another reason for the high efficiency of red light. Only that fraction of the quantum of absorbed light which can be reemitted as fluorescent light is available for photosynthesis. In the case of a quantum of red light of 684 nm, all the energy can be reemitted and is therefore available for photosynthesis. In the case of a quantum of blue light, only about two-thirds can be reemitted; the remaining one-third is given off as heat and is therefore wasted.

ENZYMES

As in the case of respiration, the dozens of reactions that comprise the process of photosynthesis are all controlled by as many enzymes, each one specific for each reaction. Many of the reactions are the same as those involved in respiration, although the direction of the reactions is reversed. This is particularly true of the Calvin cycle, which consists very largely of a reversal of the pentose phosphate pathway, the alternate path for glycolysis. Some reactions, however, are unique to the Calvin cycle, for example, the first reaction in which carbon dioxide is combined with ribulose diphosphate to yield two molecules of phosphoglycerate. The enzyme controlling this reaction has the trivial name of carboxydismutase, or ribulose diphosphate carboxylase. It has a molecular weight of 550,000 and is indistinguishable from the "protochlorophyll protein holochrome complex," or the "Fraction I protein," which has been isolated from leaves (Gibbs, 1967). There is also a series of enzymes and cofactors involved in the electron transport from chlorophyll to NADP and from water to chlorophyll. These include Mn^{++}, several plastoquinones, at least two cytochromes (b and f), a copper protein (plastocyanin), the pigment P-700, an iron protein (ferredoxin), NADP reductase, and NADP (Bishop, 1966). Their probable interrelations are shown in Fig. 15-7.

All the photosynthetic enzymes are present in the chloroplasts (carboxydismutase in large quantities), since this is where the whole process of photosynthesis occurs. The chloroplast, therefore (like the mitochondrion), is a powerhouse of enzymes. Because some of these enzymes are specific for photosynthetic reactions, there are also substances that are specific inhibitors of photosynthesis. This has proved to be of great practical importance since some of these substances can be used as herbicides (weed killers) that act specifically on autotrophic photosynthesizing organisms and have no effect on heterotrophic (nonphotosynthesizing) organisms. Two examples of these are the phenylureas (CMU and DCMU) and the aminotriazines (simazine and atrazine). These herbicides have also proved to be of great value in investigating the individual photosynthetic processes that they inhibit (Krogmann and Powers, 1965).

They specifically inhibit oxygen evolution (i.e., the Hill reaction). Several other herbicides also appear to inhibit photosynthetc reactions closely connected to the evolution of oxygen (Nishimura, 1967). Others (e.g., phlorizin) uncouple photophosphorylation (Izawa et al., 1967).

The division of labor between the two powerhouses (the chloroplast and the mitochondrion) is sharp. The mitochondrion is unable to photosynthesize, and the chloroplast is unable to respire. The chloroplast does not possess (1) a cytochrome c oxidase system, (2) a citric acid cycle, (3) pathways of hexose breakdown, or (4) as great an ability to reduce acetyl coenzyme A to fatty acids (Gibbs, 1967).

RATE OF PHOTOSYNTHESIS

The rate of photosynthesis (Table 15-1) is controlled by many factors, both internal and external. When several factors are investigated, it is usually found that only one limits the rate of photosynthesis; that is, this one factor when increased, will greatly increase the rate of photosynthesis, whereas the others when increased, have little effect on the rate of the process (Figs. 15-12 and

Table 15-1. Rate of photosynthesis in some plants at temperatures of 18° to 20° C at maximum or near-maximum light intensities and with normal carbon dioxide content of the air (0.03% or 0.56 mg per liter)*

Plant	Milligrams CO_2 per 50 cm² leaf surface per hr	Carbohydrate ($C_6H_{12}O_6$) synthesized in g/m² leaf surface per hr
Potato	9.57	1.30
Tomato	8.42	1.15
Sugar beet	9.26	1.26
Spinach	9.78	1.33
Vicia faba	8.83	1.20
Phaseolus vulgaris	9.27	1.26

*Adapted from Lundegårdh, H. 1925. Gustav Fischer Verlag, Stuttgart.

15-13). This relation has been called the *law of limiting factors.*

Thus, although light intensity, temperature, and carbon dioxide concentration are all important factors, if the temperature on a normal day is around 0° C, no increase in light intensity or carbon dioxide concentration will affect the rate of the process. Since temperature is the limiting factor, a small increase in temperature will produce a corresponding increase in the rate of photosynthesis (Fig. 15-13). As in the case of respiration, the effect of temperature is pronounced. When temperature is the limiting factor, a rise of 10° C may increase the rate of photosynthesis 2 or 3 times or sometimes even more. This is typical of ordinary chemical processes. Since physical processes are less affected by temperature, photochemical reactions show little change; for example, a 10° C rise in temperature usually increases the rate of a photochemical reaction by only about one tenth. This fact was long taken as evidence that the dark chemical reactions of photosynthesis are slower than the photochemical reactions; this is the reason why the former are limiting. Indeed, the significant temperature effect was the earliest evidence that not all of the photosynthetic process requires light. In memory of the man who first observed this, the term *Blackman reaction* was used for the dark chemical reactions. Another early proof of the existence of dark chemical reactions was the fact that intermittent light can permit just as rapid a rate of photosynthesis as continuous light if the light and dark periods are short enough, even though the dark periods between flashes reduce the total light received to about half as much per hour. However, if the length of the alternating light and dark periods is increased to about 1 min each, almost no photosynthesis takes place in some plants. This is because of the so-called *induction period*—a lag that occurs after exposure to darkness for a minute or more. On exposure to the light it may take the

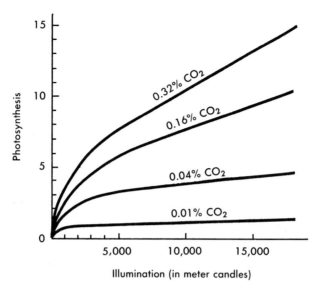

Fig. 15-12. Photosynthesis of aquatic plant *(Fontinalis)* at different light intensities and carbon dioxide concentrations. (From Harder, R. 1921. Jb. Wiss. Bot. **60**:531-571.)

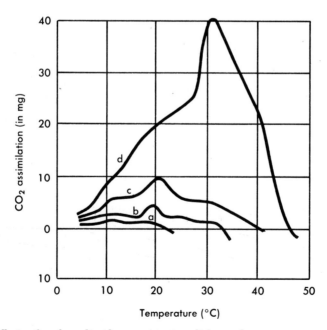

Fig. 15-13. Effects of carbon dioxide concentration, light, and temperature on photosynthesis of potato leaves. Curve **a,** a very weak light and very low carbon dioxide concentration. Curve **b,** 20% normal light and 0.03% carbon dioxide. Curve **c,** normal light and 0.03% carbon dioxide. Curve **d,** normal light and 1.22% carbon dioxide. (Redrawn from Lundegårdh, H. 1925. Klima und Boden. Gustav Fischer Verlag, Stuttgart.)

plant a full minute (the induction period) to reach its normal rate of photosynthesis.

Other external factors beside light intensity, temperature, and carbon dioxide concentration may affect the rate of photosynthesis. Low moisture supply may lead to a cessation of photosynthesis, but this is really because of the low carbon dioxide concentration after stomatal closure. Higher than normal oxygen concentrations have long been known to inhibit photosynthesis. This is called the Warburg effect. In the case of land plants (but not in algae) even the oxygen of normal air strongly inhibits photosynthesis (Björkman, 1966). Thus, when compared to normal air, absence of oxygen resulted in a 44% enhancement of photosynthesis in seven species at 30° C and an 85% enhancement in three species at 40° C (Hesketh, 1967). No enhancement occurred in true tropical grasses and in one dicotyledon. In maize there was no enhancement, even in light of low intensity.

Modern man's population explosion has led to continuous attempts to increase the food supply. This, of course, cannot exceed the amount of photosynthate produced by green plants. It is therefore of practical importance to know what are the limits to the production of photosynthate, because any attempt to exceed these limits is useless. In spite of the high efficiency of modern agriculture, only about 0.2% of the annual sunshine is stored by photosynthesizing plants (Daniels, 1967). In sunny climates, solar radiation averages about 500 cal/cm^2 per day. An idealized photochemical reaction effective at a wavelength of 500 nm, which absorbed and converted into chemical energy one tenth of the sunlight, could therefore produce 3.6 tons per day per acre (Daniels, 1967). This is, of course, as much as most crops produce in a year.

The efficiency of photosynthesis is therefore low under normal conditions. In fact, it is commonly stated that the animal is more efficient than the plant since only 1% to 2% of the sun's energy (during the grow-

ing season) is used for photosynthesis. But this is equivalent to saying that if the animal is fed wood, its efficiency would be less than 1% because just as the chemical substances in wood are mostly indigestible, so also most of the radiant energy from the sun is in the nonvisible wavelengths and, therefore, not usable by the plant. Even the usable wavelengths are not normally supplied in quantities that can be used efficiently. Therefore, only about 2% of the incident light energy is ordinarily used in photosynthesis, or 3% of the absorbed light energy. However, there are plants that under normal conditions have efficiencies considerably higher than this. Bulrush millet may accumulate 54 g/m^2 per day, storing as dry matter 4.2% of the total radiation or 9.5% of the visible radiation (Begg, 1965). In weak light and with an ample supply of carbon dioxide the efficiency may rise to 30% of the absorbed radiation. Many other examples of high efficiency have been found (Wassink, 1964). Under optimal conditions in small cultures of algae, an energy yield of 15% to 20% can be reached. Even in the case of seedlings of higher plants, under light limitations, yields of 10% to 15% have been obtained. One factor leading to a higher efficiency in algae than in higher plants is the response to increased carbon dioxide supply. In the case of the alga *Chlorella*, the concentration of carbon dioxide in the nutrient medium that produces maximal photosynthesis is in the range of 1.5% to 2.5% by volume (Ammann and Lynch, 1967). When higher plants are exposed to concentrations of carbon dioxide above that normally found in air, this may lead to stomatal closure, bringing photosynthesis to a stop. Therefore, the higher plant cannot take advantage of the higher carbon dioxide concentrations as can the alga.

The measured carbon dioxide absorption (or oxygen evolution) is really a result of the excess of photosynthesis over respiration and is therefore a measurement of the *net rate of photosynthesis.* To get the true, or "actual," rate of photosynthesis it would be necessary to add to the measured carbon dioxide absorption the carbon dioxide evolution (i.e, respiration) of the same plant when kept in the dark. This correction, however, does not always give the true value for photosynthesis since light itself may alter the rate of respiration.

By supplying the plant with atmospheric oxygen containing heavy oxygen (O^{18}) and with water with the normal isotope (O^{16}), it is possible to measure respiration directly while photosynthesis is taking place (photorespiration). This method of measurement may show (1) an inhibition of respiratory oxygen evolution by light; (2) an acceleration; or (3) no effect, depending on the plant, or even on the light intensity (Hoch et al., 1963). Low light intensities apparently inhibit the normal dark respiration, and higher light intensities induce a new and higher oxygen uptake. The inhibition may conceivably be caused by a competition between chloroplasts and mitochondria for phosphate and ADP (Jagendorf and Kok, 1964). There is a considerable variation among species in their rates of photorespiration. Some plants, like maize, have a negligible photorespiration and are therefore more efficient photosynthetically (Zelitch, 1966). The metabolic path is different from that of ordinary, dark respiration because instead of carbohydrate, the substrate is glycolic acid. In the presence of high carbon dioxide concentrations, however, this path may also be followed in the dark (McNaughton, 1966). The glycolic acid is oxidized in a reaction involving the enzyme glycolic acid oxidase. It is, in fact, possible to induce tobacco to photosynthesize as efficiently as maize by inhibiting the oxidation of glycolate. This stimulates the carbon dioxide uptake three-fold at 35° C, although not at 25° C (Zelitch, 1966), presumably because the rate of photorespiration is greatly increased at higher temperatures.

These newer results also explain the long-known Warburg effect—the inhibition of photosynthesis by oxygen (see earlier). This can logically be explained as an inhibition only of *net* photosynthesis resulting from an increase in photorespiration by the oxygen-induced production of glycolate. The oxygen would, therefore, increase the loss from the sugar phosphate pool rather than specifically inhibiting an enzyme system (Gibbs, 1967).

Whatever the effect of light on respiration, the fact that there is normally a net photosynthesis during daylight proves, of course, that the rate of photosynthesis exceeds the rate of respiration. If, however, the light intensity is decreased steadily, a point will be reached at which the actual photosynthetic rate just equals the respiration rate, and the net rate of photosynthesis is zero. This is called the *compensation point*. The plant, however, is able to adapt to changes in light intensity and other factors. For instance, when white clover is exposed to different levels of constant light and other factors, the plants adapt their respiration rates so as to maintain a constant proportion (about 20%) of the photosynthetic rate (McCree and Troughton, 1966). Consequently, even when the light level drops below the compensation point, this adaptation occurs within about 24 hr, the compensation point drops, and the plants are able to gain in weight.

The yield of a plant depends on the total gain in carbohydrate over a 24 hr period. This, in turn, depends not only on the measured rate of photosynthesis during daylight but also on the night respiration, since this uses up some of the accumulated carbohydrate. Thus the yield of a crop may sometimes be greatly increased by lowering the night temperature since this will increase the net assimilate per 24 hr period.

It should be pointed out that beside the ordinary green plants there are bacteria which photosynthesize. They also possess chlorophyll and assimilate carbon com-

$$CO_2 + H_2O \xrightarrow[\text{Chlorophyll}]{\text{Light}} \text{Intermediates of Calvin cycle}$$

Lipids ↑

Amino acids and proteins ↗ → Hexose phosphates → Sucrose

Hexoses ↗

Cellulose ↓

Starch ↘

Pentosans, pectins ↘

pounds at the expense of carbon dioxide in the light but do not evolve oxygen. In fact, they are able to photosynthesize only under anaerobic conditions.

Although photosynthesis is usually represented as leading to the formation of hexose sugar, it may also lead to the accumulation of many other substances, for example, starch, fats, and indirectly even proteins. This may be illustrated as shown above.

QUESTIONS

1. What must be measured in order to determine the rate of photosynthesis?
2. What factors must be available in order for photosynthesis to occur?
3. Where does the evolved oxygen come from?
4. How do we know where it comes from?
5. What, then, would be the overall equation for photosynthesis?
6. What method (technique) has been most useful in recent years for investigating the intermediate of photosynthesis?
7. What actual isotopes have been used?
8. According to the evidence, what is the first substance formed as a result of carbon dioxide fixation?
9. What other substances are formed soon after carbon dioxide fixation, before the release of free carbohydrates?
10. Can the photosynthetic conversion (fixation) of carbon dioxide to carbohydrates occur in the dark?
11. What, then, is the difference between photosynthetic and dark carbon dioxide assimilation?
12. What is the Hill reaction?

13. What is meant by "photolysis of water"?
14. What is the difference between photophosphorylation and oxidative phosphorylation?
15. What is the difference between cyclic and noncyclic photophosphorylation?
16. What is photoelectron transport?
17. What is the Emerson enhancement effect?
18. How many photochemical reactions participate in the photosynthetic process?
19. What is a photosynthetic unit?
20. What molecule is excited during photosynthesis?
21. What is fluorescence?
22. If chlorophyll absorbs blue light, what color does it fluoresce?
23. If it absorbs green light, what color does it fluoresce?
24. What is the role of chlorophyll in photosynthesis?
25. Describe the absorption spectrum of chlorophyll.
26. What is meant by the action spectrum of photosynthesis?
27. How do the two spectra compare?
28. What are the plastid pigments?
29. Are the other plastid pigments of any importance in photosynthesis?
30. What factors affect the rate of photosynthesis?
31. Which factor is the most important?
32. Is temperature as important in photosynthesis as in respiration?
33. What is meant by net photosynthesis?
34. What is meant by actual photosynthesis?

35. How is the actual photosynthesis determined?
36. Does respiration occur in the light?
37. How do we know whether or not it occurs in the light?
38. How is the yield of the plant related to photosynthesis and respiration?

SPECIFIC REFERENCES

Ammann, E. C. B., and V. H. Lynch. 1967. Gas exchange of algae. III. Relation between the concentration of carbon dioxide in the nutrient medium and the oxygen production of *Chlorella pyrenoidosa.* Appl. Microbiol. 15:487-491.

Begg, J. E. 1965. High photosynthetic efficiency in a low latitude environment. Nature (London) 205:1025-1026.

Björkman, O. 1966. The effect of oxygen concentration on photosynthesis in higher plants. Physiol. Plant. 19:618-633.

Harder, R. 1921. Kritische Versuche zu Blackmans Theorie der "begrenzenden Faktoren" bei der Kohlensäure-assimilation. Jb. Wiss. Bot. 60:531-571.

Harrison, A. 1965. Carbon dioxide effects on the extension in length of *Avena* coleoptiles. Physiol. Plant. 18:208-218.

Hatch, M. D., and C. R. Slack. 1966. Photosynthesis by sugar-cane leaves; a new carboxylation reaction and the pathway of sugar formation. Biochem. J. 101:103-111.

Hatch, M. D., C. R. Slack, and H. S. Johnson. 1967. Further studies on a new pathway of photosynthetic carbon dioxide fixation in sugarcane and its occurrence in other plant species. Biochem. J. 102:417-422.

Hesketh, J. 1967. Enhancement of photosynthetic CO_2 assimilation in the absence of oxygen, as dependent upon species and temperature. Planta 76:371-374.

Hess, J. L., and N. E. Tolbert. 1967. Changes in chlorophyll *a/b* ratio and products of $C^{14}O_2$ fixation by algae grown in blue or red light. Plant Physiol. 42:1123-1130.

Hind, G., and A. T. Jagendorf. 1963. Separation of light and dark stages in photophosphorylation. Proc. Nat. Acad. Sci. 49:715-722.

Hoch, G., O. v. H. Owens, and B. Kok. 1963. Photosynthesis and respiration. Arch. Biochem. Biophys. 101:171-180.

Howell, S. H., and E. N. Moudrianakis. 1967. Function of the "quantasome" in photosynthesis; structure and properties of membrane-bound particles active in the dark reactions of photophosphorylation. Proc. Nat. Acad. Sci. 58:1261-1268.

Izawa, S., T. N. Connolly, G. D. Winget, and N. E. Good. 1967. Inhibition and uncoupling of photophosphorylation in chloroplasts, p. 169-187. In J. M. Olson, G. Hind, H. Lyman, and H. W. Siegelman. Energy conversion by the photosynthetic apparatus. Brookhaven Symposium in Biology, Pub. No. 19. Brookhaven National Laboratory, Upton, N. Y.

Jagendorf, A. T., and B. Kok. 1964. Photosynthesis. Science 143:388-395.

Junge, W., E. Reinwald, B. Rumberg, U. Siggel, and H. T. Witt. 1968. Further evidence for a new function unit of photosynthesis. Naturwissenschaften 55:36-37.

Kortschak, H., C. E. Hartt, and G. O. Burr. 1965. Carbon dioxide fixation in sugar-cane leaves. Plant Physiol. 40:209-213.

Krotkov, G., V. C. Runeckles, and K. V. Thimann. 1958. Effect of light on the CO_2 absorption and evolution by *Kalanchoe*, wheat, and pea leaves. Plant Physiol. 33:289-292.

Mantai, K. E., and N. I. Bishop. 1967. Studies on the effects of ultraviolet irradiation on photosynthesis and on 520 nm light-dark difference spectra in green algae and isolated chloroplasts. Biochem. Biophys. Acta 131:350-356.

McCree, K. J., and J. H. Troughton. 1966. Prediction of growth rates at different light levels from measured photosynthesis and respiration rates. Plant Physiol. 41:559-566.

McNaughton, S. J. 1966. Light-stimulated oxygen uptake and glycolic acid oxidase in *Typha latifolia* L. leaf disks. Nature (London) 211:1197-1198.

Mühlethaler, K. 1966. Der Feinbau des Photosynthese-Apparates. Umschau Wiss. Tech. 66:659-662.

Nishimura, M. 1967. Oxidation-reduction reactions of cytochromes in red algae. In J. M. Olson, G. Hind, H. Lyman, and H. W. Siegelman. Energy conversion by the photosynthetic apparatus. Brookhaven Symposium in Biology, Pub. No. 19. Brookhaven National Laboratory, Upton, N. Y.

Osmond, C. B. 1967. B-carboxylation during photosynthesis in *Atriplex.* Biochim. Biophys. Acta 141:197-199.

Plamondon, J. E., and J. A. Bassham. 1966. Glycolic acid labeling during photosynthesis with $C^{14}O_2$ and tritiated water. Plant Physiol. 41:1272-1275.

Ramirez, J. M., F. F. del Campo, and D. I. Arnon. 1968. Photosynthetic phosphorylation as energy source for protein synthesis and carbon dioxide assimilated by chloroplasts. Proc. Nat. Acad. Sci. 59:606-612.

Schwartz, M. 1967. Quantum yield of ferrocytochrome C photo-oxidation in chloroplast particles. Nature (London) **213**:1187-1189.

Splittstoesser, W. E. 1966. Dark CO_2 fixation and its role in the growth of plant tissue. Plant Physiol. **41**:755-759.

Thornber, J. P., R. P. F. Gregory, C. A. Smith, and J. L. Bailey. 1967. Studies on the nature of the chloroplast lamellae. I. Preparation and some properties of two chlorophyll-protein complexes. Biochemistry **6**:391-396.

Vishniac, W., and S. Ochoa. 1952. Fixation of carbon dioxide coupled to photochemical reduction of pyridine nucleotides by chloroplast preparations. J. Biol. Chem. **195**:75-93.

Walker, D. A. 1966. Carboxylation in plants. Endeavour **25**:21-26.

Wassink, E. C. 1964. Efficiency of light energy conversion in laboratory experiments and crop growth. [Transl. title.] Meded. Landboewhogesch. Wageningen **64**:1-33.

Zelitch, I. 1965. The relation of glycolic acid synthesis to the primary photosynthetic carboxylation reaction in leaves. J. Biol. Chem. **240**:1869-1876.

Zelitch, I. 1966. Increased rate of net photosynthetic carbon dioxide uptake caused by the inhibition of glycolate oxidase. Plant Physiol. **41**:1623-1631.

Zscheile, F. P., and C. L. Comar. 1941. Influence of preoperative procedure on the purity of chlorophyll components as shown by absorption spectra. Bot. Gaz. **102**:463-481.

GENERAL REFERENCES

Arnon, D. I. 1961. Cell free photosynthesis and the energy conversion. *In* D. W. McElroy, and B. Glass. Light and life. The Johns Hopkins Press, Baltimore.

Bishop, N. J. 1966. Partial reactions of photosynthesis and photoreduction. Ann. Rev. Plant Physiol. **17**:185-208.

Brown, A. H., and A. W. Frenkel. 1953. Photosynthesis. Ann. Rev. Plant Physiol. **4**:23-58.

Calvin, M., and J. A. Bassham. 1962. The photosynthesis of carbon compounds. W. A. Benjamin, Inc., New York.

Clayton, R. K. 1965. Molecular physics in photosynthesis. Blaisdell Publishing Co., Division of Ginn & Co., Waltham, Mass.

Daniels, F. 1967. Direct use of the sun's energy. Amer. Sci. **55**:15-47.

Franck, J., and W. E. Loomis. 1949. Photosynthesis in plants. Monogr. Amer. Soc. Plant Physiol. Iowa State University Press, Ames, Iowa.

Gibbs, M. 1967. Photosynthesis. Ann. Rev. Biochem. **36**:757-784.

Kamen, M. D. 1963. Primary processes in photosynthesis. Academic Press, Inc., New York.

Kirk, J. T., and R. A. E. Tilney-Bassett. 1967. The plastids. W. H. Freeman & Co., Publishers, San Francisco.

Krogmann, D. W., and W. H. Powers. 1965. Biochemical dimensions of photosynthesis. Wayne State University Press, Detroit.

Lundegårdh, H. 1925. Klima und Boden. Gustav Fischer Verlag, Stuttgart.

Olson, J. M., G. Hind, H. Lyman, and H. W. Siegelman, 1967. Energy conversion by the photosynthetic apparatus. Brookhaven Symposium in Biology, Pub. No. 19. Brookhaven National Laboratory, Upton, N. Y.

Olson, J. M. 1967. Energy conversion in photosynthesis. Brookhaven Lect. Ser., Pub. No. 62, Brookhaven National Laboratory, Upton, N. Y.

Rabinowitch, E. I. 1945. Photosynthesis and related processes. Interscience Publishers, New York.

Rabinowitch, E. I. 1952. Photosynthesis. Ann. Rev. Plant Physiol. **3**:229-265.

Smith, J. H. C., and C. S. French. 1963. The major and accessory pigments in photosynthesis. Ann. Rev. Plant Physiol. **14**:181-224.

Stiller, M. 1962. The path of carbon in photosynthesis. Ann. Rev. Plant Physiol. **13**:151-170.

Thomas, J. B. 1965. Primary photoprocesses in biology. Interscience Publishers, Inc., Division John Wiley & Sons, Inc., New York.

van Niel, C. B. 1962. The present status of the comparative study of photosynthesis. Ann. Rev. Plant Physiol. **13**:1-26.

Vernon, L. P., and M. Avron. 1965. Photosynthesis. Ann. Rev. Biochem. **34**:269-296.

Vernon, L. P., and G. R. Seely. 1966. The chlorophylls. Academic Press, Inc., New York.

chapter 16

Other metabolic paths

A. Dependence on respiration and photosynthesis
B. N-metabolism
   1. Proteins and amino acids
   2. N-assimilation
   3. Protein synthesis
   4. Other nitrogen-containing substances
C. Mineral metabolism
D. Control of metabolism
E. Secondary metabolites

DEPENDENCE ON RESPIRATION AND PHOTOSYNTHESIS

Respiration and photosynthesis involve breakdown and synthesis, respectively, of carbon-containing substances, and they are the two basic processes of all C-metabolism. But they are also the two basic processes for *all* plant metabolism (e.g., N-metabolism) for two reasons:

1. They are the only sources of high energy phosphates. All other syntheses in the plant, therefore, depend on them for the energy needed to drive their endergonic reactions; and since catabolism is possible only if substances have first been synthesized, catabolism is also dependent on them, at least indirectly as shown in Fig. 16-1.

2. Respiration and photosynthesis also supply the basic organic substances that are the starting points for the synthesis of all other organic substances in the plant. Thus, although the beginning of respiration and the end of photosynthesis are represented as hexose, other carbohydrates may also be formed. In the case of photosynthesis, as already indicated, the hexose phosphates may lead to formation of any one of a large number of different carbohydrates. However, the metabolic path may also branch off at some earlier point in the cycle,

leading to synthesis of fats, proteins, etc.:

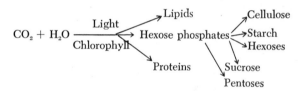

The metabolism of other carbon substances (lipids, glycosides, terpenes, lignins, etc.) is not as well understood as that of the carbohydrates, although great advances in our knowledge concerning it are being made daily. Each of these groups includes a large number of substances that must be synthesized step by step from simpler substances in the plant. The pivot substance in many cases seems to be acetic acid, which links the anaerobic phase of respiration with the aerobic Krebs cycle. We may therefore visualize many metabolic pathways radiating from this point like the spokes of a wheel, each involving its own series of enzymes and each leading to a step-by-step biosynthesis of the substances at the end of the spoke. These interrela-

191

Glycosides

Lignins

Fatty acids   Phenols   Tannins

Carbohydrate ⟶ Acetic acid   Krebs cycle

Other organic acids,   Terpenes, etc.
alcohols, esters, etc.

tions with respiration may be represented as shown above.

There is no basic difference between the metabolism of these substances and that of the carbohydrates, so it will not be considered here. N-metabolism, however, does involve some basic differences, which are fundamental to all life.

N-METABOLISM
Proteins and amino acids

The mechanism of protein synthesis is of fundamental importance to many biological disciplines—genetics, evolution, morphology, taxonomy, etc. In physiology it must be understood before the more complex mechanisms of growth and development can be explained. In fact, the true explanations of these processes are only now beginning to unfold as a result of the recent discoveries of the fundamentals of protein synthesis. It is now known that the process of protein breakdown is taking place constantly in all living protoplasm, although the total protein content may remain unchanged because of a compensating synthesis. In tobacco leaves, half of the proteins may be broken down and resynthesized in about 7 days (Holmsen and Koch, 1964). Both protein catabolism and anabolism are therefore basic to all life. Although the two processes take place at the same rate in the steady state, the rates in the nonsteady state may change independently of each other. In growing cells the rate of breakdown is very low and the rate of synthesis is high. In nongrowing cells the rate of breakdown is high and the rate of synthesis may decrease to a very low value. However, the rates may not be the same for all proteins. In tobacco leaves the so-called fraction I protein is synthesized more rapidly than other proteins during formation of the leaves and is broken down more rapidly during aging (Dorner et al., 1957). When the leaves are beginning to develop, it accounts for 17% of the protein, after 17 days for 55% of the protein, and after 45 days for only 26% of the protein. Furthermore, its synthesis is affected by light and it is not found in albino leaves or roots.

Protein breakdown in plants was early investigated in the case of leguminous seeds (Chibnall, 1939). When these high protein seeds germinate, the proteins largely disappear from the nongrowing storage tissues and are replaced by amino acids and amides, which can then be translocated to the growing regions where proteins are resynthesized from them. The greater part of the protein loss can usually be accounted

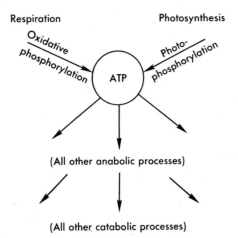

Fig. 16-1. Dependence of all other metabolic processes on respiration and photosynthesis—the only two processes that generate ATP.

for by the two amides, *asparagine* and *glutamine;* the remainder can be accounted for by a smaller quantity of several amino acids. This would appear to be a digestion process similar to the hydrolysis of carbohydrates to sugars and of fats to fatty acids and glycerol:

Digestion of storage substances

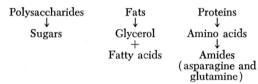

Polysaccharides   Fats   Proteins
↓   ↓   ↓
Sugars   Glycerol   Amino acids
    +   ↓
    Fatty acids   Amides
    (asparagine and glutamine)

But proteins are far more complex than these other reserves because the polysaccharides are digested to one or a few sugars, each of the fats to no more than three fatty acids, whereas even the simplest protein gives rise on hydrolysis to about twenty different amino acids. Some dozen proteins have had their amino acid sequence determined, and they all contain from fifty to 250 amino acid residues per molecule (Lienier, 1966). This is illustrated by the structure of a small protein molecule (Fig. 16-2). The complexity of the molecule is indicated by the three levels of structure: the primary (the order of the amino acids in the chain), the secondary (the spiral or helical structure of the chain, not shown here), and the tertiary (the folds in the spiral molecules). Most proteins contain more than one peptide chain, and this gives rise to the quaternary structure of the molecule. The protein molecule of enzymes is usually globular but other proteins may be rod shaped or a random coil.

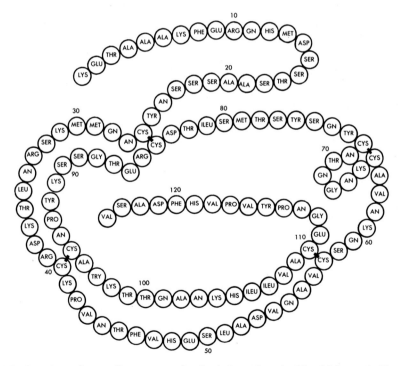

Fig. 16-2. Structure of a small protein molecule (ribonuclease). The folds are held in place by four covalent disulfide bonds joining the eight paired cysteine residues. For symbols, see p. 195. (From Anfinsen, C. R. 1967. *In* J. M. Allen. Molecular organization and biological function. Harper & Row, Publishers, New York.)

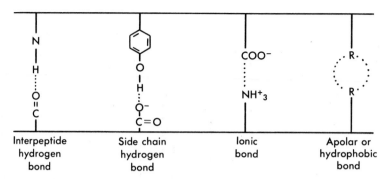

Fig. 16-3. Intramolecular noncovalent bonds responsible for the secondary and tertiary structure of proteins. (From Schachman, H. K. 1963. Sympos. Quant. Biol. **28**:409-430.)

The primary structure of a protein molecule results from the strong covalent peptide bonds. The secondary and tertiary structures are sometimes fixed, partly by a small number of strong covalent, disulfide bonds (Fig. 16-2) but in most cases by a much larger number of weak noncovalent bonds (Fig. 16-3). It was formerly thought that the latter were primarily H-bonds, but modern evidence indicates that they are mainly the result of the weak attraction between the hydrophobic side groups of the protein molecules. The polypeptide chains may be linked either covalently or noncovalently in the quaternary structure.

It is only when the protein possesses its normal, three-dimensional (secondary, tertiary, quaternary) folded structure that it is in the *native* state. Normally, only native proteins are enzymatically active, although there are some exceptions. When it unfolds, it is said to be in the *denatured* state. Not only is the denatured protein inactive enzymatically but many of its other properties may change. This denaturation may, however, be reversible, for example, when induced by low temperature (Brandts, 1967). If, however, two or more denatured protein molecules combine with each other, even by weak hydrogen or hydrophobic bonds, they form *aggregates* and the denaturation is then irreversible. Denatured proteins are, however, more readily hydrolyzed to peptides and amino acids, which can then serve as raw materials for the resynthesis of native proteins (p. 145). Denaturation may therefore be a normal component of plant metabolism because although the enzyme must be stable under a precise set of conditions, it must also be able to disintegrate when no longer required (Gutfreund, 1967).

According to modern concepts, both the secondary and tertiary structures (i.e., the folding of the molecule) are a direct consequence of the order and chemical nature of the amino acids in the chain (Schachman, 1963; Epstein et al., 1963). The template (see later) is therefore needed only for the two-dimensional structure of the polypeptide. An understanding of protein breakdown and synthesis therefore depends on a knowledge of amino acid structure and synthesis. All the amino acids found in proteins are of the α-amino type (i.e., the amino group is on the so-called α-carbon atom, which is next to the carboxyl group):

Generalized structure	*Specific amino acids*
$R\cdot CH\cdot COOH$	$CH_2\cdot NH_2\cdot COOH$ (glycine)
$\quad\vert$	
NH_2	$CH_3\cdot CHNH_2\cdot COOH$ (alanine) etc.

Some amino acids have more than one carboxyl group and are therefore acidic; others have more than one amino group and are basic. Those with one of each group are nearly neutral. In the formation of proteins the amino acids link together to form a peptide as follows:

$$\underset{\substack{\text{Amino acid}}}{NH_2 \cdot \overset{R_1}{C}H \cdot COOH} + \underset{\substack{\text{Amino acid}}}{NH_2 \overset{R_2}{C}H \cdot COOH} \rightarrow \underset{\substack{\text{Dipeptide}}}{NH_2 \cdot \overset{R_1}{C}H \cdot CO\text{-}NH \cdot \overset{R_2}{C}H \cdot COOH} + H_2O$$

The protein chain therefore has the following makeup:

The average molecular weight of an individual amino acid is about 100. The smallest protein molecules contain about 100 such amino acids (less in a few cases) and therefore have molecular weights of approximately 10,000. All proteins contain essentially the same twenty amino acids, but each one differs in the proportion of each amino acid and in the order in which they occur in the molecule (the primary structure of the molecule). If a protein consists of 200 amino acid residues consisting of the twenty different amino acids, there are 20^{200} possible combinations. The number of different proteins possible in nature, therefore, staggers the imagination.

Calculations from the amino acid analyses of eighty separate proteins have led to the following numbers of molecules in an "average protein" of 233 amino acid residues and a molecular weight of 25,700 (Smith, 1966):

Asparagine°	(An)	25
Glutamine°	(Gn)	24
Leucine	(Leu)	20
Alanine	(Ala)	19
Glycine	(Gly)	17
Serine	(Ser)	17
Valine	(Val)	16
Lysine	(Lys)	15

°Aspartic acid (ASP) and glutamic acid (Glu) presumably included in asparagine and glutamine, respectively.

Proline	(Pro)	13
Threonine	(Thr)	13
Isoleucine	(Ileu)	11
Arginine	(Arg)	10
Phenylalanine	(Phe)	10
Tyrosine	(Tyr)	8
Histamine	(His)	5
Methionine	(Met)	4
Cysteine	(Cys)	3
Tryptophan	(Try)	3

N-assimilation

One characteristic of the higher plant is its ability to synthesize from inorganic nitrogen all the twenty-odd amino acids necessary for the formation of its proteins. Animals, on the other hand, must be supplied with the ten *essential amino-acids,* that is, the ten that they are unable to synthesize. The others they can synthesize from these ten. They are unable to synthesize any from inorganic nitrogen. Nitrogen is absorbed by the plant in the inorganic form, either as NO_3^- or as NH_4^+. The NO_3^- must first be reduced to NH_4^+ in a series of steps, probably as follows:

$$NO_3^- \longrightarrow NO_2^- \longrightarrow N_2O_2^= \longrightarrow NH_2OH \longrightarrow NH_4^+$$

where NO_3^- = nitrate
NO_2^- = nitrite
$N_2O_2^=$ = hyponitrite
NH_2OH = hydroxylamine
NH_4^+ = ammonium

This would seem to indicate that the NH_4^+ form is a more efficient source of nitrogen since respiratory energy must be used to reduce NO_3^- to this form. However, under suitable conditions both forms are equally satisfactory sources of nitrogen, and the NO_3^- form is suitable over a larger range of conditions (e.g., over a wider pH range).

The NH_4^+ is then converted to organic nitrogen by a reductive combination with carboxylic acids of the Krebs cycle (and pyruvic acid) to give amino acids. The reaction, in simplified form, is as follows:

$$NH_3 + CH_3 \cdot CO \cdot COOH + 2H \rightleftharpoons$$
Pyruvic acid
$$CH_3 \cdot CHNH_2 \cdot COOH + H_2O$$
Alanine

Specific enzymes are, of course, required for these reactions, for example, NO_3^- reductase for the reduction of NO_3^- and aminases and transaminases for the synthesis of amino acids from NH_4^+. The enzymes that control NO_3^- reduction require molybdenum as a cofactor.

In many cases the amino acid synthesis occurs in the roots of a plant (e.g., the apple), and the amino acids are translocated to the growing shoots, where they are built up to form proteins. In other plants (e.g., wheat), however, the inorganic nitrogen is translocated directly to the leaves and synthesized there to amino acids and proteins. The synthesis in the roots occurs, of course, in the dark and is dependent on respiration for energy and carbon chains. The synthesis in the leaves, on the other hand, may occur only in the light, at least in the case of certain plants. The tracer work on photosynthesis has shown that some amino acids are synthesized within about 5 min after photosynthesis has started. Apparently, there is therefore a close connection between photosynthesis and amino acid synthesis in green cells. It is interesting that the more advanced species (e.g., the grasses) synthesize their amino acids in the leaves and only in the light. This would be a more

efficient process since the endergonic syntheses would use light energy more directly than when dependent on the stored reserves of the root.

Protein synthesis

The synthesis of proteins is dependent on nucleotides, which are also nitrogen-containing substances. They consist of three components:

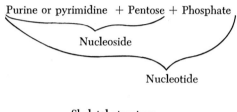

Purine or pyrimidine + Pentose + Phosphate

Nucleoside

Nucleotide

Skeletal structure

Pyrimidines   Purines

Thus, the purine *adenine* combines with the pentose ribose to form the nucleoside *adenosine,* which then combines with phosphoric acid to form the nucleotides *adenosine monophosphate* (AMP), *adenosine diphosphate* (ADP), or *adenosine triphosphate* (ATP). These simple nucleotides then may be combined in a long chain to form the complex nucleotides or nucleic acids, just as the amino acids are combined to form the proteins. When nucleotides contain ribose, they form the polymeric molecules of ribonucleic acid (RNA); when they contain deoxyribose, they form deoxyribonucleic acid (DNA). These nucleic acids resemble the proteins in several ways. They have large molecular weights (e.g., 10^6), they occur in the folded form, and they can be denatured (unfolded) just as in the case of proteins, for example, by heating. However, their relation to proteins is much closer than this similarity since they are required for protein synthesis.

Unlike the carbohydrates, fats, and most

other high molecular weight substances that are mainly chemically identical in different plants, proteins differ from species to species and even from variety to variety. Since the hundreds of known proteins differ from each other in the order of arrangement of the twenty amino acids, there must be some proce ⬚ ⬚ mple enzymatic catalysi ⬚ ⬚ rder. Indeed there c ⬚ ⬚ rent, specific enzyn ⬚ ⬚ each of the 100 to ⬚ ⬚ acid linkages requ ⬚ ⬚ ven a single protei ⬚ ⬚ zymes are themselv ⬚ ⬚ would multiply ge ⬚ ⬚ process would be ir ⬚ ⬚ litions. The metho⬚ ⬚ fore be basically d ⬚ ⬚ other substances ⬚ ⬚ em has been the ⬚ ⬚ igations in recent ⬚ ⬚ as been done witl ⬚ ⬚ mechanisms wc ⬚ ⬚ bly also

apply in general to higher plants since this is true of metabolism as a whole. That this assumption is basically correct is indicated by the newer evidence accumulating from direct experiment with higher plants.

For decades there has been a *template* theory of protein synthesis. According to this concept, a template or pattern for each specific protein must first be made, and the amino acids are laid down on the pattern according to the order specified by it. In recent years direct evidence for this theory has been accumulating. The more specific template concept (or "dogma") of today is as follows.

The function of nucleic acids is the storage and transcription of biological information (Spiegelman and Hayashi, 1963):

Transcription   Translation

Duplication DNA ⟶ RNA ⟶ Protein

The function of the protein is the expression of this information. Protein synthesis is controlled initially by a nucleic acid (DNA) in the nucleus or, more specifically, in the

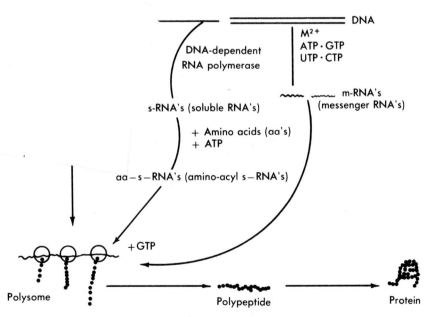

Fig. 16-4. Schematic model of protein synthesis (translation process). M^{2+} = divalent cation activator. (From Cohen, N. R. 1966. Biol. Rev. **41**:503-560.)

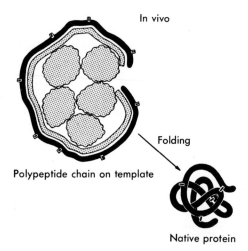

In vivo

Folding

Polypeptide chain on template

Native protein

Fig. 16-5. Schematic diagram of the conversion of a newly synthesized polypeptide chain to a folded "native" protein. The eight protruding attachments on the chain represent chemical side groups which later combine to form four bonds holding the folds in place. (From Epstein, C. J., et al. 1963. Sympos. Quant. Biol. **28**:439-449.)

genes of the chromosomes. The nuclear DNA itself is a double polymer of purines (guanine and adenine) and pyrimidines (cytosine and thymine). Adenine is always paired with thymine, cytosine with guanine. Longitudinal splitting of the DNA yields reciprocal half polymers, each of which regenerates the missing half, thus multiplying the DNA. However, these half polymers can also serve as templates, on which a different nucleic acid (messenger RNA, or mRNA) is formed, with a complementary sequence of units. RNA differs from DNA by having uracil instead of thymine and ribose instead of deoxyribose. The mRNA then peels off, passes from the nucleus into the cytoplasm, and becomes associated with a polyribosome. It then acts in turn as a template on which is deposited a series of small RNA molecules—the transfer or soluble RNA (tRNA or sRNA). Beginning at one end of a messenger molecule, a series of these small tRNA molecules attach successively to it. In each case the tRNA has a structure complementary to that of the

mRNA and holds a specific amino acid attached to it. In this way, as amino acids are brought in, they attach to each other by peptide linkages and finally peel off as a polypeptide chain, that is, a protein molecule (Fig. 16-4). The polypeptide chain then folds to form the native, folded protein (Fig. 16-5). The released proteins do not necessarily remain in the soluble molecular form. They may link up with other proteins or with other substances (lipids, carbohydrates, nucleic acids, etc.) forming *conjugated proteins.* Protein synthesis has been stated to require up to 90% of the biosynthetic energy of the cell (Lehninger, 1965). This energy is required not only for the synthesis of the amino acids but also for their incorporation into the protein molecule. It is required, for instance, for attachment of the amino acid to its specific tRNA (Novelli, 1967):

$$E + ATP + AA_1 \rightleftharpoons E - AA_1 - AMP + PP_1$$
$$E - AA_1 - AMP + tRNA^1 \rightleftharpoons AA_1 - tRNA^1 + E + AMP$$

where E = enzyme aminoacyl—RNA synthetase specific for AA_1
AA_1 = a specific amino acid
$tRNA^1$ = its specific transfer RNA

Although most of the work on which the preceding concepts are based has been done with microorganisms and animal cells, recent investigations have shown that the ribosomes of higher plants have characteristics similar to those of bacteria and mammalian cells and are therefore capable of the same mechanism of protein synthesis (Marcus and Feeley, 1965).

It is now known that protein synthesis occurs not only on the ribosomes which are free in the cytoplasm but also in the mitochondria and chloroplasts, presumably on the ribosomes which have been found within these organelles. Just as in the case of the nucleus, protein synthesis in these organelles is under the control of their own DNA (Shephard, 1965). About 9% of the total leaf DNA is in the chloroplasts (Tewari and Wildman, 1966). It differs both physically and chemically from nuclear DNA. Chloroplast DNA is also capable of replication more rapidly than nuclear DNA in the case of tobacco (Green and Gordon, 1966). Chloroplast DNA is just as capable of synthesizing RNA as is nuclear DNA (Spencer and Whitfeld, 1967; Gibbs, 1967), and about 15% to 25% of the total leaf RNA is located in the chloroplasts (Wollgiehn et al., 1966).

Other nitrogen-containing substances

There are, of course, many nitrogen-containing substances in the plant beside proteins. Some 4000 alkaloids have been found in 3600 species of plants (Leete, 1967), and new ones are being discovered at a rapid rate. These plants obviously must have a special metabolism not found in all species and even differing within those species that synthesize them (Willaman and Li, 1963).

Since all these nitrogen-containing substances in the plant (amines, betaines, alkaloids, purines, pyrimidines) are derived indirectly from the inorganic nitrogen (NO_3^- or NH_4^+) obtained from the soil, they must be related in some way to the first-formed organic substances—the amino acids. Just as acetic acid and pyruvic acid may be considered as pivot substances for C-metabolism, the amino acids must act as the pivot for N-metabolism as follows:

$$
\begin{array}{c}
\text{Amides Betaines Amines} \\
NO_3^- \qquad \qquad \nwarrow \ \uparrow \ \nearrow \\
\text{or } + \ R \cdot CO \cdot COOH \longrightarrow \text{Amino acids} \rightleftharpoons \text{Proteins} \\
NH_4^+ \qquad \qquad \qquad \downarrow \\
\text{Alkaloids} \longleftarrow \left\{ \begin{array}{c} \text{Purines} \\ \text{and} \\ \text{pyrimidines} \end{array} \right\} \longrightarrow \begin{array}{c} \text{Nucleo-} \\ \text{tides} \end{array} \\
\text{where } R \cdot CO \cdot COOH = \text{an } \alpha\text{-keto acid}
\end{array}
$$

MINERAL METABOLISM

Although all the mineral elements play some role in the metabolism of the plant, it does not necessarily follow that they are all themselves metabolized. Manganese, for instance, is active in the ionic form since it is an enzyme activator. Iron, on the other hand, becomes a component of organic substances (e.g., of the cytochrome molecule) and, therefore, is metabolized. The metabolism of the mineral elements is not as well worked out as the metabolism of most of the carbon- and nitrogen-containing compounds, even though a much smaller number of substances is involved. The two mineral elements with the most complex metabolism are phosphorus and sulfur. P-metabolism is, of course, an essential component of both respiration and photosynthesis and of both C-metabolism and N-metabolism. Yet its metabolism is relatively simple since the plant is unable to change the oxidation state of phosphorus. It is absorbed in the most highly oxidized form (+5) as the inorganic phosphate, which is merely transferred to a vast number of organic substances, forming organic phosphates. Sulfur, on the other hand, is also absorbed in the most highly oxidized form (+6, inorganic sulfate), but it can be reduced in the plant to all possible levels. S-metabolism is therefore in itself complex. The most important sulfur-containing substances in the plant are the three amino acids (*cysteine, cystine,* and *methionine),* the peptide *glutathione* (glutamyl-cysteinyl-glycine), and certain vitamins or enzyme cofactors (*thiamine, biotin,* and *thioctic acid).* It is as the mercaptan, containing the thiol (SH, or sulfhydryl) group, and as the corresponding oxidized disulfide form that sulfur is most reactive in the plant.

The interrelations between respiration and the other phases of plant metabolism are shown in Fig. 16-6.

CONTROL OF METABOLISM

The chemical reactions in the various metabolic pathways are all controlled by enzymes and, therefore, by proteins. The discovery of the enzymes that catalyze the individual reactions of respiration, photosynthesis, and other metabolic paths have therefore revealed the specific proteins that enable these processes to take place. However, there is a vast number of these chemical reactions taking place in the plant, and

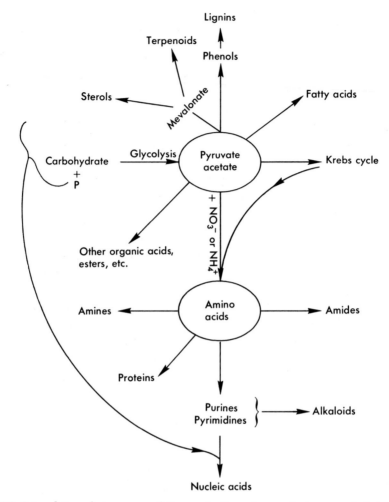

Fig. 16-6. Interrelations between respiration and other phases of plant metabolism.

indeed in each cell or even each organelle at any one time, and many of these are interdependent. How does the plant succeed in producing each of these enzymes in the right quantity, in the right place, and at the right time in order to ensure the continuation of each process at the speed and in the order needed for the normal metabolism of the plant?

The importance of this question goes far beyond an understanding of metabolism itself. Most physiological processes, including the very growth and development of the plant (see Chapters 19 and 20), are dependent on metabolic energy. The physiology of the plant is therefore largely controlled by its metabolism. The factors that control the metabolism of the plant also control all the physiological processes that are dependent on metabolic energy. Hence, no true understanding of plant physiology is possible without an understanding of the control of metabolism.

Since the metabolic reactions are all catalyzed by enzymes, it is obvious that any control of enzyme activity will control metabolism. There are two basic kinds of enzyme control that may, however, be interdependent:

1. All enzymes can be inhibited (or conversely, deinhibited or activated) in various ways, and these inhibiting factors can, therefore, control metabolism.

2. Since enzymes are proteins, any factor that affects the rates of synthesis and breakdown of proteins will also control metabolism.

It is now obvious why the mechanism of protein synthesis and its relation to DNA and RNA are of such fundamental importance to the physiologist. As a result of the tremendous research explosion in this area in recent years, the mechanism of the control of metabolism is gradually unfolding.

The first mechanism, enzyme inhibition or activation, is believed to be a part of all control devices (Perutz, 1967) since the synthesis or breakdown of inactive enzymes can have no effect by itself on metabolism. The protein molecules, comprising the enzymes, are able to change their structure reversibly from an active to an inactive form in response to chemical stimuli. Thus, sulfhydryl enzymes (e.g., many dehydrogenases) are inactivated reversibly by oxidation of the sulfhydryl group to disulfide. Conversely, disulfide enzymes (e.g., ribonuclease) are reversibly inactivated by reduction of their disulfide group to sulfhydryl. There are, of course, other activation and inactivation processes not involving sulfhydryl. The chemical stimuli that induce these reversible changes are known as *effectors* or *modifiers*. Effectors or modifiers are therefore substances that modify (positively or negatively) the affinity of an enzyme for its substrate, presumably by changing it from the inactive to the active form, or vice versa.

The second mechanism must be the primary step in enzyme control since it involves either *repression* or *derepression* of the synthesis of the enzyme protein. The importance of this control is obvious from the fact that although each cell contains at least one copy of all of the organism's genetic information, only a small part is usually active in synthesizing proteins at any one time. Thus, RNA from pea buds does not synthesize pea seed globulin; RNA from developing pea cotyledons does (Bonner et al., 1967). The repressors are believed to be basic proteins known as *histones*, which prevent chromosomal DNA from acting as a template. In agreement with this concept the vegetative nucleus of pollen grains shows high RNA and protein synthesis and no nucleohistone; the generative nucleus shows no detectable synthesis and a high degree of nucleohistone (Sauter and Marquardt, 1967). Furthermore, this protein synthesis controls even the replication of the genetic material itself. In principle, this second mechanism of control may be exerted at different levels: (1) DNA → mRNA, or the transcription level; (2) stability of mRNA; and (3) mRNA → protein, or the translation level. In bacteria, at least, the major control is at the third level—the level of translation (Vogel and Vogel, 1967). The first-described mechanism (inhibition by reversible structural change) can only follow the preceding three control levels and may therefore be considered a fourth level of control.

However, there is also a fifth level of control, by the substrate and the product, whose reaction is catalyzed by the enzyme. Synthesis of catabolic enzymes is induced by the substrate, and synthesis of anabolic enzymes is repressed by the product. Even in a series of metabolic steps the activity of the first enzyme may be inhibited by the product of the last step. The formation of ATP may therefore inhibit a metabolic process, and conversely, ADP or AMP may stimulate it (Atkinson, 1966). An example of this control is phosphofructokinase in carrot plants, which is controlled by citrate and ATP concentration (Dennis and Coultate, 1966). The whole course of metabolism may therefore be altered by changes in intracellular concentrations of these substances.

Both repression and derepression of the synthesis of a single enzyme have been identified. Nitrate reductase of cultured tobacco cells is induced by the presence of nitrate in the medium (Filner, 1966). The enzyme disappears as a result of nitrate starvation. It is not formed in cells grown on casein hydrolysate as the sole nitrogen source. Some amino acids repress synthesis of the enzyme, others derepress it in the presence of the repressor amino acids. Such enzymes, whose formation is selectively provoked by the substrate are called *adaptive enzymes.* Most of the plant's enzymes are probably *constitutive*—their synthesis is not provoked by the substrate (Bell, 1965).

Artificial control is also possible by supplying antibiotics. Actinomycin D prevents the formation of mRNA from DNA (i.e, it inhibits transcription), puromycin appears to destroy the activated amino acid–tRNA complex, and chloramphenicol inhibits the transfer of activated amino acids to mRNA (i.e., they inhibit the translation process). Other antibiotics inhibit other steps in protein synthesis.

It must be realized, of course, that the control of enzyme activity or synthesis is merely a means to an end, not an end in itself. The actual goal of regulatory control is twofold (Atkinson, 1966): (1) to maintain an essentially constant level of building blocks (ATP, NADPH, etc.), in spite of wide fluctuation in demand; and (2) to facilitate future maintenance by storage of excess energy.

SECONDARY METABOLITES

The aforementioned control obviously cannot be perfect in all cases, and an escape valve must be available if the plant is to survive. If, for instance, one metabolic chain of reactions is too active, leading to the accumulation of toxic quantities of a substance, the plant must have a mechanism of getting rid of the excess. In the case of simple substances, an enzyme may be present whose function is merely to destroy this toxic substance as rapidly as it is formed. This is apparently the function of the ubiquitous (in aerobic organisms) catalase. It destroys hydrogen peroxide at the rate of 2000 molecules per second, thus preventing the accumulation of toxic quantities from aerobic respiration. But most metabolites are too complex to be broken down into such simple substances as water and oxygen. If they are toxic, they must be excreted. Even before excretion they may have to be converted first into a less toxic waste substance. Different species possess different enzymes and metabolic paths and therefore may produce different waste metabolites.

Unlike the animal, the plant does not possess a system for excreting toxic or waste metabolites into the surrounding environment. Instead, each cell has its own internal garbage can—the vacuole. This has one advantage over the other system because if conditions change, so that what was formerly waste or toxic is now needed, it can be reabsorbed from the vacuole. This system is not suited to the animal, which produces so much more waste material than the plant, and unlike the plant, cannot reutilize its waste.

It is usually stated that the function of such waste products is unknown since many of them remain unaltered in the cell until its death. Some, however, are now known to be reutilized to a degree and therefore may be considered as storage metabolites. But even the vast number of metabolites that are not reutilized may be considered as playing a role in plant metabolism since in many cases at least, they are the end products of detoxification mechanisms in the plant. It is also possible, however, that many of these waste metabolites are simply the end products of metabolic pathways that may be considered blind alleys. The plant happens to have a series of enzymes leading to these end products but no enzymes to convert them once they are formed.

Since this is a polyglot group of metabolites, including toxic substances, waste substances, substances possibly sometimes serving as reserves, and just accidental end products, they are all grouped together as *secondary metabolites.* They are distinguished from the *primary metabolites,* substances that (1) are found in essentially all plants and (2) play reasonably well understood roles in the physiology of the plant. It is, of course, difficult if not impossible to draw a sharp line between the two groups.

The major secondary metabolites are the alkaloids, terpenes (essential oils, resins, rubber), many phenolic substances (such as tannins), glycosides, etc. Some of the more complex secondary metabolites are found in only one or a few taxonomic groups; the simpler ones are usually more widespread. In the case of the latter, the synthesis may depend on enzymes whose normal function is in association with the primary metabolites. Thus, phenolic substances are probably formed from acetylcoenzyme A in fungi and from aromatic amino acids in higher plants (Neish, 1964). The number of secondary metabolites in plants is very large. Some 4000 alkaloids alone are known. Aside from these and the macromolecules, nearly 1000 simpler organic substances have been isolated from plants and chemically identified (Mentzer, 1964). Some of the secondary metabolites accumulate in extraordinarily large amounts; for example, tannins account for 14.5% of the dry matter of some leaves (El Sissi and Abd Alla, 1966). The metabolic pathways involved in the synthesis of these secondary metabolites are gradually being discovered, for example, for rubber (Archer and Audley, 1967). In many cases, however, little is known as yet.

QUESTIONS

1. What chemical change occurs when high protein seeds germinate?

2. What substances accumulate in large quantity when high protein seeds germinate?

3. What substances accumulate in smaller quantity?

4. When a plant is supplied with inorganic nitrogen, what must it synthesize before proteins?

5. Are any amino acids essential for higher plants?

6. Why are (are not) amino acids essential for higher plants?

7. What is the difference between higher plants and animals in this respect?

8. Where are the amino acids synthesized in the plant?

9. Where are the proteins synthesized?

10. What structures are involved in protein synthesis?

11. Is there any relation between protein synthesis and respiration?

12. Is there any relation between protein synthesis and photosynthesis?

13. When lipids disappear in higher plants, what other substances often disappear?

14. When lipids accumulate, what other substances disappear?

15. What are glycosides?

16. Give an example of a glycoside.

17. On what chemical substances does protein synthesis depend?

18. What basic difference exists between the mechanism of protein synthesis and the synthesis of other substances in the plant?

19. What are effectors?

20. What is repression?

21. What are secondary metabolites?

SPECIFIC REFERENCES

Anfinsen, C. R. 1967. Molecular structure and the function of proteins, p. 1-19. *In* J. M. Allen. Molecular organization and biological function. Harper & Row, Publishers, New York.

Archer, B. L., and B. G. Audley. 1967. Biosynthesis of rubber. Adv. Enzymol. 29:221-257.

Atkinson, D. E. 1966. Regulation of enzyme activity. Ann. Rev. Biochem. 35:85-124.

Bell, E. 1965. Molecular and cellular aspects of

development. Harper & Row, Publishers, New York.

Bonner, J., M. E. Dahmus, D. Fambrough, R. C. Huang, K. Marushige, and D. Y. H. Tuan. 1967. The biology of isolated chromatin. Science **159**: 47-56.

Brandts, J. F. 1967. Heat effects on proteins and enzymes, p. 25-72. *In* A. H. Rose. Thermobiology. Academic Press, Inc., New York.

Dennis, D. T., and T. P. Coultate. 1966. Phosphofructokinase, a regulatory enzyme in plants. Biochem. Biophys. Res. Commun. **25**:187-191.

Dorner, R. W., A. Kahn, and S. G. Wildman. 1957. The proteins of green leaves. VII. Synthesis and decay of the cytoplasmic proteins during the life of the tobacco leaf. J. Biol. Chem. **229**: 945-952.

El Sissi, H. I., and M. F. Abd Alla. 1966. Polyphenolics of the leaves of *Catha edulis* (L.). Planta Med. **14**:76-83.

Epstein, C. J., R. F. Goldberger, and C. B. Anfinsen. 1963. The genetic control of tertiary protein structure; studies with model systems. Sympos. Quant. Biol. **28**:439-449.

Filner, P. 1966. Regulation of nitrate reductase in cultured tobacco cells. Biochim. Biophys. Acta **118**:299-310.

Fowden, L. 1965. The chemical approach to plants. Sci. Progr. **53**:583-599.

Gibbs, S. P. 1967. Synthesis of chloroplast RNA at the site of chloroplast DNA. Biochem. Biophys. Res. Commun. **28**:653-657.

Green, B. R., and M. P. Gordon. 1966. Replication of chloroplast DNA of tobacco. Science **152**: 1071-1074.

Gutfreund, H. 1967. The dynamic behavior of proteins during catalysis. Naturwissenschaften **54**: 402-406.

Holmsen, T. W., and A. L. Koch. 1964. An estimate of protein turnover in growing tobacco plants. Phytochemistry **3**:165-172.

Marcus, A., and J. Feeley. 1965. Protein synthesis in imbibed seeds. II. Polysome formation during imbibition. J. Biol. Chem. **240**:1675-1680.

Mothes, K. 1966. Zur Problematik der metabolischen Exkretion bei Pflanzen. Naturwissenschaften **53**:317-323.

Novelli, G. D. 1967. Amino acid activation for protein synthesis. Ann. Rev. Biochem. **36**:449-484.

Perutz, M. F. 1967. Some molecular controls in biology. Endeavour **26**:3-8.

Rich, A. 1967. The structural basis of protein synthesis, p. 20-36. *In* J. M. Allen. Molecular organization and biological function. Harper & Row, Publishers, New York.

Sauter, J. J., and H. Marquardt. 1967. Die Rolle des Nukleohistons bei der RNS -und Proteinsynthese während der Mikrosporogenese von *Paeonia tenuifolia* L. Z. Pflanzenphysiol. **58**:126-137.

Schachman, H. K. 1963. Considerations on the tertiary structure of proteins. Sympos. Quant. Biol. **28**:409-430.

Shephard, D. C. 1965. An autoradiographic comparison of the effects of enucleation and actinomycin D on the incorporation of nucleic acid and protein precursors by *Acetabularia* chloroplasts. Biochim. Biophys. Acta **108**:635-643.

Smith, M. 1966. The amino acid composition of proteins. J. Theor. Biol. **13**:261-282.

Spencer, D., and P. R. Whitfeld. 1967. DNA synthesis in isolated chloroplasts. Biochem. Biophys. Res. Commun. **28**:538-542.

Spiegelman, S., and M. Hayashi, 1963. The present status of the transfer of genetic information and its control. Sympos. Quant. Biol. **28**:161-181.

Taylor, W. I. 1966. The source of indole alkaloids. Science **153**:954-956.

Tewari, K. K., and S. G. Wildman. 1966. Chloroplast DNA from tobacco leaves. Science **153**: 1269-1271.

Vogel, H. J., and R. H. Vogel. 1967. Regulation of protein synthesis Ann. Rev. Biochem. **36**:519-538.

Willaman, J. J., and H. L. Li. 1963. General relationships among plants and their alkaloids. Econ. Bot. **17**:180-185.

Wollgiehn, R., M. Ruess, and D. Munsche. 1966. Ribonucleinsäuren in Chloroplasten. Flora **157**: 92-108.

GENERAL REFERENCES

Bonner, J. F., and J. E. Varner. 1965. Plant biochemistry. Academic Press, Inc., New York.

Chibnall, A. C. 1939. Protein metabolism in the plant. Yale University Press, New Haven, Conn.

Cohen, N. R. 1966. The control of protein biosynthesis. Biol. Rev. **41**:503-560.

Davies, D. D., J. Giovanelli, and T. ap Rees. 1964. Plant biochemistry. F. A. Davis Co., Philadelphia.

Goodwin, T. W. 1965. Chemistry and biochemistry of plant pigments. Academic Press, Inc., New York.

Haurowitz, F. 1963. The chemistry and functions of proteins. Academic Press, Inc., New York.

Leete, E. 1967. Alkaloid biosynthesis. Ann. Rev. Plant Physiol. **18**:179-196.

Lehninger, A. L. 1965. Bioenergetics; the molecular basis of biological energy transformations. W. A. Benjamin, Inc., New York.

Lienier, I. E. 1966. Organic and biological chemistry. The Ronald Press Co., New York.

McKee, H. S. 1962. Nitrogen metabolism in plants. Clarendon Press, London.

Mentzer, C. 1964. Actualités de phytochimie fondamentale. Masson et Cie, Paris.

Morton, R. A. 1965. Biochemistry of quinones. Academic Press, Inc., New York.

Neish, A. C. 1964. Major pathways of biosynthesis of phenols, p. 293-359. *In* J. B. Harborne. Biochemistry of phenolic compounds. Academic Press, Inc., New York.

Pridham, J. B. 1967. Terpenoids in plants. Academic Press, Inc., New York.

Schubert, W. J. 1965. Lignin biochemistry. Academic Press, Inc., New York.

part IV

GROWTH AND DEVELOPMENT

chapter 17

Growth

A. Quantitative relations
B. Cellular basis of growth
C. Stages of growth
D. Influence of external factors on growth
   1. Temperature
   2. Light
   3. Water
   4. Chemical stimulants and inhibitors
E. Dormancy or rest period

The preceding treatment of the fundamental biophysics and biochemistry of the plant is, in a sense, merely a preliminary to the remaining aspects of the physiology of the plant. The ultimate goal of the plant physiologist is to discover and explain in terms of physics and chemistry the complete mechanism of plant growth and development. Today, this requires a utilization of the most modern principles of molecular biology as a basis for experimental attack. The result is an explosion of evidence that is daily flooding the literature of plant physiology. Although it is impossible to do justice to the available experimental evidence, a slightly different approach will be necessary from that in the preceding chapters. Because of the many different lines of attack and today's far from complete understanding of the physiology of growth and development, it is necessary to adopt a kind of "recent advances" approach (e.g., Chapters 19 and 20). In this way the student will be given a glimpse of the tremendous ferment of activity in these plant physiology fields and of some of the principles that are slowly but surely beginning to appear. At the same time the student must realize that when he has mastered the following information, it will be already out of date, and he will have to supplement it with the more current publications.

QUANTITATIVE RELATIONS

Of all the phases of physiology, growth and development are the most complex since they depend on all other physiological processes. From the chemical point of view, growth has been defined as the transformation of simple, inorganic foodstuffs into new chemical entities that form the organized protoplasm of the plant (Robertson, 1923). However, this cannot be a complete definition of growth since there are also physical aspects of the process. It is therefore impossible to understand plant growth without a knowledge of the basic facts of cell physiology and of the physiology of the plant as a whole—transfer of materials, mineral nutrition, and metabolism. Beside these processes, other factors specific to growth and development must be introduced. Foremost among these are the substances known as growth regulators (see Chapter 19). Yet it must be remembered that these substances are essentially switch mechanisms for turning growth or development on or off. No understanding of the switch can

209

substitute for an understanding of the growth mechanism itself.

The first requirement for studying growth is a suitable method of measuring it. The simplest definition of growth is an increase in size. Since size is a synonym for volume, this means that:

$$Growth = V_t - V_o$$
$$and\ growth\ rate = \frac{V_t - V_o}{t}$$

where V_t = volume at end of time t
V_o = original volume at time zero

The ideal method of measuring growth would therefore be to determine the volume of the plant or plant part, but this is usually difficult to do with any degree of accuracy. In the case of growing leaves the area is more readily determined:

$$Growth = S_t - S_o$$

where S_t = surface area at end of time t
S_o = original surface area

In the case of stems and roots, a simpler measurement is length:

$$Growth = L_t - L_o$$

where L_t = length at end of time t
L_o = original length

The easiest and most accurate measurement, however, is usually the weight of the plant or plant part and:

$$Growth = W_t - W_o$$

where W_t = weight at end of time t
W_o = original weight

This leads to some complicating factors, however. It must be decided whether fresh or dry weight is a more nearly correct measurement of growth. In the case of germinating seeds, growth occurs rapidly, yet dry weight decreases steadily for some time because of rapid loss of reserves. During this period only the fresh weight reveals the amount of growth. The same is true in the case of a growing tree in the spring. On the other hand, during midsummer when the tree has stopped growing, there is a steady increase in dry weight because of an accumulation of reserves. Dry weight

can therefore serve as a measurement of growth only when it does not include any significant changes (increases or decreases) in the plant's food reserves. It is a satisfactory measurement of growth of seedlings only if the endosperm or seed leaves are not included. In other words, weight is a measure of growth only if $W \propto V$.

CELLULAR BASIS OF GROWTH

The mere increase in size of a plant or a plant part is not necessarily the result of growth. The first uptake of water by dry seeds is purely an imbibition process and may occur equally in living and dead seeds. Since true plant growth occurs only in living cells, volume or fresh weight measurements in this case would reveal a spurious growth, and increased volume is therefore not an adequate definition of growth. Obviously, then, growth must be defined in another way, preferably on a cellular basis. Since the growth of a plant is initiated in special tissues (meristems), the cellular changes in these tissues should throw some light on the process. The three main changes in the complete development of a cell are cell division, cell enlargement (or elongation) and cell differentiation (or maturation). That growth of a plant part may result from either cell division or cell enlargement is shown in Table 17-1. In this case both phosphorus and nitrogen increased root growth—phosphorus because of increased cell division, nitrogen mainly because of increased cell enlargement.

However, it must be realized that even growth resulting from cell division is possible only because the daughter cells enlarge to the size of the mother cell before they again divide. Furthermore, whether cell enlargement goes only this far or much farther, it must not include any simple turgor-induced change in cell volume if it is a true measure of cell growth. Consequently, *the growth of a plant or plant part is the increase in size caused by cell en-*

Table 17-1. Growth of isolated wheat roots with low and high phosphate and nitrate[*]

Nutrient solution	Increase in root length (mm)	Cell length (μ)	Increase in cell number (longitudinally)
Low P, low N	13.7 ± 1.7	177 ± 3	35 ± 4
High P, low N	25.5 ± 2.0	170 ± 3	120 ± 5
Low P, high N	25.2 ± 1.3	269 ± 4	51 ± 3

[*]From Burström, H. 1951. *In* F. Skoog (Ed.). Plant growth substances. University of Wisconsin Press Madison, Wis.

Table 17-2. Wall (or turgor) pressures of developing cells in the hypocotyl of *Helianthus* seedlings; maximum rate of cell enlargement in zone 2, and beyond this, increasing zone number corresponds to increasing maturity[*]

Zone	Distance from cotyledons (mm)	Wall pressure (atm) Epidermis	Cortical cells
1	0-5	2.0	0.235
2	5-10	1.44	0.210
3	10-15	1.39	0.120
4	15-20	1.14	0.120
5	20-25	2.05	0.42
6	25-30	2.28	
7	30-35	2.58	0.42

[*]Adapted from Beck, W. A., and B. Andrus. 1943. Bull. Torrey Bot. Club **70:**563-598.

largement when measured at constant positive cell turgor.

According to the classical concept of cell growth, the motivating force is turgor pressure. Yet it is a fact that mature, nongrowing cells usually have higher turgor (and wall) pressures than growing cells (Table 17-2). Therefore, the difference between growing and nongrowing cells does not result from a difference in turgor pressure. There is obviously another factor that must be considered: the resistance to the force. The mature cells have more rigid cell walls than the growing cells. The latter, therefore, stretch more readily; and during cell growth, the decrease in rigidity (i.e., the increase in elastic and plastic stretch) of the cell walls

is directly related to the rate of growth (Fig. 17-1). Consequently, a small turgor pressure is sufficient to stretch the walls of young, growing cells; but in order for cells to continue growth, other changes must also take place. In the simplest cases of cell growth there is nothing more than an uptake of water, producing a turgor pressure sufficient to overcome the force of attraction between the particles in the cell wall. As a result, the wall stretches and becomes thinner, and the osmotic pressure drops because of uptake of water, which dilutes the cell contents. The consequent increase in the cell's water potential soon brings water uptake (and therefore growth) to a stop. This simplest type of growth can be induced artificially in some cases. Yet under normal conditions, cells are able to increase in size more than fifteen-fold (Fig. 17-1). In these cases, as water enters the cell during enlargement (accounting for about 90% of the increase in cell size), solutes are also absorbed (Table 17-3). In this way the cell maintains a high enough turgor pressure to continue stretching its wall.

Besides water and solute absorption, other changes occur during cell growth. New wall material must be laid down by *intussusception*, an insertion of new particles between the old wall particles that have become separated. The exact mechanism is not fully understood, although it has been suggested that the cell wall material is synthesized in the Golgi bodies and ejected through the plasma membrane into the cell wall (Wil-

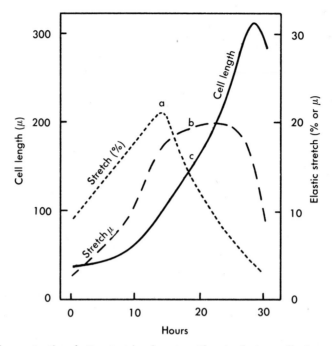

Fig. 17-1. Change in the elastic stretch of root epidermis during cell elongation. Between 10 and 25 hr the ability to stretch elastically is at a maximum (curves **a** and **b**). During this same period, the growth rate (curve **c**) is most rapid. (From Burström, H. 1951. *In* F. Skogg (Ed.). Plant growth substances. University of Wisconsin Press, Madison, Wis.)

son, 1964). Wall thickening by *apposition* (a deposit of new particles on the inner surface of the wall) occurs during cell maturation (after enlargement has ceased).

As mentioned earlier, under artificial conditions, it is possible for cell enlargement to occur without cell wall synthesis. In pea stem segments, longitudinal growth is not correlated with the formation of cellulose, yet the segments may become very long (Winter, 1966). It may therefore be concluded that during this longitudinal growth

Table 17-3. Growth and solute content of epidermal cells of *Helianthus annuus;* explanation of zones as in Table 17-2*

Zone	Average distance from cotyledons (mm)	Average cell length (μ)	Average equivalent concentration (moles sucrose)	Total solutes per cell (arbitrary units)
1	2.5	25.6	0.270	7.18
2	7.5	96.1	0.248	23.90
3	12.5	144.3	0.243	34.60
4	17.5	173.2	0.228	39.60
5	22.5	203.6	0.225	44.90
6	27.5	212.8	0.226	48.00
7	32.5	221.4	0.226	50.00

*Adapted from Beck, W. A. 1941. Plant Physiol. **16:**637-642.

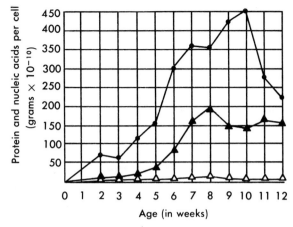

Fig. 17-2. Changes in level of protein, ribonucleic acid, and deoxyribonucleic acid per cell. November fruit (tomato), natural set. • = protein per cell; ▲ = RNA per cell; △ = DNA per cell. (From Davies, J. W., and E. C. Cocking. 1965. Planta **67**:242-253.)

there is neither intussusception nor apposition of cellulose microfibrils. Lateral growth, on the other hand, was positively correlated with the formation of cellulose. During normal plant cell growth, however, intussusception probably always occurs.

The meristematic cell is filled with cytoplasm and has few small vacuoles. As enlargement progresses, the vacuole occupies a larger and larger fraction of the cell, and the cytoplasm becomes a thin and almost invisible layer around it. Plant cell enlargement is therefore mostly vacuole enlargement. This might lead one to suspect that no new cytoplasm is formed during cell growth. Such is not the case. In spite of the smaller fraction of the cell occupied by the cytoplasm, as a result of cell enlargement, the actual amount per cell increases as much as fourfold, as judged by the protein-nitrogen content of the cell (Blank and Frey-Wyssling, 1944). More detailed analyses have since revealed nearly tenfold increases in protein and twentyfold increases in RNA during cell enlargement in growing fruit (Fig. 17-2). There was, however, no increase in DNA. The protein content of the nucleus also increases during cell enlargement (Lyndon, 1963), but nuclear

RNA decreases. The nucleus even increases in volume and hydration. That protein synthesis is an essential component of cell growth is further supported by recent evidence that inhibitors of protein synthesis (e.g., chloramphenicol) also stop cell enlargement (see Chapter 19).

Recent evidence indicates a possible role of cell wall protein in cell enlargement, and it has therefore been given the name "extensin." It is rich in hydroxyproline—an amino acid found in only a few proteins, such as animal collagen. Free hydroxyproline inhibits cell enlargement at as low a concentration as 1 mM, supposedly because it blocks the utilization of this protein (Cleland, 1967). It is obvious that phospholipids must also increase during cell enlargement since the plasma membranes increase tremendously in area.

The increase in RNA per cell during cell enlargement is easy to understand since the RNA is required for protein synthesis, and the increased metabolic rate accompanying cell enlargement is possible only if there is an increase in rate of protein (i.e, enzyme) synthesis. In spite of the lack of increase during cell enlargement, changes in DNA content have been associated with the

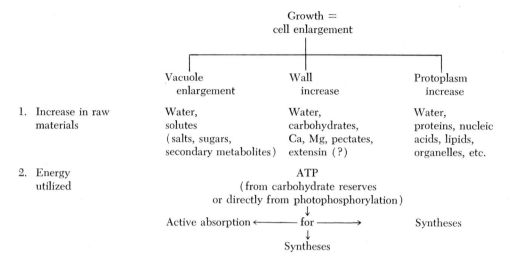

Growth =
cell enlargement

	Vacuole enlargement	Wall increase	Protoplasm increase
1. Increase in raw materials	Water, solutes (salts, sugars, secondary metabolites)	Water, carbohydrates, Ca, Mg, pectates, extensin (?)	Water, proteins, nucleic acids, lipids, organelles, etc.
2. Energy utilized		ATP (from carbohydrate reserves or directly from photophosphorylation)	

Active absorption ←——— for ——→ Syntheses
↓
Syntheses

growth process. In the case of germinating peas, cell elongation consists of two phases: a lag phase during which the DNA content doubles, and a true elongation phase, which begins only after the DNA doubling (Van Parijs and Vandendriessche, 1966*a*). Even during this true cell elongation phase, a 20% increase in DNA occurs (Van Parijs and Vandendriessche, 1966*b*).

It is apparent that during normal cell elongation all parts of the cell increase; but all the increases do not occur at the same rate (Williams and Rijven, 1965). Therefore, the *proportions* change as the cell expands. During 5 days prior to leaf emergence in wheat plants, proteins decreased from 69% to 45% of the dry weight, and cell wall materials increased from 12% to 40%. Yet the protein per cell increased eightfold, and the RNA threefold, reaching a maximum 7 days after emergence.

The increase in cytoplasm during cell enlargement has now been shown by a more direct method—electron microscopy. It has been found that all the cell organelles in the cytoplasm increase in cell number (Table 17-4).

Thus, although the simplest type of growth involves nothing more than a stretching of the wall caused by an uptake

of water, nevertheless, under normal conditions the following processes all take place:

1. Absorption of water
2. Absorption of solutes, maintaining the osmotic potential and turgor pressure at an almost constant level
3. Deposit of new wall material within the wall (intussusception)
4. Formation of new protoplasm (synthesis of proteins, nucleic acids, phospholipids, etc.) and multiplication of organelles

The bases of cellular growth can therefore be represented as shown above.

Growth is therefore a complex process,

Table 17-4. Increase in protoplasmic organelles per cell during cell enlargement in a root cap*

	Endoplasmic reticulum (area)	Mitochondria (no.)	Golgi bodies (no.)	Plastids (no.)
Original quantity	300	200	30	15
Final quantity	10,000	2000	300	25

*From Juniper, B. E., and F. A. L. Clowes. 1965. Nature (London) **208**:864-865.

involving all or nearly all the other physiological processes occurring in the plant. It requires large quantities of energy since (1) absorption of solutes is probably active and therefore dependent on respiratory energy and (2) syntheses of cell wall and protoplasmic and vacuole substances are endergonic processes. It is therefore obviously dependent on respiration or photosynthesis or both since these are the plant's only sources of reserve energy (carbohydrates) and available energy (ATP).

Although the preceding classical concept of cell growth resulting from turgor pressure is logical enough and is based on considerable sound evidence, it may not always be an exact picture of the order of events. That cell enlargement can occur under artificial conditions according to the classical turgor concept is easily proved. When potato slices are aerated in water, their turgor pressure becomes so great that the cell walls are stretched beyond their elastic limits, and a permanent increase in cell volume of as much as 50% occurs. But this is not a normally growing tissue. Roots grow with such a low turgor pressure (about 1.5 atm according to Burström) that the cell wall is not even stretched elastically, let alone plastically. The root cells therefore grow despite the fact that the turgor pressure is too small to produce growth. Instead of being the initiating force in growth, turgor is here only a secondary factor. The initial cause of growth must be the insertion of new material within the cell wall (intussusception), leading to an increase in the perimeter of the cell. The turgor pressure would then serve simply to keep the wall taut and to prevent it from wrinkling as its perimeter increases. This would still be an essential role because in the absence of turgor pressure, the wall particles would soon become too close together to permit any further intussusception.

In the case of many types of cells (e.g., development of vascular cells from cam-

bium cells), growth appears to be uniform. In others (e.g., elongation of epidermal cells) growth seems to be localized at the tips (Frey-Wyssling, 1952). The protoplasm penetrates the loose cellulose framework in their tips, and even moves ahead of it, laying down a new framework as the cell elongates. The new cellulose framework is in the form of long strands that comprise the weft (parallel to the elongating cell). When the extending tip pushes further ahead, the warp (the cross strands) is laid down in the wall, reinforcing it. All this represents only the formation of the primary wall. The secondary wall is later formed by apposition.

STAGES OF GROWTH

The growth of a plant or plant part characteristically passes through stages represented by an S-shaped curve (Fig. 17-3). The time during which this occurs has been called by Sachs the *grand period of growth*. Several attempts have been made to develop a mathematical expression for the

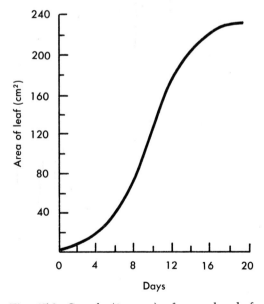

Fig. 17-3. Growth (in area) of cucumber leaf. (From Gregory, F. G. 1921. Ann. Bot. **35**:93-123.)

growth curve. Blackman (1919) assumed that the shape of the curve results from the fact that the rate of production of new materials is proportional to the size of the plant. It has long been recognized that when growth results from cell division (and enlargement of the daughter cells to the size of the mother cell), each cell is replaced by two cells of the same size as the mother cell, and this process is repeated so that there is a geometric progression (2, 4, 8, 16, 32, etc.). Blackman applied the compound interest formula to explain growth:

$$A = ae^{rt}$$

or in logarithmic form:

$$\log_e \frac{A}{a} = rt, \text{ or } 2.3026 \log A/a = rt$$

where A = final size
　　a = initial size
　　e = base of natural logarithm
　　r = rate of interest (or growth)
　　t = time interval

Thus, if a plant has doubled itself in a time (t) of 10 days, A/a = 2, and since $\log_e 2 = 0.69315$, the rate of increase, or growth, was therefore:

$$r = \frac{\log_e (A/a)}{t} = \frac{0.69315}{10} = 0.0693,$$
$$\text{or } 6.93\% \text{ per day}$$

If the period of doubling were 5 days, the rate would be 13.8% per day. But this constant rate of growth can obviously apply only up to the end of the increasing slope of the growth curve.

Robertson assumed that growth is based on a chemical transformation of assimilated or reserve substances into living protoplasm. As soon as it is formed, the new protoplasm begins to participate in these transformations, thus catalyzing its own synthesis. In this respect, growth resembles autocatalyzed chemical reactions (Fig. 17-4)—reactions in which one of the products of the chemical change has the property of accelerating or catalyzing the further progress of the change. Since growth depends on many chemical reactions, Robertson (1923) assumes that the slowest reaction sets the pace for the rest and becomes the

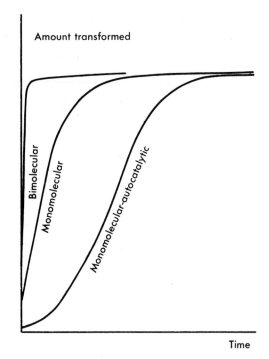

Amount transformed

Time

Fig. 17-4. Comparison of the relationship of extent of transformation to time in monomolecular, bimolecular, and autocatalyzed monomolecular reactions. (From Robertson, T. B. 1923. The chemical basis of growth and senescence. J. B. Lippincott Co., Philadelphia.)

"master reaction." Although he admits that this reaction is probably multimolecular, for purposes of simplification he applies the formula for monomolecular, autocatalytic reactions, assuming that a single master reaction limits the growth rate:

$$\frac{dx}{dt} = K(A-x), \text{ or } \log \frac{x}{A-x} = k(t-t_1)$$

where K = empirical constant
　　x = size reached in t days from beginning of growth
　　A = final size
　　t_1 = time to reach one half final size

According to Robertson, there are two possible reasons why the chemical reaction (or the growth process) comes to a stop: (1) by exhaustion of substrate and (2) by accumulation of the products of the reaction, resulting in acceleration of the reverse reaction. The theoretical curve obtained from

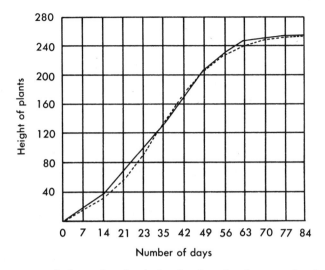

Fig. 17-5. Comparison of observed and calculated values for the mean height of *Helianthus*. Full line indicates the observed heights, and the discontinuous line indicates the calculated heights. (From Robertson, T. B. 1923. The chemical basis of growth and senescence. J. B. Lippincott Co., Philadelphia.)

Robertson's equation agrees very well with the actual growth curve (Fig. 17-5). However, it must be admitted that not all growth curves follow the simple S-shape, and therefore the more complex cannot be fully described by either of the preceding equations.

Other formulas have been proposed by various workers, but all fail to express adequately the progress of growth since they do not take into account all the processes and variables involved, which will be discussed next.

INFLUENCE OF EXTERNAL FACTORS ON GROWTH
Temperature

Growth of higher plants occurs in the range of about 0° to 35° C. Within most (but not all) of this range, raising the temperature 10° C increases the growth rate 2 to 3 times. There are three temperatures known as the *cardinal points* for growth: the *minimum,* or lowest temperature at which growth can be detected; the *optimum,* or temperature of maximum rate of growth; and the *maximum,* or highest tem-

Table 17-5. Cardinal points (° C) for growth of different plants*

Plant	Mini-mum	Opti-mum	Maxi-mum
Barley	5	29	38
White mustard	0	21	28
Scarlet runner bean	10	34	46
Maize	10	34	46
Gourd, squash	14	34	46

*From Palladin, V. I. 1923. Plant physiology. B. E. Livingston (Ed.). Blakiston Division, McGraw-Hill Book Co., New York.

perature at which growth can be detected. These are not sharp temperatures, and they vary from species to species (Table 17-5).

It should be pointed out that the optimum temperature for growth as determined in these tests over a short period of time is not necessarily the optimum for the general development and yield of the plant over long periods of time. Furthermore, the cardinal temperatures may change with the stage of development of the plant. In the case of cocksfoot *(Dactylis glomerata)*, relative growth rates increased with temperature from 14° to 26° C during the early stages but not during a later stage (Davidson and Milthorpe, 1965). It must also be remembered that growth depends on other processes, (e.g., photosynthesis, respiration, etc.), and these processes also possess cardinal temperatures.

The cardinal temperatures for growth are of great practical importance since maximum growth and yield would require a climate with a maximum time at or near the optimum temperature. But the existence of cardinal temperatures is also of theoretical importance, and an understanding of their cause might also lead to an ability to control them.

Chemical reactions do not show minimum, optimum, and maximum temperatures but yield straight line relations between rates and temperature. Enzymatically controlled reactions, on the other hand, show a similar straight line relation only within a narrow temperature range—between the two temperatures slightly above the minimum and slightly below the optimum. The rate drops off steeply at both extremes. This, therefore, explains the existence of minimum and maximum temperatures for all plant processes since all chemical reactions in the plant are enzymatically controlled. A maximum temperature must occur as soon as the enzyme protein is denatured and therefore inactivated. This usually occurs at a temperature slightly above the maximum for a plant process, probably because the denaturation temperature for a protein is lower in vivo than in vitro. The minimum temperature may be explained in the same way. Although an enzyme is not irreversibly denatured at low temperatures as it is at high temperatures, it may be *reversibly* denatured (Brandts, 1967). The cardinal temperatures for growth, therefore, must be controlled by the denaturation temperatures of a plant's enzymes.

Light

Although the growth of higher plants eventually depends on photosynthesis, nevertheless light is not essential for the growth process itself, so long as sufficient quantities of organic substances are available. Some tuberous or bulbous plants (e.g., jack-in-the-pulpit) can complete their whole life cycle in the dark at the expense of their large food reserves, but the kind of growth is different when light is absent. In the dark, higher plants show a weak, spindly growth known as *etiolation*. In the case of most dicotyledons, the stem is excessively elongated, and the leaves are underdeveloped. Little differentiation occurs, the tissue remaining mostly parenchymal. Usually the leaves remain free of chlorophyll, and the color is therefore pale yellow, although exceptions are found among gymnosperm seedlings, some ferns, and many algae. These may form chlorophyll in the dark (Vernon and Seeley, 1966, see Chapter 15). Some nongreen plants show a similar stretching in darkness. Monocotyledons may show excessive elongation of the first internode and normal or excessive leaf development in the dark.

Relatively short, daily exposures to light prevent etiolation. Consequently, light must retard this excessive growth. An extreme case is the dwarfing of alpine plants by the intense light they are subjected to. This light is richer in violet and ultraviolet radiations, which appear to have more of a stunting effect than other wavelengths.

In spite of this commonly observed retardation of growth by light, the effect in many cases is just the opposite. Although the stems of etiolated plants are excessively long, the leaves may fail to grow altogether (e.g., potato sprouts). According to Thomson (1951), the primary effect of light is to *accelerate* growth when cells are in the division stage. It is only after cell enlargement has progressed to near its normal limit that light retards any further excessive enlargement which would occur in the dark.

The relation of root growth to light is somewhat complex (Burström, 1965). Weak illumination may be indispensable for continuous growth of excised roots. It appears to be in some way related to tryptophan requirement—perhaps for synthesis of some growth regulator. Roots of intact plants, however, are comparatively insensitive to light and show only a growth inhibition. Yet if roots respond phototropically at all, they are negatively phototropic (see Chapter 18), indicating greater growth on the illuminated side. Even excised roots are inhibited by light if it exceeds a very low level. As in the case of the stunting effect (see earlier), the enhancing effect of light on growth may depend on light quality. Short wavelengths (blue light) lead to an increase in protein synthesis in fern gametophytes, making possible the formation of normal, two-dimensional prothallia (Drumm and Mohr, 1967). Long wavelengths (red light) lead to formation of filaments that are much lower in protein content, even when the rate of dry matter accumulation is the same as in the blue light. The RNA content is also higher in the blue light. These results are interpreted to indicate that the blue light initiates differential gene (i.e., DNA) activation.

Other aspects of the effects of light will be discussed in Chapter 20.

Water

Since all growth depends on a hydrostatic turgor pressure, a water deficiency will, of course, retard or completely stop it. On the other hand, an excess of water may result in an abnormal type of growth. Thus, in a saturated atmosphere the development of leaves is poor and the differentiation of the tissues is retarded. This is undoubtedly the result of excessive stretching of the cell walls because of the abnormally high turgor pressure. Plants adapted to such conditions (e.g., aquatics) have low osmotic pressures and therefore cannot develop such excessive turgor pressures even when their tissues are saturated.

Chemical stimulants and inhibitors

Even nutrient salts required by plants for normal growth may inhibit growth or actually kill the plants if applied in unbalanced solutions. On the other hand, they stimulate growth when applied in suitable quantities and in balanced solutions. Nonnutrient mineral elements frequently inhibit growth. Salts of the heavy metals (copper, lead, silver, mercury, etc.) are particularly toxic. Even the metabolic products of a plant (e.g., oxalic acid) may be poisonous when supplied to the protoplasm instead of being stored in the vacuole. In very weak doses many poisons actually stimulate growth. Thus phenol is poisonous in 1:1000 concentration but stimulates when used in 4 to 8:100,000; ethyl alcohol checks growth in 25 to 75:1000 and stimulates it in 25 to 75:100,000 (Maximov, 1938). Mercury compounds used to disinfect seeds sometimes stimulate growth. The subject of chemical stimulants and inhibitors will be returned to later when considering so-called growth regulators (see Chapter 19). Inhibitors of enzymes and of protein synthesis have already been mentioned (see Chapters 13 and 16).

DORMANCY OR REST PERIOD

Some of the preceding environmental factors may completely stop growth. This is most commonly the case in nature when the plant temperature is below the minimum for growth (e.g., in winter) or when its water content is too low (e.g., in seeds). But even when the environmental conditions are favorable for growth, the plant may still fail to grow because of some internal factors. Twigs collected in the fall, newly harvested potato tubers, or Dutch bulbs (tulips, hyacinths, daffodils, etc.) fail to grow although exposed to temperature, water, and other environmental conditions that are optimum for their later spring growth. In some cases, part of the plant may grow while another part is arrested; for example, in the early growth of grasses the leaves grow but the internodes do not. Similarly, the lateral buds of a plant may fail to grow while the terminal bud is growing; and roots of at least some species may grow appreciably while the aboveground parts are unable to grow (Meyer and Tukey, 1967).

In the past the attempt has been made to differentiate rigorously between these two kinds of suspended growth. Horticulturists have called a nongrowing plant or plant part *dormant* when growth can be readily induced by transferring it to environmental conditions optimum for growth of nondormant plants. When these conditions failed to induce growth, the plant was said to be in its *rest period*. However, it was early

recognized that the ability to resume growth is a quantitative character (Table 17-6). More recent results have shown that even for a specific degree of suspended growth, the time required for growth to recommence will depend on the particular environmental conditions to which it is exposed. The degree of rest or dormancy therefore depends on an intimate interaction between internal and external factors. Furthermore (Table 17-6), the plant or its organ passes successively through phases that differ in degree of suspended growth. It is, therefore, now common to refer to the degree of suspended growth in terms of either dormancy or rest period (Table 17-7).

It is only in phase III—the true dormancy, or main rest period—that immediate normal growth does not occur despite the external conditions. But even this phase may not be as rigid as formerly believed. It is now becoming clear that many plants previously believed to have such a theoretical extreme dormancy may actually be induced to grow within a narrow range of environmental conditions (Vegis, 1964c). Under natural conditions, of course, the plant may be in the completely nongrowing state, even though it is not in true dormancy, simply because the external conditions do not fall within this narrow range. Many potatoes, for instance, have no true dormancy because even when at their maximum rest, they can sprout at high temperatures (e.g., above 30° C); but since they are normally at temperatures well below this, they re-

Table 17-6. Growth of excised twigs of *Prunus avium* on transfer to the greenhouse[*]

Date removed from outdoors to greenhouse	Date of first flower opening	Number of days for opening of flowers
Dec. 11	Jan. 10	27
Jan. 10	Jan. 28	18
Feb. 2	Feb. 19	17
March 2	March 14	12
March 11	March 21-22	10½
March 23	March 31	8
April 3	April 8	5

[*]From Vegis, A. 1964a. *In* W. Ruhland. Encyclopedia of plant physiology. Vol. XV. Springer-Verlag, New York, Inc., New York.

Table 17-7. The phases of suspended growth in terms of dormancy and rest period[*]

Phase	Growth	Dormancy	Rest period
I	Fully normal growth	Nondormancy	Active period
II	Ability to continue normal growth not completely lost	Predormancy	Early rest
III	Immediate normal growth cannot occur	True dormancy	Main rest
IV	Able to start growth within narrow limit of external conditions	Post dormancy	After-rest

[*]Adapted from Vegis, A. 1964c. Ann. Rev. Plant Physiol. **15:**185-224.

main dormant. Conversely, wheat seed has a high temperature dormancy (George, 1967). At a germination temperature of 20° C, all twelve varieties tested were dormant at harvest, and this dormancy persisted for 20 to 60 days. At 10° C, germination was normal; at 30° C, all varieties showed a persistent dormancy from which only one recovered after 80 days and only two more after a year. Thus in dormant seed the speed and completeness of germination was inversely related to the germination temperature; in nondormant seed the relation was direct.

Dormancy therefore can be best described in terms of the cardinal points for growth. In the actively growing phase a plant possesses the widest range from the

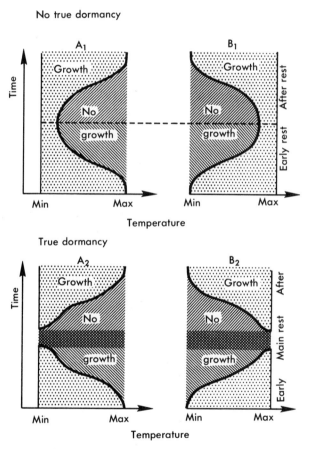

Fig. 17-6. *Top:* No true dormancy occurs, only an early rest and an after-rest. During early rest the temperature range for growth gradually narrows to a minimum; during after-rest it gradually widens until the original maximum range is attained. A_1, During early rest the maximum temperature for growth gradually decreases to a point near the minimum temperature, which remains constant. During after-rest the process is reversed. B_1, During early rest the minimum temperature for growth gradually rises to a point near the maximum temperature, which remains constant. During after-rest the process is reversed. *Bottom:* True dormancy occurs because early rest progresses until the minimum and maximum temperatures for growth are identical. After-rest again mirrors early rest. A_2 and B_2 are the projections of A_1 and B_1. (From Vegis, A. 1964c. Ann. Rev. Plant Physiol. **15**:185-224.)

minimum to the maximum temperature for growth. As it begins to enter the dormant state (i.e., in the predormant state), the range from minimum to maximum narrows (Fig. 17-6). The smallest range for growth occurs at the phase of true dormancy, or main rest. In some plants this range may be zero (Fig. 17-6, A_2 and B_2). In others it may still have a positive although small value (Fig. 17-6, A_1 and B_1).

Both entrance into dormancy and passage through it to the active state depend on the external factors. It has long been known that growth may be "started by such diverse agencies as heat, cold, increased and decreased water content, anesthetics, narcotics, gases and liquids, bases and acids, salts singly or combined, electricity, mechanical injury, and in vacuo" (Howard, 1915). Under natural conditions, dormancy is broken by "after-ripening." This may also be accomplished artificially by *stratification,* for example, by burying the plant with dormant buds (or the dormant seed) in moist soil or sand and by maintaining it at a low temperature (e.g., 0° C) for some days, weeks, or months. Conversely, the postdormant plant may be given a secondary true dormancy by exposing it to high temperature. Some plants show the opposite reaction to temperature—low temperature induces dormancy and high temperature breaks it. Paradoxically, some plants that normally pass through their rest periods at low temperatures may have their rest broken by warm-water baths (30 to 35° C for 9 to 12 hr). Moderate temperature and moisture may be the deciding factors in the case of the potato tuber. When stored under moist conditions (20° to 25° C), its dormant period is terminated in 7 days as opposed to 7 weeks in the case of dry-stored tubers (Goodwin, 1966).

In addition to temperature and water, light may also be an important factor. This is particularly true in the case of many seeds. Some (e.g., Poa, tobacco, carrot, lettuce, *Oenothera*) will not germinate under normal conditions unless exposed to light. Others (e.g., *Crataegus*) are prevented from germinating by exposure to light. In the case of many plants, all three factors interact in producing or breaking dormancy. For instance, secondary dormancy caused by high temperature may be reversed by light. The effect of photoperiod will be considered in Chapter 20.

The inability of nonripened seed to germinate is often dependent on the presence of intact seed coats, supposedly because they limit the supply of oxygen to the embryo (Vegis, 1964b). Dormancy resulting from an impermeable seed coat may be overcome by filing the coat or by immersion in high concentrations of sulfuric acid (e.g., 60% to 70% for 24 hr). Even the pressure exerted by such coverings may sometimes be important. However, embryos that are truly dormant are not usually induced to germinate by breaking their coats.

In addition to oxygen, other chemicals have long been known to affect dormancy. Anesthetics such as ether (0.5 ml per liter of air for 24 to 48 hr) are sometimes effective. Nitrates stimulate the germination of light-requiring seeds. Thiourea is often effective both in the case of seeds and buds (e.g., of potato tubers). It may break secondary dormancy induced by high temperatures, or it may substitute for low temperature after-ripening. It does not seem to be effective against true dormancy. Thiocyanates and ethylene chlorohydrin are effective in breaking the rest period of potato tubers. The Boyce Thompson Institute (Denny, 1945), has developed a particularly effective mixture called Rindite (7 parts ethylene chlorohydrin:3 parts ethylene dichloride:1 part carbon tetrachloride) for breaking dormancy of potato tubers (0.8 ml/kg for 48 hr). Over a hundred substances have been reported to have some effect on the breaking of dormancy (Vegis, 1964b). Some (e.g., maleic hydrazide) are also able to induce dormancy. The effects of growth regulators will be considered in Chapter 19.

It should be emphasized that only the specific part of the plant treated by any of the aforementioned methods has its dormancy broken. The effect is not transmitted to the rest of the plant. Thus if a single branch is exposed to low temperature, immersed in warm water, exposed to ether, etc., only this branch will have its dormancy broken.

According to Vegis (1964*b*), the onset and termination of dormancy is to some extent connected with the synthesis and decomposition of fats and lipids. The most deeply dormant seeds, for instance, commonly contain large amounts of lipids. It has been suggested that this is connected with a restricted oxygen supply, which may favor the formation of lipids. On the other hand, if the restriction of the oxygen supply is too extreme, anaerobic respiration may lead to the accumulation of ethyl alcohol and acetaldehyde, which may break the dormancy. Some of the older methods of breaking the rest period may be explained in this way; for example, immersion in warm water may induce a sufficient oxygen deficiency to stimulate an active anaerobic respiration, leading to the accumulation of substances that break dormancy.

As has already been indicated (see Chapter 4), dormancy is often associated with minimum surface and, therefore, with minimum surface energy. It is obviously a mechanism by means of which the plant can conserve its stored energy until conditions are suitable for growth. But how does the dormant plant succeed in preventing its growth, even when external conditions are favorable, and how does it later contrive to overcome its dormancy, unlocking its energy reserves and making them available for growth? All the foregoing relations indicate that there are two possible mechanisms of dormancy (Wareing, 1965). Growth can be prevented (1) by the enclosing structure (seed coats, bud scales, etc.) restricting oxygen uptake and, therefore, metabolic activity; or (2) by the production of inhibitors. This will be discussed after considering growth regulators (see Chapter 19). Recent evidence indicates that enzyme activity may be a factor. The activity of nitrate reductase, extracted from embryos of *Agrostemma* seeds, is closely connected with their ability to germinate (Borriss and Schulze, 1966). Dormant embryos showed little or no activity. Embryos from after-ripened (i.e, nondormant) seeds possessed substantial activity even without induction and developed strong activity when induced by potassium nitrate. The change in enzyme activity during seed activation must be controlled by the nucleic acid metabolism. Direct evidence of this has been produced in several cases. Messenger RNA is formed during the imbibition of seeds, prior to germination (Marcus and Feeley, 1964). Presumably, the seeds cannot germinate until this RNA leads to synthesis of the proteins required by the new cells. DNA also increases, both prior to and during cell division (Holdgate and Goodwin, 1965). Inhibitors of the synthesis of nucleic acid (e.g., uracil derivatives) can actually prevent the release of lettuce seeds from dormancy (Smith and Frankland, 1966). The reverse changes have been found to occur during seed development on the plant. From the time of maximum fresh weight of the seed to the time it is completely air dry, the RNA of the nuclear fraction of castor bean endosperm decreases to about 50% (Cocucci and Sturani, 1965). Similarly, the mitochondrial RNA almost disappears, but the RNA of the soluble fraction increases. The ribosomes completely disappear. Barker and Rieber (1967) have also found that the concentration of functional polysomes is low in dry pea seeds and that this limits their protein synthesizing system. All these changes are interpreted as leading to the progressive metabolic and enzymatic inactivation characteristic of the ripening process. The decrease in RNA seems to be caused by both inactivation of synthesis and acceleration of breakdown.

The control of all these processes, and therefore of dormancy or rest period, is presumably possible because of the synthesis and breakdown by the plant of specific inhibitors (e.g., dormin, see Chapter 19).

QUESTIONS

1. How is growth measured?
2. Are fresh or dry weights satisfactory measurements?
3. Where does growth occur in the plant?
4. What cellular changes occur in these regions?
5. Which cellular changes really involve growth?
6. Define growth on a cellular basis.
7. What is the motivating force for growth?
8. How does turgor pressure compare in growing as opposed to nongrowing cells?
9. What cell property permits growth of cells at the same turgor as fails to produce it in mature cells?
10. What processes accompany cell enlargement?
11. Why is energy required for cell enlargement?
12. Is it conceivable that cell enlargement may be initiated by something other than turgor pressure?
13. What is the initiating process in this case?
14. Do all cells grow uniformly?
15. Is there any relation between the protoplasm and the growing cell wall?
16. What is the shape of the growth curve?
17. What is the grand period of growth?
18. What concept of growth is used to develope an equation for growth?
19. What is meant by the cardinal points for growth?
20. Are these cardinal points exact?
21. What effect does a 10° C temperature rise have on growth?

22. Is growth possible in the absence of light?
23. What is etiolation?
24. Does light enhance or retard growth?
25. How does a water deficit affect growth?
26. How does excess water affect growth?
27. What is meant by the rest period?
28. What is meant by dormancy?
29. List the phases of the rest period and of dormancy.
30. How may dormancy be overcome artificially?
31. What changes occur in cells entering dormancy?
32. What changes occur in cells emerging from dormancy?
33. What temperature relation changes during the entrance into and emergence from dormancy?

SPECIFIC REFERENCES

Barker, G. R., and M. Rieber. 1967. The development of polysomes in the seed of *Pisum arvense*. Biochem. J. **105**:1195-1202.

Beck, W. A. 1941. Production of solutes in growing epidermal cells. Plant Physiol. **16**:637-642.

Beck, W. A., and B. Andrus. 1943. The osmotic quantities of the cells in the hypocotyl of *Helianthus annuus* seedlings. Bull. Torrey Bot. Club **70**:563-598.

Blackman, V. H. 1919. The compound interest law and plant growth. Ann. Bot. **33**:353-360.

Blank, F., and A. Frey-Wyssling. 1944. Protoplasmic growth in the hypanthium of *Oenothera acaulis* during cell elongation. Ann. Bot. **8**:71-78.

Borriss, H., and J. Schulze. 1966. Die Nitratreductase-Aktivität der Embryonen ruhender und nachgereifter *Agrostemma*-Samen. Z. Pflanzenphysiol. **55**:449-457.

Brandts, J. F. 1967. Heat effects on proteins and enzymes, p. 25-72. *In* A. H. Rose. Thermobiology. Academic Press, Inc., New York.

Burström, H. 1953. Studies on the growth and metabolism of roots. IX. Cell elongation and water absorption. Physiol. Plant. **6**:262-276.

Burström, H. 1965. Light in the regulation of root growth, p. 45-60. *In* P. R. White, and A. R. Grove. Proceedings of the International Conference of Plant Tissue Culture. McCutchan Publishing Corp., Berkeley.

Cleland, R. 1967. Inhibition of formation of protein-bound hydroxyproline by free hydroxyproline in *Avena* coleoptiles. Plant Physiol. **42**:1165-1170.

Cocucci, S., and E. P. Sturani. 1965. Nucleic acids in ripening castor bean seed endosperm. II. Changes of the RNA from different fractions. [Transl. title.] G. Bot. Ital. **72**:357-358.

Davidson, J. L., and F. L. Milthorpe. 1965. The effect of temperature on the growth of cocksfoot (*Dactylis glomerata* L.). Ann. Bot. **29**:407-417.

Davies, J. W., and E. C. Cocking. 1965. Changes in carbohydrates, proteins, and nucleic acids during cellular development in tomato fruit locule tissue. Planta **67**:242-253.

Denny, F. E. 1945. Synergistic effects of three chemicals in the treatment of dormant potato tubers to hasten germination. Contrib. Boyce Thompson Inst. Plant Res. **14**:1-14.

Drumm, H., and H. Mohr. 1967. Die Regulation der RNS-Synthese in Farngametophyten durch Licht. Planta **72**:232-246.

Frey-Wyssling, A. 1952. Growth of plant cell walls. Sympos. Soc. Exp. Biol. **6**:320-328.

George, D. W. 1967. High temperature seed dormancy in wheat (*Triticum aestivum* L.). Crop Sci. **7**:249-253.

Goodwin, P. B. 1966. The effect of water on dormancy in the potato. Eur. Potato J. **9**:53-63.

Gregory, F. G. 1921. The increase of leaves and leaf surface of *Cucumis sativus*. Ann Bot. **35**:93-123.

Holdgate, D. P., and T. W. Goodwin. 1965. Metabolism of nucleic acids during early stages of the germination process in rye (*Secale cereale*). Phytochemistry **4**:845-850.

Howard, W. L. 1915. An experimental study of the rest period of plants. Res. Bull. Mo. Agric. Exp. Sta. Pub. No. 21. Columbia, Mo.

Juniper, B. E., and F. A. L. Clowes. 1965. Cytoplasmic organelles and cell growth in root caps. Nature (London) **208**:864-865.

Lyndon, R. F. 1963. Changes in the nucleus during cellular development in the pea seedling. J. Exp. Bot. **14**:418-430.

Marcus, A., and J. Feeley. 1964. Activation of protein synthesis in the imbibition phase of seed germination. Proc. Nat. Acad. Sci. U. S. A. **51**:1075-1079.

Meyer, M. M., Jr., and H. B. Tukey, Jr. 1967. Influence of root temperature and nutrient applications on root growth and mineral nutrient content of *Taxus* and *Forsythia* plants during the dormant season. Proc. Amer. Soc. Hort. Sci. **90**:440-446.

Smith, H., and H. Frankland. 1966. Specific inhibition by uracil derivatives of the mechanism of dormancy release in light-sensitive lettuce seeds. Nature (London) **211**:1323-1324.

Thomson, B. F. 1951. The effect of light on the rate of development of *Avena* seedlings. Amer. J. Bot. **37**:284-291, 635-638.

Van Parijs, R., and L. Vandendriessche. 1966a. Changes of the DNA content of nuclei during the process of cell elongation in plants (*Pisum sativum*). I. The formation of polytene chromosomes. Arch. Int. Physiol. Biochim. **74**:579-586.

Van Parijs, R., and L. Vandendriessche. 1966b. Changes of the DNA content of nuclei during the process of cell elongation in plants (*Pisum sativum*). II. Variations, other than doublings, of the amount of Feulgen stain in maturing plant cells. Arch. Int. Physiol. Biochim. **74**:587-591.

Wareing, P. E. 1965. Dormancy in plants. Sci. Progr. **53**:529-537.

Williams, R. F., and A. H. G. C. Rijven. 1965. The physiology of growth in the wheat plant. II. The dynamics of leaf growth. Aust. J. Biol. Sci. **18**:721-743.

Wilson, K. 1964. The growth of plant cell walls. Int. Rev. Cytol. **17**:1-49.

Winter, H. 1966. Effect of auxin and sugar on cell wall synthesis and growth of pea stem segments. Proc. Kon. Nederl. Akad. Wet. [Biol. Med.] **69**:64-72.

GENERAL REFERENCES

Burström, H. 1951. Mechanism of cell elongation. *In* F. Skoog (Ed.). Plant growth substances. University of Wisconsin Press, Madison, Wis.

Crocker, W. 1948. Growth of plants. Reinhold Publishing Corp., New York.

Leopold, A. C. 1964. Plant growth and development. McGraw-Hill Book Co., New York.

Maximov, N. A. 1938. Plant physiology. McGraw-Hill Book Co., New York.

Palladin, V. I. 1923. Plant physiology. B. E. Livingston (Ed.). Blakiston Division, McGraw-Hill Book Co., New York.

Robertson, T. B. 1923. The chemical basis of growth and senescence. J. B. Lippincott Co., Philadelphia.

Thompson, D. W. 1942. On growth and form. Cambridge University Press, New York.

Vegis, A. 1964a. Ruhezustände bei höheren Pflanzen, Induktion, Verlauf und Beendigung; Übersicht, Terminologie, allgemeine Probleme, p. 499-533. *In* W. Ruhland. Encyclopedia of plant physiology. Vol. XV. Springer-Verlag, New York, Inc., New York.

Vegis, A. 1964b. Die Bedeutung von physikalichen und chemischen Aussenfaktoren bei der Induktion und Beendigung von Ruhezustanden bei Organen und Geweben höherer Pflanzen, p. 534-668. *In* W. Ruhland. Encyclopedia of plant physiology. Vol. XV. Springer-Verlag, New York, Inc., New York.

Vegis, A. 1964c. Dormancy in higher plants. Ann. Rev. Plant Physiol. **15**:185-224.

chapter **18**

Irritability and movement

IRRITABILITY

Like the animal the plant shows the property of *irritability*—the ability to respond physically to an environmental factor. For instance, its leaves or flowers may close in response to darkness and open in the light; its stem may bend toward the light, or grow upward against the force of gravity; its leaves may close in response to touch by an insect, trapping the insect. Even the organelles within the cell may show irritability responses, for example, plastid movements in response to light and streaming movement, which may arise or speed up on exposure to light or because of a rise in temperature. The specific environmental factor that gives rise to the irritability reaction is called the *stimulus,* and the reaction of the plant is called the *response.* The earlier physiologists were fascinated with the phenomenon, but more modern investigators seem to have lost interest in it. It is, in fact, rare nowadays to find reference to the term *plant irritability.* Recently, however, interest in outer space has given rise to attempts to grow

plants in the complete absence of polarization of environmental factors, that is, with all the environmental factors either reduced to a minimum or of equal strength on all sides of the plant (Gordon et al., 1964). This tends to remove the stimuli and, therefore, presumably also the response. In the complete absence of polarization in the environment, the plant has been found to grow as an undifferentiated mass of tissue instead of differentiating into organs. Thus, the phenomenon of *polarity* in plant growth (e.g., the formation of a stem tip at one pole, a root tip at the other) is an example of irritability. In fact, the whole process of organ differentiation may be considered to be the result of irritability. This is, perhaps, one reason why the concept has been dropped by many physiologists because, in its broadest sense, it may include nearly the whole field of growth and development.

MOVEMENTS
Cell movements

When used in its narrower sense, the main irritability reactions of the plant that have been intensively studied are plant movements. Although the plant is popularly thought of as differing from animals by its lack of movement, it can be seen (Fig. 18-1) that there are six distinct kinds of

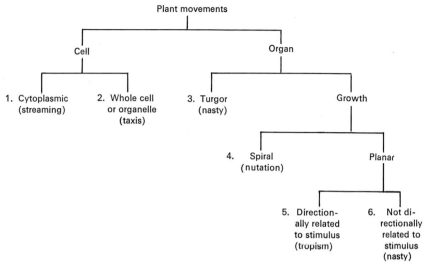

Fig. 18-1. Classification of plant movements.

plant movement. But these are all either cell or organ movements and, unlike the animal, the higher plant does not move as a whole but remains anchored in place. Cytoplasmic streaming can also be found in animal cells, but it is usually much more pronounced in plant cells (Ambrose, 1965). Rates as high as $107\mu/sec$ have been measured (see Chapter 10). It occurs in the inner cytoplasm, or *endoplasm,* which is in the sol, or liquid, state. Between the endoplasm and the outer plasma membrane is the *ectoplasm,* which is stationary because of its gel, or solid, state. Although the organelles may be carried along in the streaming cytoplasm, they may move at independent speeds and even in opposite directions within the same stream. Chloroplasts even rotate about their own axes. The direction of the stream may be reversed, and in the case of myxomycete plasmodia, this reversal occurs regularly at about 60 sec intervals (Kamiya, 1942).

The driving force has been sought for years by many investigators. Hydrostatic forces cannot be involved since the streaming can continue in a *Nitella* cell that has been cut open and is, therefore, at atmo-

spheric pressure. In this case the cytoplasm may flow right out of one side of the cut surface while flowing up the opposite side (Ambrose, 1965). The evidence indicates that the driving force originates at the contact between the streaming endoplasm and the stationary ectoplasm. The key to the movement may be found in the microtubules, which frequently can be seen under the electron microscope near the plasmalemma (see Fig. 2-6). Kamiya found that the movement is dependent on adenosine triphosphate (ATP). He therefore suggests that there may be contractile proteins localized near the cell membrane which are responsible for the movement. An electrical mechanism has also been proposed.

Although taxis (motility) is more common among lower orgnaisms (blue-green algae, bacteria), the chloroplasts of higher plants are also capable of this kind of movement (Haupt, 1966). In *Vallisneria spiralis* the negative phototaxis of the chloroplasts may be accompanied by a rotation (Seitz, 1967). Other organelles also show mass movements within the cytoplasmic stream but at velocities differing from that of the protoplasmic mass (Mahlberg, 1965). This may also be considered a kind of taxis.

Organ movements

Organ movements are of two main kinds: (1) those caused by turgor changes and (2) those caused by asymmetrical growth. Loss of turgor is frequently a rapid process, leading to rapid movements, for example, the closing of the leaves of the sensitive plant *(Mimosa pudica)* when touched or the closing of the traps of insectivorous plants (Lloyd, 1942). Such rapid turgor changes are believed to be the result of a rapid loss in semipermeability. Slower turgor changes caused by changes in solute content lead to the slower turgor movements of stomata. The slow sleep movements of plant leaves, however, are believed to be caused by the aforementioned permeability changes. The slowest movements are those resulting from differential (i.e., asymmetrical) rates of growth. Investigations of these so-called growth movements have led to some of the most important information as to the nature of the growth process.

There are three main types of growth movements: (1) *circumnutation*, (2) *tropisms*, and (3) *nastic movements*. Unfortunately, the latter two may be caused by turgor changes as well as by growth, but those caused by growth movements are far more common and have been more intensively studied. All or nearly all plants show circumnutation. This is a spiral type of growth of the plant apex caused by a different rate of growth on opposite sides of the growing tip. The more rapid rate of growth travels around the tip, which, as it grows upward (in the case of the stem), must therefore rotate. The cause of this cyclic change in rates of growth of cells on different sides of the tip is not known. In some plants, circumnutation may come to a stop when the gravitational force has been compensated for on a clinostat (see later) and the plant is kept in the dark; but in other cases the circumnutation has continued under these conditions (Johnsson, 1966). The latter result has led to the conclusion that the movement is autonomous (independent of the environment). If so, circumnutation could not truly be classified as an irritability since this is a response to the environment. This problem has been circumvented in the past by dividing all plant movements into *autonomic* (controlled by internal stimuli) and *paratonic* (controlled by external stimuli). However, there is no basis for such a classification until an internal stimulus is identified and proved to induce a plant movement. If the relation to gravity is confirmed, it could conceivably be an overshoot in response to gravity. In support of this explanation the ratio between the periodic time for circumnutation and the reaction time for geotropic curvature (see later) has been found to remain constant with temperature changes from 15° to 40° C (Israelsson and Johnsson, 1967).

The other two growth movements, however, are definitely a response to an environmental stimulus. The length of time the plant must be exposed to the stimulus in order for a reaction to occur is the *presentation time*. The length of time required for it to react to the stimulus is the *reaction time*. In other words, the plant may be exposed to the stimulus for the presentation time without showing any response. Later, however, in the absence of the stimulus the reaction may appear. The time it takes for the plant to recover its original position after the stimulus has been removed is the *relaxation time*. In the case of tropisms the direction of movement shows a definite relation toward, away from, or at a definite angle to the direction of application of the stimulus. It is actually the *plane* of the response that is controlled by the spatial relation between the stimulus and the tropically responding organ (Wilkins, 1966). A movement toward a stimulus is called a positive movement and away from the stim-

Table 18-1. Different kinds of tropisms and nasties and their respective stimuli

Stimulus	Movement directionally related to stimulus (tropisms)	Movement not directionally related to stimulus (nasties)
Gravity	Geotropism	
Light (sun)	Phototropism (heliotropism)	Photonasty
Temperature	Thermotropism	Thermonasty
Night		Nyctinasty
Touch	Thigmotropism (haptotropism)	
Chemicals	Chemotropism	
Water	Hydrotropism	

ulus, a negative movement. Nastic movements, on the other hand, show no definite directional relation to the stimulus. Since movements can be caused by different stimuli, the name of the movement consists of a prefix that indicates the stimulus and a suffix that states the kind of movement.

There is some evidence of other growth movements, such as electrotropism and magnetotropism (Pittman, 1963). A greater root concentration on the north and south sides of wheat seedlings has been explained as a geomagnetotropic response (Woolley and Pittman, 1966). Electrotropism has been demonstrated in a number of plants (Murr, 1965). The electrostatic or electrokinetic field can either stimulate or retard their growth, although in all cases leaf damage occurs.

Although chemotropism is usually an artificial phenomenon, it is also found in nature. The pistils of many higher plants contain substances that induce chemotropism of pollen tubes. One substance has been found to induce positive chemotropism and another, negative chemotropism (Miki-Hirosige, 1964). But the only two tropisms that have been studied in considerable detail are geotropism and phototropism.

GEOTROPISM

Primary roots are *positively* geotropic (i.e., bend toward the center of the earth), and primary stems are *negatively* geotropic. Secondary lateral roots and shoots show a weaker response and take up a position at an angle to the gravitational force. They are said to be *plagiogeotropic* as opposed to the primaries, which are *orthogeotropic* (parallel to the gravitational force). Lateral roots and shoots of a higher order are almost insensitive to the stimulus of gravity (*apogeotropic*). Rhizomes grow horizontally; that is, they take up a position at right angles to the gravitational stimulus and are called *diageotropic*. The response of an organ to the gravitational stimulus may change with the stage of development. The peduncle of the poppy bud is positively geotropic, but it gradually becomes negative as the flower opens. A change may also be induced artificially; if the primary root or stem tip is cut off, the lateral nearest to the wound becomes orthogeotropic.

In dicotyledons the first sign of a geotropic response occurs a short distance back from the apex. In grasses and other monocotyledons it occurs near the nodes where the meristems are found. If roots or shoots are *decapitated* (the apex is removed), they fail to show a geotropic response. If, however, they are exposed to the stimulus and then decapitated before curvature can occur (after the presentation time but before the reaction time), the stumps will nevertheless show a geotropic response. Thus the tip is necessary for the *perception* of the stimulus (and is called the region of perception) but not for the *response*. The region of response is some distance back from the tip. This control of the responses of one part of a plant by another part is known as *correlation*.

The exact way in which the gravitational stimulus is transmitted to the protoplasm is not known. According to the *statolith theory*, starch grains fall under the gravitational force, exerting a pressure on the protoplasm, that leads in some way to the geotropic response. In the case of the rhizoids of *Chara foetida,* a recent explanation (Sievers, 1967*b*) is that the starch grains block the normal distribution of the Golgi vesicles on the lower side, preventing the growth on this side and stimulating growth on the upper side. On the other hand, in the case of the normal, vertical rhizoid (Sievers, 1967*a*) the statoliths block the central portion of the tip and pack the Golgi vesicles and microvesicles in the region adjacent to the apical wall (Fig. 18-2). This

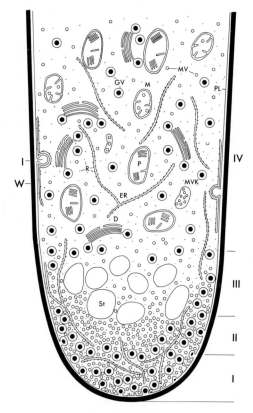

Fig. 18-2. Schematic representation of a longitudinal section through zones I to IV of a normal growing rhizoid of *Chara foetida.* **D** = dictyosome; **ER** = endoplasmic reticulum; **GV** = Golgi vesicle; **I** = invagination of plasmalemma with parallel ER cisterna; **M** = mitochondrion; **MV** = microvesicle; **MVK** = multivesicular body; **P** = plastid; **PL** = plasmalemma; **R** = ribosome; **St** = statolith; **W** = cell wall. (From Sievers, A. 1967. Protoplasma **64**:225-253.)

leads to enhanced growth at the tip. The Golgi vesicles appear to resemble the inner part of the cell wall. The plasmalemma is very irregular, with invaginations that may be caused by the incorporation of Golgi vesicles or microvesicles. Sievers concludes that the Golgi vesicles and microvesicles each supply the apical wall with growth material. During the process, their membranes fuse with the plasmalemma.

According to the geoelectric effect theory, the gravitational force is converted into an electric charge that leads to the response. The geoelectric effect definitely does occur, but the evidence now indicates that it is the *result* of a redistribution of substances (auxins, see Chapter 19) and not the *cause* of the tropism (Wilkins, 1966). Most of the evidence seems to favor the statolith theory (Audus, 1962), although there is some recent evidence of response in the absence of starch grains. The dependence of geotropism on metabolism has been demonstrated by its absence under anaerobic conditions (Wilkins and Shaw, 1967). Since the plant is constantly subjected to the force of gravity, its response to this force must be of fundamental importance in its normal growth and development. Until now, terrestrial plants have been subjected to a constant gravitational force. With the interest in outer space research, it is now becoming important to know the effects on the plant of varying the gravitational force, and experiments have been initiated to determine this. Thus, it has long been known that the polarity of the gravitational force can be eliminated by rotating the plant axis while maintaining it in a horizontal position. The instrument that rotates the plant is called a *clinostat.* Recent improvements in construction of the clinostat have permitted exposure of the plant to gravitational forces all the way down to zero (Dedolph et al., 1966). No threshold in gravity perception by the plant was detected. In fact, the geotropic curvature response of oat seedlings increased with the gravitational force,

PHOTOTROPISM

As a rule, stems are positively phototropic and leaves are plagiophototropic. The majority of roots are insensitive. The peduncle of *Lynaria cymbalaria* is positively phototropic when the flower opens and negatively phototropic after fertilization. As a result, the capsule is pressed into cracks between rocks, where the seed may later germinate.

Etiolated coleoptiles of grasses (e.g., oats) have been most commonly used for experimental work on phototropism. As in the case of geotropism, it is easily shown that the region of perception and the region of response to the light stimulus are not the same. The sensitive zone is the first 1 to 5 mm from the tip.

Phototropism is in accord with what has been said earlier about the retarding effect of light on cell elongation. The darkened side grows more rapidly, and as a result, curvature is toward the light. This, however, is a superficial and not uniformly correct explanation.

As in the case of photosynthesis, the action spectrum for phototropism is known. Maximum sensitivity is in the blue, and essentially no response occurs at the long wave end of the spectrum (Fig. 18-3). This action spectrum resembles the absorption spectra of both carotenes and flavins, and there has therefore been considerable controversy as to which is responsible for the absorption of the light that stimulates the phototropic response. Unfortunately, the absorption spectrum of neither pigment quite matches the action spectrum of phototropism (Thimann, 1964), and analysis of the pigments of *Avena* coleoptiles demonstrates the presence of both flavins and carotenoids.

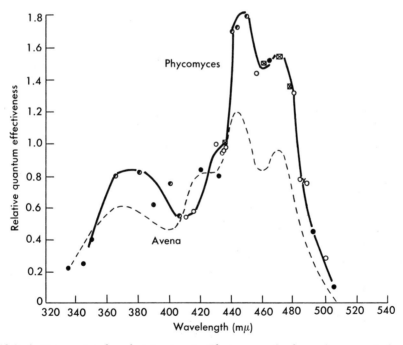

Fig. 18-3. Action spectra for phototropism in *Phycomyces* (a fungus) sporangiophores and *Avena* coleoptiles at low light intensities. (From Curry, G. M., and K. V. Thimann. 1961. *In* B. C. Christensen and B. Buchmann. Progress in photobiology. Elsevier Publishing Co., Amsterdam.)

Since light of the wavelength 450 mμ was absorbed to only 13% by the carotenoids, 87% by the flavones, Zenk (1967) concluded that carotenoids can play only a minor role in creating a transverse light gradient in the coleoptile.

NASTIC MOVEMENTS

The commonest of nastic movements are *nyctinastic* movements, the day and night movements of leaves and flowers. Flower perianths and many compound leaves may open during the day and close at night. If these are growth movements, they are, of course, possible only while the organ is capable of growth. The amplitude of movement therefore decreases with age, but some retain the ability to open and close even when mature. This is because some nastic movements depend on turgor changes rather than growth differences. Families typically showing nyctinastic leaf movements are *Leguminosae, Oxalidaceae, Euphorbiaceae, Marantaceae.*

Nyctinastic movements may be controlled by temperature, light, etc. The perianths of tulip, crocus, and other similar flowers open at a constant temperature when illuminated and close when darkened. If, on the other hand, the light is kept constant, they open at high temperatures and close at low temperatures. They are therefore both *photonastic* and *thermonastic*. In each case, the opening is the result of a more rapid growth of the upper surface *(epinasty)*. Such nastic movements may also be caused by the presence of certain substances in small quantity. One of the most sensitive tests for ethylene is the epinasty of tomato petioles, which occurs if the ethylene is present in 1:500,000 parts of air.

QUESTIONS

1. What is irritability?
2. What is a stimulus?
3. What is a response?
4. Where exactly does cytoplasmic streaming occur?
5. What is a taxis?
6. Does taxis occur in higher plants?
7. Name the three main classes of growth movements.
8. In each case what is the immediate, direct cause of movement?
9. In what way do tropic and nastic movements differ from circumnutation?
10. What is the difference between tropic and nastic movements?
11. What is the presentation time?
12. What is the reaction time?
13. What is the region of perception?
14. What is the region of response?
15. What is meant by correlation?
16. Name some tropisms.
17. Name some nastic movements.
18. What is epinasty?
19. In what direction does a plant move geotropically?
20. Where is the region of perception for geotropism?
21. Where is the region of response for geotropism?
22. What is a statolith?
23. How may the Golgi apparatus be related to geotropism?
24. What is a clinostat?
25. What is the action spectrum for phototropism?

SPECIFIC REFERENCES

Ambrose, E. J. 1965. Cell movements. Endeavour **24**:27-32.

Audus, L. J. 1962. The mechanism of the perception of gravity by plants. Sympos. Soc. Exp. Biol. **16**:197-226.

Audus, L. J., and J. C. Whish. 1964. Magnetotropism, p. 170-182. *In* M. F. Barnothy. Biological effects of magnetic fields. Plenum Publishing Corp., New York.

Curry, G. M., and K. V. Thimann. 1961. Phototropism; the nature of the photoreceptor in higher and lower plants, p. 127-134. *In* B. C. Christensen and B. Buchmann. Progress in photobiology. Proceedings International Congress on Photobiology, 3rd. Elsevier Publishing Co., Amsterdam.

Davies, J. W., and E. C. Cocking. 1965. Changes in carbohydrates, proteins, and nucleic acids during cellular development in tomato fruit locule tissue. Planta **67**:242-253.

Dedolph, R. R., S. A. Gordon, and D. A. Oemick.

1966. Geotropism in simulated low-gravity environments. Amer. J. Bot. **53**:530-533.

Gordon, S. A., et al. 1964. Growth and development in compensated fields: Introduction and design. USAEC Argonne Nat. Lab. Rep. **6971**: 166-71.

Graham, L., and C. H. Hertz. 1964. Measurements of the geoelectric effect in coleoptiles. Physiol. Plant. **17**:186-201.

Haupt, W. 1966. Phototaxis in plants. Int. Rev. Cytol. **19**:267-299.

Israelsson, D., and A. Johnsson. 1967. A theory for circumnutations in *Helianthus annuus*. Physiol. Plant. **20**:957-976.

Johnsson, A. 1966. Spontaneous movements in plants studied as a random walk process. Physiol. Plant. **19**:1125-1137.

Kamiya, N. 1942. Physical aspects of protoplasmic streaming, p. 199-244. *In* W. Seifriz. The structure of protoplasm. Monogr. Amer. Soc. Plant Physiol. Iowa State Univeresity Press, Ames, Iowa.

Mahlberg, P. G. 1965. Phase cinemicrographic observations on cultured cells. II. Mass movement of cytoplasm in *Euphorbia marginata*. Amer. J. Bot. **52**:438-443.

Miki-Hirosige, H. 1964. Tropism of pollen tubes to the pistils, p. 152-158. *In* H. F. Linskens. Pollen physiology and fertilization. A symposium. North-Holland Publishing Co., Amsterdam.

Murr, L. E. 1965. The biophysics of plant electrotropism. Trans. N. Y. Acad. Sci. **27**:759-771.

Pickard, B. G., and K. V. Thimann. 1966. Geotropic response of wheat coleoptiles in absence of amyloplast starch. J. Gen. Physiol. **49**:1065-1086.

Pittman, U. J. 1963. Magnetism and plant growth. Canad. J. Plant Sci. **43**:513-518.

Seitz, K. 1967. Eine Analyse der für die lichtabhängigen Bewegungen der Chloroplasten verantwortlichen Photorezeptor-systeme bei *Vallisneria spiralis* ssp. torta. Z. Pflanzenphysiol. **57**: 96-104.

Sievers, A. 1967a. Elektronenmikroskopische Untersuchungen zur geotropischen Reaktion. II. Protoplasma **64**:225-253.

Sievers, A. 1967b. Elektronenmikroskopische Untersuchungen zur geotropischen Reaktion. III. Z. Pflanzenphysiol. **57**:462-473.

Sievers, A. 1967c. Zum Wirkungsmechanismus der Statolithen in der pflanzlicher Zelle. Naturwissenschaften **54**:252-253.

Thimann, K. V. 1964. Phototropism. Photochem. Photobiol. **3**:463-469.

Wilkins, M. B. 1966. Geotropism. Ann. Rev. Plant. Physiol. **17**:379-408.

Wilkins, M. B., and S. Shaw. 1967. Geotropic response of coleoptiles under anaerobic conditions. Plant Physiol. **42**:1111-1113.

Woolley, D. G., and U. J. Pittman. 1966. P^{32} detection of geomagnetotropism in winter wheat roots. Agron. J. **58**:561-562.

Zenk, M. H. 1967. Untersuchungen zum Phototropismus der Avena-Koleoptile. II. Pigmente. Z. Pflanzenphysiol. **56**:122-140.

GENERAL REFERENCES

Lloyd, F. E. 1942. The carnivorous plants. Chronica Botanica Co., Waltham, Mass.

Weevers, T. 1949. Fifty years of plant physiology. Scheltema & Holkema, Amsterdam.

chapter 19

Growth regulators

TERMINOLOGY

In order for a plant to grow normally, many substances must be available.

1. The raw materials: all the essential mineral elements, water, carbon dioxide, and oxygen
2. The building blocks for cell production and enlargement: the carbohydrates, proteins, nucleic acids, lipids
3. The sources of energy: both chemical (carbohydrates and high energy phosphates) and pressure (water for turgor pressure)
4. The catalysts to maintain adequate rates of metabolism (enzymes plus their cofactors)

Since the green plants are fully autotrophic, the only substances they must be supplied with are, therefore, the raw materials—the mineral elements, water, carbon dioxide, and oxygen. From these they can synthesize the other three groups of substances. In the presence of two other essential environmental factors—adequate light and suitable temperature—these raw materials are all that need be supplied to the plant. Under certain conditions, of course, the plant may

not be fully autotrophic, for example, albino mutants, immature embryos, etc. In such cases the plant is heterotrophic and must be supplied with some of the other three groups of substances in order to grow normally.

If any of the aforementioned substances is present in suboptimal quantities, growth will be decreased. In the broadest sense of the term, all of these substances may therefore be called growth substances since they can alter the growth rate of the plant. Yet, even if all the raw materials are available to the autotrophic green plant in optimum quantities, the plant may fail to grow in spite of favorable light and temperature. This will occur when the plant is in its rest period. However, even in the case of a rapidly growing plant, some parts grow continuously (e.g., promeristems), and others grow until a certain point is reached and then stop; yet both the raw materials and the environmental conditions are adequate for the whole plant.

It is obvious, then, that the plant must possess some mechanism for regulating its growth. It does this by means of an additional group of substances called *growth regulators*. These are substances that the plant synthesizes in small quantities, which do not serve as raw materials, building blocks, sources of energy, or enzyme com-

Table 19-1. Comparison of activities of different kinds of growth regulators*

	Natural quantities		Artificial quantities	
Regulator	Quantity found in plant	Plant	Applied concentration	Effect
Auxin	6 μg/kg (or ppb)	Pineapple	10^{-2} mg/l (10 ppb)	Increase in length of oat coleoptile
Gibberellin	0.001 g/100 buds	Sunflower	2×10^{-11} mole/l (1 ppt)	Initiation of starch digestion
Kinin	50-100 ppb in bleeding sap	Grapevine	0.04 mg/l (40 ppb)	Increase in fresh weight of tobacco callus

*Adapted from van Overbeek, J. 1966. Science **152:**721-731.

ponents; but which act as keys capable of unlocking or locking up the growth process.

Growth regulators are usually *hormones*—substances synthesized in one part of the plant and translocated to another part where they exert their effects. They are therefore frequently called *growth hormones* to distinguish them from other hormones that do not specifically control growth (e.g., flowering hormones, see Chapter 20). However, it is conceivable for a growth regulator to act where it is synthesized, in which case it would not be a hormone. Furthermore, there are many synthetic growth regulators, and since these are not synthesized in the plant, they cannot be hormones if the preceding definition is accepted. Conversely, substances may be growth hormones without belonging to the group of growth regulators. Thiamine and niacin and probably other vitamins may act as growth hormones (Thimann, 1965), yet the mechanism of their activity (as enzyme cofactors) is fundamentally different from the mechanism of action of growth regulators (see later), and they therefore do not belong in this group. The term *phytohormones* includes all plant hormones—the growth regulators, vitamins, flowering hormones, etc.

The four main groups of naturally occurring growth regulators are the auxins, cytokinins, gibberellins, and a miscellaneous group including dormins (or abscisins) and other inhibitors. The high activities of the regulators are illustrated by the low concentrations required to produce an effect on the plant (Table 19-1).

AUXINS

Auxins were the first substances recognized to be normal growth regulators. They were discovered as a result of investigations of growth movements, which they were soon found to control. However, beside controlling the asymmetrical growth that results in growth movements, they were later found to control the normal, symmetrical ("straight") growth of the plant. This led to the principle that "without auxin, there is no growth" (Went and Thimann, 1937; Avery et al., 1947).

The evidence for the role of auxins is mainly based on studies of tropisms—particularly phototropism. As indicated in Chapter 18, both geotropic and phototropic reactions are characterized by one region of perception and another of response. This phenomenon is known as correlation. It has been shown conclusively by various methods that correlation results from the transfer of a chemical substance from the region of perception to the region of response (Fig. 19-1). When the coleoptile tips of oat seedlings are exposed to light, a substance is released in the tip and is translocated downward to the region of response. One method of showing this is to cut off the tip after the presentation time, but before the reaction time, to place it on a small square of agar gel or gelatin gel for a while, then to transfer the agar to the decapitated coleoptile or, better, to one that has not been exposed to unilateral light. Curvature always occurs toward the side of the agar under the illuminated side of the tip. An electric wave also travels down the coleoptile at speeds up to 14 mm/hr (Newman, 1963), but this appears to be caused by the stream of translocated substance.

The substances responsible for this correlation are known as *auxins,* a term originally coined for any chemical capable of inducing a curvature in the oat coleoptile. Three such substances were originally isolated from plant and animal material. They were called auxin *a,* auxin *b* (auxin *a* minus a molecule of water), and heteroauxin.

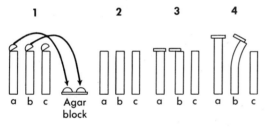

Fig. 19-1. Effects of decapitation and replacement of substances from the tip (in agar blocks) on the growth of oat coleoptiles. **1,** Three coleoptiles decapitated. **2,** Tips from **a** and **b** transferred to agar blocks. **3,** One block placed symmetrically on **a,** one asymmetrically on **b,** none on **c. 4, a** shows straight growth, **b** growth curvature, **c** no growth.

Later evidence led to the conclusion that only the last of these actually occurs in plant tissue. It is the chemical substance β-indoleacetic acid, or indole-3-acetic acid (IAA), which is closely related to, and therefore assumed to be formed from, the ubiquitous amino acid tryptophan (Fig. 19-2). Direct evidence of the synthesis of IAA in the plant from tryptophan has not been obtained (Wightman, 1964); but although tryptophan itself has no auxin activity, some presumed intermediates between it and IAA (e.g., indoleacetaldehyde, indoleacetonitrile) are active. When applied uniformly to decapitated coleoptiles, these substances cause increased growth. When applied to one side, they cause curvature toward the other side.

Phototropic curvature is caused by a higher concentration of auxin on the side of the coleoptile further from the light. Two theories have been proposed to explain this asymmetry. According to the Went-Cholodny theory, it is caused by translocation of auxin from the illuminated to the dark side of the coleoptile. Other investigators have shown that IAA can be destroyed in vitro by light and explain the lower concentration on the illuminated side by this destruction. Recent and thorough investigations appear to have conclusively established the Went-Cholodny explanation (Thimann, 1964, see Chapter 18). This means that phototropism is not simply caused by an inhibition of growth by light since the redistribution of auxin must result in both a reduced rate on the lighted side and an enhanced rate on the darker side.

So much of the fundamental work on

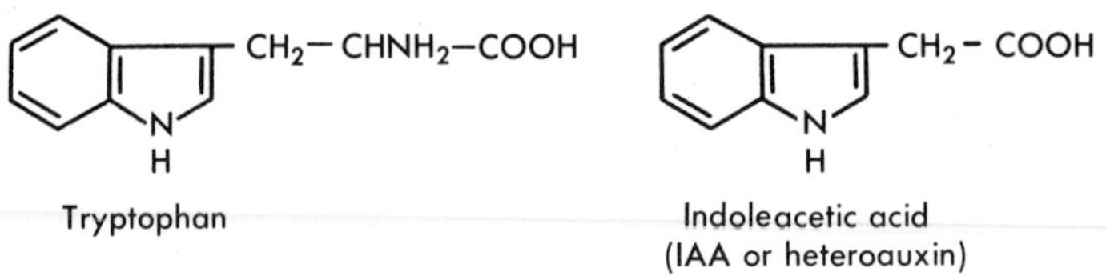

Fig. 19-2. Chemical structure of indoleacetic acid (IAA) and its chemical relation to tryptophan.

phototropism is based on the oat coleoptile that there is some danger in generalizing from these results. In the case of sunflower seedlings (Lam and Leopold, 1966) an asymmetry of auxin content can be produced without the involvement of lateral auxin transport from the lighted side of the stem to the shaded side. Darkening one *cotyledon* will cause curvature of the stem toward the lighted cotyledon. More diffusible auxin is obtained from the stem below the darkened cotyledon than below the lighted one. The relationship is basically similar to that in the oat coleoptile, but in this case the *cotyledons* induce the stem curvature by producing an asymmetrical auxin concentration in the stem. The cotyledons are also the region of perception in this case.

Geotropism can be similarly explained. The gravitational stimulus causes a greater concentration of auxin on the lower side, but it must be remembered that this causes negative geotropism in stems and positive geotropism in roots. This has been explained by the difference in sensitivity of roots and stems to the concentration of auxin (Fig. 19-3). All cells may be either stimulated or inhibited, depending on the concentration of auxin; but the concentration (10^{-6} M) causing increased growth of stem cells de-

creases the growth of root cells. Consequently, if the gravitational stimulus increased the auxin concentration on the lower side to 10^{-6} M, this would lead to an upward curvature of the stem and a downward curvature of the root. Although this is an adequate explanation of geotropism, recent investigations point to the involvement of gibberellins (see later).

Some of the apparent exceptions, when investigated more thoroughly, only help to support the generalization that all plant growth movements are controlled by auxins. Ethylene gas can produce tropisms, but this is a result of the immediate cessation of lateral auxin transport induced by ethylene, at least in pea stems (Burg and Burg, 1966 and 1967). Paradoxically, IAA itself induces ethylene formation. Even nastic growth movements have been shown to be controlled by auxins (Brauner and Arslan, 1951).

This dual effect of auxin gives the plant a switch mechanism for turning the growth process on or off. The inhibition, for instance, has been proposed to explain the *polarity* of growth (such as in apical dominance), the rapid growth of the terminal bud and the relative or absolute inhibition of growth of the lateral buds. The auxin is synthesized or released in the terminal bud;

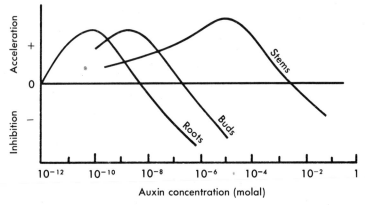

Fig. 19-3. Inhibition and acceleration of growth of different organs as a function of auxin concentration. (From Thimann, K. V. 1937. Amer. J. Bot. **24:**407-412.)

and since the movement of IAA is strongly polar in a basipetal direction, it is translocated down the stem. The concentration in the terminal bud apparently remains optimum for its growth. The concentration produced in the lateral buds by the downward translocation of the excess is high enough to inhibit growth, and they are kept in the dormant state. This fact is used in practice because potato tubers can be prevented from sprouting by soaking them in a concentrated solution of auxin. However, the difference in sensitivities of different organs is also apparent here. Auxin can promote the growth of inhibited pea *shoots*, although it cannot promote the growth of the inhibited pea *buds* (Sachs and Thimann, 1967). Similarly, once a lateral bud is induced to begin growing, it becomes less sensitive to inhibition by the terminal apex.

In all the cases just mentioned, the effects of auxins on growth result from an increased

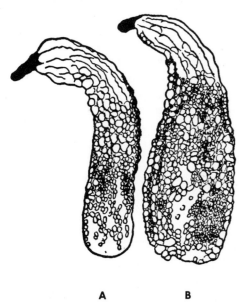

A   **B**

Fig. 19-4. Fruit formation in squash by treatment with 2% indolebutyric acid in lanolin, **A,** as compared with normal fruit by pollination, **B.** (From Gustafson, F. G. 1951. *In* F. Skoog. Plant growth substances. University of Wisconsin Press, Madison, Wis.)

or decreased cell enlargement at low or high concentration, respectively. It has long been known, however, that auxins may also affect cell division. Callus tissue may be formed when IAA is applied to a cut stem or petiole surface. Similarly, root and fruit initiation (Fig. 19-4) is favored by the same substance, and fruit abscission can be prevented or induced. Renewed activity of cambium and apical meristems in spring seems to be dependent on auxins. In summary, IAA may produce the following effects on the plant (Linser, 1966): (1) promotion or inhibition of cell enlargement, (2) promotion of cell division, (3) promotion of root formation, (4) inhibition of root growth, (5) inhibition of bud growth or formation, (6) enhancement or inhibition of abscission (see Chapter 20), (7) enhancement or inhibition of flowering (see Chapter 20), (8) formation of parthenocarpic fruit, (9) promotion of respiration, and (10) promotion of protein synthesis. Optimum concentrations for growth are 10^{-10} M for roots, 10^{-8} M for buds, 10^{-5} M for stems, and 10^{-4} M for flowers.

Any complete explanation of auxin action must therefore include the effects on cell division and cell enlargement; yet most theories have attempted primarily to explain the effect on cell enlargement. This is perhaps understandable since auxins apparently control the growth of only those organisms that show significant cell enlargement—the higher plants. They generally have no effect on the growth of bacteria or fungi; neither do they induce any growth response in organisms without cellulose cell walls, as for example, animals. These facts favor the view that auxins alter the cellulose cell walls in such a way as to lead to cell enlargement (Heyn, 1940). For a time it was thought that the effect might be on the pectins that are present in small quantities in the walls of growing plant cells; but more recent evidence has eliminated this possibility (Cleland, 1963). As mentioned in Chapter 17, cell enlargement can occur either if the force exerted on the cell wall by the turgor pressure is greater than the force of attraction between the particles of the cell wall or if particles are inserted into the wall by intussusception. In either case a cell may begin to enlarge as the result of either increased turgor pressure or decreased rigidity of the wall, or both. It has been amply proved that auxins do not cause any increased turgor pressure, but they do increase the plasticity of the wall. This, then, is at least part of the explanation for the increased cell enlargement. In those cells that enlarge primarily as the result of intussusception, auxins in some way must be involved in the process. In both cases it is an effect on the wall.

Such an explanation, however, is by no means complete. The mode of action of the auxin on the cell wall is not at all clear. It has, however, been shown that auxins greatly affect respiration rate. Those concentrations of auxin that increase growth usually cause increased respiration and those that decrease growth usually cause decreased respiration, although there are exceptions. The way in which this effect on respiration is linked to the effects on the cell wall is impossible, as yet, to say. Perhaps the auxin effect on respiration releases the energy needed to break the bonds between the micelles of the cell wall and to synthesize cellulose molecules (Ray and Baker, 1962), thus both increasing wall plasticity and permitting the intussusception of new wall material. According to Cleland (1967), the covalent bonds that render the cell wall rigid are broken by a combined action of auxin and turgor pressure. It has been suggested that IAA activates the formation of one or more new enzymes, which act on cell walls to increase plasticity (Nooden and Thimann, 1965). One of these may be cellulase, an enzyme that splits cellulose hydrolytically and whose synthesis in pea seedlings may be controlled by auxin concentration (Fan and MacLachlan, 1966).

This relation of auxin action to protein synthesis has stimulated an intensive search for evidence of a relation to nucleic acids, which control protein synthesis (see Chapter 16). Growth-promoting concentrations of IAA were soon found to enhance RNA synthesis and growth-inhibitory concentrations, to decrease it (Key, 1964; Key and Shannon, 1964). Some evidence seemed able to separate this from the growth effect since cell wall extensibility was apparently increased even by auxin concentrations that inhibited RNA synthesis more than 90% (Cleland, 1965). However, later investigations of a number of different plants have demonstrated an inhibition of auxin-induced cell wall extensibility in all cases when RNA synthesis was inhibited (Coartney et al., 1967). Similarly, specific inhibitors of both RNA and protein synthesis suppress IAA-induced growth. On the basis of these results, Knypl (1966) suggests that IAA is a specific inductor of messenger RNA (mRNA) synthesis. This conclusion is supported by the incorporation of labeled uracil and orthophosphate into RNA in the presence of IAA (Masuda et al., 1967). Trewavas (1968), however, has concluded that ribosomal RNA is the major species whose synthesis is increased by IAA, although small changes occurred also in mRNA.

In addition to the natural auxins, there are many synthetic substances capable of producing the same effects on plant growth. By definition they are also auxins (since they induce curvature of oat coleoptiles), although not naturally occurring. Examples of these are indolebutyric acid, naphthaleneacetic acid, and the important weed killer 2,4-D (2,4-dichlorophenoxyacetic acid). Quantitatively, some of these substances are more effective than the naturally occurring IAA and some are less effective, depending on the test that is used, since there are many effects of both the natural and the synthetic substances. The substance that is most effective for one may be least effective for another of these plant responses (Avery et al., 1947). Some may be more effective than the native auxin because they are more persistent in the plant, perhaps because they resist oxidation by IAA oxidase (see later). Just as in the case of IAA, the synthetic auxin 2,4-D may increase enzyme activity—a 3-day treatment of chicory roots with 10^{-5} M increased invertase 60 times (Flood et al., 1967).

Although IAA (and its immediate indolic precursors) is usually considered to be *the* natural auxin, recent evidence has thrown some doubt on the generality of this conclusion. A nonindolic "citrus auxin" has recently been found in several plant families, but its chemical nature is as yet unknown (Lewis et al., 1965). Native, nonindolic auxins that are steroid in nature have also been identified in *Coleus* (Vendrig, 1967). p-Hydroxybenzoic acid has recently been isolated from woody cuttings of *Ribes rubrum* (Vieitez et al., 1966) and has been found to act as an auxin in the oat coleoptile straight growth test, producing both growth inhibition at high concentration and growth stimulation at low concentration. It is therefore not surprising that among the synthetic substances with auxin activity, some may be found which are very different chemically from IAA (Kefford and Caso, 1966).

However, the effects of these other substances may still be secondary, either by acting as antiauxins (antagonizing the effect of IAA) or as synergists (enhancing the effect of IAA). Thus, monophenols (p-coumaric acid, dichlorophenol, kaempferol) may decrease auxin action by promoting its oxidation; diphenols (caffeic, protocatechuic, and chlorogenic acids, catechol) may increase the auxin action by preventing its oxidative destruction (Thimann, 1965).

GIBBERELLINS

At the same time as the European (especially the Dutch) investigators were proving the existence and functions of auxins,

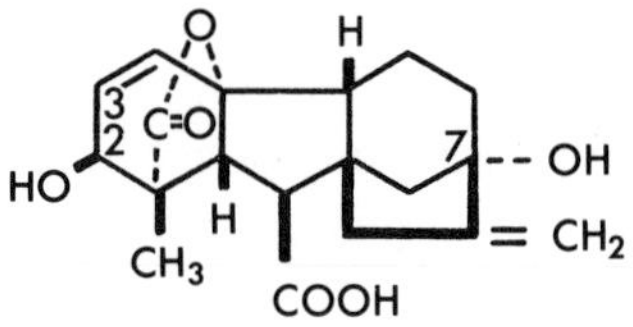

Gibberellic acid
(GA₃)

Fig. 19-5. Chemical structure of gibberellins. (From van Overbeek, J. 1966. Science 152:721-731. Copyright 1966 by the American Association for the Advancement of Science.)

Japanese investigators were discovering gibberellins in rice plants affected by the "foolish disease." However, it was not until about 30 years later that these substances were found to be present in normal plants and their importance was recognized by scientists throughout the world. Unlike auxins, of which only one (IAA) appears to be the normal hormone in most plants, there are over twenty known gibberellins (MacMillan and Takahashi, 1968), of which at least five have now been found to occur naturally in the tissues. All the gibberellins are chemically related to and therefore probably synthesized from diterpenes. Although the biosynthetic pathways for both of these groups of substances are as yet unknown, both are believed to be synthesized from mevalonic acid

$$CH_3$$
$$(COOH \cdot CH_2 \cdot C \cdot CH_2 \cdot CH_2OH)$$
$$OH$$

berellin activity may be shown by substances differing greatly in chemical structure from gibberellins. The best known is gibberellic acid, Fig. 19-5, GA_3).

Although they do not all have the same effects, gibberellins (like auxins) characteristically lead to cell enlargement. For instance, they induce internode elongation in plants during their rosette stage of growth and, therefore, lead to the phenomenon called "shooting" (or "bolting")—the significant stem elongation that precedes flowering in many plants (e.g., biennials). This internode elongation is frequently accompanied by a slight decrease in root growth, probably because of a competition for raw materials. Excised roots, on the other hand, show increased growth in the presence of gibberellin. In fact, van Overbeek believes that root growth is promoted only by gibberellin, not by auxin as indicated earlier (Fig. 19-3). It is even involved in the growth regulation of red algae (Jennings and McComb, 1967).

When applied artificially, gibberellins may also stimulate cell division in meristems, leading to flowering (Bernier et al., 1964). They may also stimulate seed germination. Leaf growth responds to gibberellin, although not to auxin; yet gibberellins may promote fruit growth, resembling the effect of auxin. It has even been suggested that gibberellins which are produced in the apple ovule after fertilization are responsible for fruit set (Dennis, 1967). The response to gibberellin may be altered by an environmental factor: The dwarf pea fails to show its dwarfness in the dark when supplied with gibberellin; light, however, renders it less responsive to gibberellic acid, and it fails to grow to full size.

In summary, gibberellins (0.05 to 0.5 ppm) can promote (1) internode growth, (2) hypocotyl growth, (3) germination, (4) sexual development of flowers, (5) flower formation (see Chapter 20), (6) parthenocarpy, (7) breaking of the rest period of buds and underground organs, (8) flower formation (see Chapter 20), (9) cell division in the cambial zone, (10) change in shape and size of leaves, and (11) activity of enzymes (Linser, 1966). They may also participate in the expression of apical dominance (Ruddat and Pharis, 1966) and may increase the storage life of fruit (Kitagawa et al., 1966). Unlike auxins, gibberellins show no polarity of movement. Any attempt to explain the mechanism of gibberellin action must be capable of explaining all of these results.

Many of these effects are explainable by the control of enzyme activity. Stimulation of seed germination is apparently the result of an activation of α-amylase, which then digests the endosperm starch. In the case of barley the series of events is as follows (van Overbeek, 1966): Gibberellin is produced naturally by the embryo after swelling in water. It then stimulates α-amylase activity in the aleurone layer. The enzyme is secreted by the aleurone layer into the endosperm, leading to the digestion of starch to sugar, which can then be used by the embryo. The preceding activation by gibberellin is apparently the result of *synthesis* of the α-amylase protein because when barley endosperm halves were incubated with phenylalanine-C^{14}, the α-amylase protein became labeled in the presence of gibberellin but not in its absence. Other enzymes are also activated by gibberellin—lipase in cottonseed (Jones et al., 1967), protease from ungerminated barley (Jacobsen and Varner, 1967), and cytolytic enzymes, which promote cell wall degradation. The latter effect could well explain cell enlargement since it would lead to greater plasticity of the cell walls. At the extreme it could explain breaking of the seed coat on germination. In the case of dwarf peas, internode expansion can be induced by gibberellic acid treatment. This results in a considerable increase in protein synthesis, which apparently increases the rate of cell wall synthesis (Broughton and McComb, 1967). In this way the dwarf form is converted into a normal tall type that does not require gibberellic acid treatment because it synthesizes the substance in sufficient quantity. In Alaska pea, for instance, growth rate is correlated with the synthesis of gibberellic acid (Moore, 1967).

Gibberellins interact greatly with auxins and may, in fact, help to explain the mechanism of auxin action. Thus, gibberellic acid does not cause coleoptile curvature as auxin does, but when added to IAA, the combination causes more curvature than IAA alone.

When auxin is added to gibberellin, it prevents the gibberellin-induced promotion of leaf growth. In the case of the oat seedling, it is not until auxin production ceases and growth of the coleoptile is completed that gibberellin-dominated leaf development can occur. High IAA and low gibberellic acid concentrations favor xylem formation in woody plants; low IAA and high gibberellic acid concentrations favor phloem production (Digby and Wareing, 1966). Seedling germination is complex but can be largely explained by an interaction between gibberellin and auxin. Gibberellin stimulates protease activity, leading to a breakdown of proteins to amino acids, including tryptophan. The tryptophan may, perhaps, then be converted to IAA. According to van Overbeek (1966) the sequence of events is as follows:

Barley seed germination

1. Embryo is activated by imbibition of water.
2. Gibberellin is produced in the embryo.
3. Gibberellin moves to aleurone layer.
4. Gibberellin induces synthesis of digestive enzymes (hydrolases): α-amylase, protease, (cellulase ?).
5. These enzymes convert starch to sugars, proteins to amino acids, including tryptophan.
6. Gibberellin induces root growth.
7. Tryptophan produced by protease activity is translocated to the coleoptile tip.
8. In the coleoptile tip it is converted to IAA.
9. IAA moves polarly from the coleoptile tip to the base.
10. The coleoptile cell walls become weakened.
11. Water is taken up and the cells enlarge.

Even this detailed speculative scheme is an oversimplification because it fails to explain any of the other changes accompanying cell enlargement, for example, increase in protoplasm and its organelles.

As in the case of auxin, the evidence now seems to indicate that gibberellin controls growth by an effect on the nucleic acids responsible for protein synthesis. Thus gibberellin is prevented from inducing α-amylase synthesis by treatment with actinomycin D, an inhibitor that forms complexes with DNA but not with RNA. Presumably, it inhibits DNA-dependent but not RNA-dependent synthesis (Khan, 1967b). The scheme is as follows:

$$DNA \xrightarrow[\text{gibberellic acid}]{\text{Action of}} mRNA \longrightarrow \alpha\text{-Amylase}$$

On this basis, gibberellic acid may conceivably lead to α-amylase synthesis by uncovering the genes responsible for synthesis of the α-amylase molecule. The same explanation may apply to the stimulation of the RNA-controlled protein synthesis in yeast, leading to sporulation (Kamisaka et al., 1967). But the foregoing scheme does not seem to apply to all tissues. In the case of lentil epicotyl, DNA synthesis was promoted by gibberellin (Nitsan and Lang, 1966). It therefore does not seem possible at present to specify the exact link in the nucleic acid chain where gibberellins exert their control.

CYTOKININS (PHYTOKININS OR KININS)

In opposition to auxins, which primarily affect cell enlargement, substances have been found that primarily affect cell division and have been called *kinins* (from kinesis, meaning division). Unfortunately, the name has priority for a chemically different (peptide) group of substances in animal physiology. The terms *phytokinin* (Osborne, 1963, see Chapter 20) and *cytokinin* (Skoog et al., 1965) have therefore been suggested, and the latter is now most commonly used. Unlike auxins and gibberellins, cytokinin activity was not discovered in plant extracts. Adenine sulfate was the first cytokinin discovered (Skoog, 1951). It greatly alters the kind of growth in tissue cultures (e.g. tissues cut from apical or cambial meristems and grown in suitable culture media). It may convert the undifferentiated tissue so formed into one that develops buds and even mature organs (Fig. 19-6). Later investigations resulted in the isolation of a more active substance

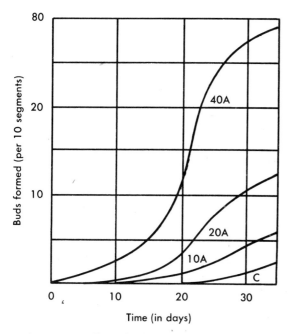

Fig. 19-6. Effect of adenine sulfate (40, 20, 10, and 0 mg per liter) on bud formation in tobacco stem segments. (From Skoog, F., and C. Tsui. 1951. *In* F. Skoog. Plant growth substances. University of Wisconsin Press, Madison, Wis.)

HN – CH₂ – CH = C – CH₃
CH₂OH

Zeatin

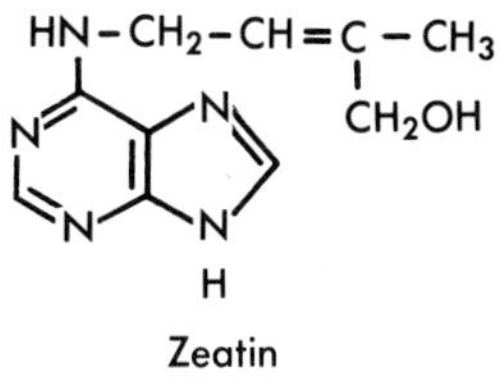

Fig. 19-7. Zeatin. (From van Overbeek, J. 1966. Science **152**:721-731. Copyright 1966 by the American Association for the Advancement of Science.)

formed by the breakdown of nucleic acid. It was named *kinetin* and is now known to be 6-furfuryl-aminopurine. Thus one of the earliest discovered characteristics of cytokinins was their interaction with auxins. The aforementioned ability of tissues to differentiate in culture and to form buds (Fig. 19-6) was found to depend on the presence of a high enough ratio of cytokinin to auxin, and the cytokinins were active only in the presence of auxins. Similarly, when lateral buds of peas are inhibited by the growing apex (because of the excess auxin transported to the lateral buds from the apex), this inhibition is overcome by applying kinetin directly to the lateral buds. This release from apical dominance must be followed by an application of auxin to the bud apices in order to obtain normal elongation (Sachs and Thimann, 1967). The fact that auxins sometimes induce cell division (although their normal effect is on

cell enlargement) and cytokinins may accelerate elongation (although their normal effect is on division) may perhaps be a result of this interrelationship. Cytokinins (in the presence of auxins) strongly promote the formation of lignin at the expense of polysaccharide synthesis (Koblitz, 1967).

Many extracts of plants are now known to have cytokinin activity. This may result from the presence of *zeatin*, which has been isolated from corn grains. It is a hydroxy derivative of 6-dimethylallyl adenine (Fig. 19-7). Cytokinins commonly occur in seeds. The best known synthetic cytokinin is benzyladenine. Even urea derivatives have cytokinin activity although not related to them chemically.

Cytokinins have the capacity to direct the flow of chemicals within the plant. This may be partly a result of the high ratio of protein nitrogen to soluble nitrogen that they produce (Banerji and Laloraya, 1967). The consequent low concentration of amino acids could induce translocation of amino acids from regions of higher concentration. However, this flow occurs even in the case of an amino acid not synthesized into a protein. The movement is therefore not simply a "source and sink" relationship (i.e., a movement to a region where the substance is used). Cytokinins are able to inhibit senescence (see Chapter 20), perhaps because of this directional flow of usable materials. However, their most striking effect is their modification of the action of other hormones. Alone, they usually have little effect.

In summary, as in the case of the other two groups of growth regulators, cytokinins have many effects on the plant (Linser, 1966). They may accelerate or promote (1) cell division and the related DNA and RNA synthesis; (2) cell enlargement in leaves (they may also inhibit it); (3) leaf bud formation (eliminating the related polarity); (4) root formation and growth (inhibition occasionally); (5) germination; (6) breaking of rest; (7) respiration; and (8) translocation of low molecular weight nitrogen compounds. They may also (9) inhibit protein degradation and (10) influence leaf shape and pigments.

Earlier observations seemed to indicate that kinins were not translocated but acted in situ, where they were synthesized. This, if true, would eliminate them from the class of hormones. More recent results, however, indicate that they must be translocated under normal conditions. It has, in fact, been concluded that they are formed in the roots of cereals and control the activity of enzymes (e.g., glycolic acid oxidase) in the leaves (Dezsi and Farkas, 1964). Even a rooted leaf may be fully supplied with a cytokinin ascending from the root (Engelbrecht, 1964). When applied to the leaf blade of several plants, kinetin is immobile (Lagerstedt and Langston, 1967), but it is not immobile when applied to petiole, vein, or root. Basipetal as well as acropetal translocation occurred in tobacco leaves. Kinetin was not taken up by cotton roots but was both absorbed by and translocated from tobacco roots.

As might be expected from its chemical relation to nucleic acids, kinetin has been found to promote RNA metabolism (e.g., in moss protonemata, Brandes, 1967). Aminoacyl-transfer-RNA synthetase activity increased in discs of tobacco leaf after treatment for 7 days with kinetin (Anderson and Rowan, 1966). A labeled kinin (N, 6-benzyladenine), when supplied to tobacco and soybean, was incorporated into several RNA components (Fox, 1966). Similarly, kinetin increased the incorporation of labeled orotate into the RNA of radish leaf discs (Burdett and Wareing, 1966). Kinetin has also been found to inhibit the breakdown of adenine in *Pelargonium* leaves (Schlee et al., 1966). A highly purified soluble RNA from roots and shoots of germinating corn *(Zea mays)* seeds has been found to possess cytokinin activity (Letham and Ralph, 1967), leading to the conclusion that it induces growth by allowing the synthesis of specific transfer RNA's (tRNA). Cytokinins can also inhibit RNA synthesis (Srivastava, 1967); but this is to be expected since they can also inhibit growth (e.g., of roots, Gaspar and Xhaufflaire, 1967).

All three groups of growth regulators are apparently interrelated. They sometimes act in sequence. Cytokinins and gibberellins appear to dominate the early phase of development from seeds, and auxins dominate the later phases. The normal pattern is for each regulator to rise in concentration quickly to a peak, then to decline quickly. When the concentration is kept artificially high for long periods, this normal rise and fall is interefered with and the normal interactions are disturbed. This may be the reason why 2,4-D acts as a weed killer (van Overbeek, 1966). It does not necessarily follow that these three groups include all the growth regulators. Isolated potato buds required both gibberellic acid and IAA for their growth but were found to cease growth when the stem was 6 to 10 mm long unless 8-hydroxyquinoline was present (Goodwin, 1966).

DORMINS AND OTHER INHIBITORS

The preceding three groups of growth regulators basically induce or initiate growth. Other regulators exist that are basically growth inhibitors and are, therefore, primarily inducers of dormancy or the rest period. For instance, the resumption of growth by dormant seeds, buds, or other organs has long been explained on the basis of auxin interactions with *antiauxins*—substances that counteract the effects of auxins. Coumarin is an antiauxin that prevents seeds from germinating. Only when it is destroyed (e.g., by light), can some seeds germinate. On the other hand, if the auxin is originally present at a level capable of causing growth inhibition, a competitive antiauxin can counteract part of the auxin and can reduce its effective concentration to the level of growth stimulation. In this case the antiauxin indirectly stimulates growth. The same result may be brought about by an auxin-destroying enzyme, indoleacetic oxidase (IAA oxidase), as by a competitive inhibitor. Even in the growing plant it is believed that IAA is constantly being destroyed by IAA oxidase. Diphenols such as caffeic acid have growth-promoting activities because they inhibit this oxidation (i.e., destruction) of IAA. They may therefore be thought of as anti-antiauxins. Monophenols, on the other hand, activate IAA oxidation.

Recent results indicate that dormancy is frequently regulated by a complicated series of interactions between gibberellins, cytokinins, auxins, and endogenous inhibitors (Cornforth et al., 1966). Potato dormancy (Hemberg, 1952) has been related to an inhibitor of α-amylase, and bud dormancy in woody plants has been related to *dormins*. In the presence of dormins, vegetative buds change to dormant winter buds by converting the developing leaf primor-

dia into bud scales. One of these dormins is particularly potent, and its inhibitory effect is overcome by gibberellin. It is identical with an independently discovered substance abscisin II (now called *abscisic acid*) which stimulates abscission (e.g., of cotton) and inhibits gibberellic acid–induced formation of α-amylase in barley aleurone (Chrispeels and Varner, 1967). It is an isoprenoid related to the half carotene vitamin A (Fig. 19-8). Light appears to stimulate its formation from the products of photooxidation of the carotenoid xanthophyll (Taylor and Smith, 1967). This relation as well as its relation to gibberellins is indicated by the biosynthetic pathway recently proposed for all three substances (Graebe, 1967, and personal communication) (Fig. 19-9).

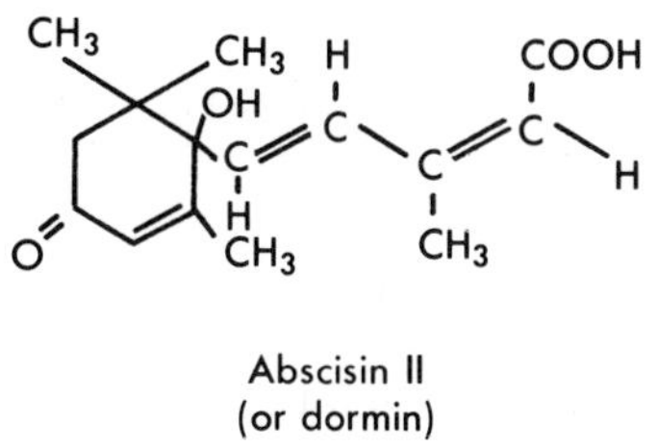

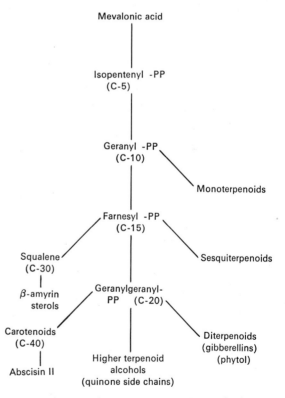

Fig. 19-8. Dormin or abscisin II or abscisic acid. (From van Overbeek, J. 1966. Science **152**:721-731. Copyright 1966 by the American Association for the Advancement of Science.)

Fig. 19-9. Proposed pathway for synthesis of gibberellins and abscisins (abscisic acid). (From Graebe, J. E. 1967. Science **157**:73-75, and personal communication.)

Since dormin (or abscisic acid) inhibits the α-amylase activity induced by gibberellin in barley endosperm, it must therefore be a gibberellin antagonist or antigibberellin. But more than one kind of interaction may be involved (Aspinall et al., 1967). Thus, dormin also interacts with other growth stimulators, and gibberellin interacts with other growth inhibitors. Cytokinin, for instance, may also counteract the dormin-induced inhibition of germination (Khan, 1967a); and gibberellin also interacts with coumarin, overcoming the inhibition of seed germination induced by this inhibitor. The complexity of the interactions is indicated by the promotion of fruit set in *Rosa* by dormin (Jackson and Blundell, 1966).

On the other hand, the breaking of dormancy has frequently been related to a rise in concentration of gibberellin without any evidence of an inhibitor. In the case of lettuce seed not enough gibberellin is produced in the dark for normal germination; on illumination with near red there is a sharp increase in gibberellin within an hour (Köhler, 1966). Gibberellic acid may also overcome the dormancy of nonlight-requiring seed, although it has not been clearly demonstrated whether it can break true dormancy; furthermore, not all plants respond to gibberellic acid (Vegis, 1964a, see Chapter 17). In some seeds (e.g., oats, Barralis, 1965) a double dormancy mechanism appears to exist—the caryopsis dormancy and that caused by the hulls. Gibberellic acid can break the dormancy of some seeds and of excised embryos, but it has no effect on intact seeds containing pericarp and endosperm inhibitors (e.g., *Fraxinus excelsior*, Szalai, 1965b) and therefore seems to be inactive on the second mechanism. Furthermore, if short of oxygen, such seeds may enter a second dormancy. In the case of dormant buds of *Syringa vulgaris*, GA_3 and kinetin were unable to replace the chilling that breaks the bud dormancy under natural conditions.

They promoted bud growth only during predormancy and postdormancy (Leike and Lau, 1967). Potato dormancy also seems to be related to gibberellin content. As the nondormant new tubers enter their rest period, the gibberellin level drops (Szalai, 1965a), and during sprouting, it increases to 3 times the dormant quantity. Nevertheless, when artificially stimulated to sprout with Rindite, the gibberellin level drops. Perhaps the Rindite destroys dormin. This may also indicate that dormins were possibly involved in the other cases where gibberellins overcame dormancy.

As in the case of the other growth regulators, the dormins appear to control nucleic acid metabolism. Cultures of *Lemna minor* become nearly completely dormant when dormin is added (3.8×10^{-6} M). In this case only the cytokinin benzyladenine (not auxin or gibberellin) was able to counteract the effect of dormin (van Overbeek et al., 1967); and even this was possible only if the dormin concentration did not exceed a critical level. The dormin suppressed nucleic acid synthesis. Inhibition of DNA synthesis seemed to precede that of RNA. Cytokinins reversed this process. It is therefore suggested that dormin is an inhibitor of DNA synthesis and that cytokinin is an activator of the process. This would seem to imply that cytokinin and gibberellin both act in the same way, although presumably on different DNA molecules.

A number of synthetic growth retardants have been produced that shorten the internodes of plants. Amo 1618, chlorocholine chloride (CCC), and Phosfon D (Fig. 19-10) are the best known; they also inhibit α-amylase production in germinating barley seed. The evidence indicates that they do this by inhibiting the biosynthesis of gibberellins (Paleg et al., 1965; Khan and Faust, 1967). Other antigibberellins may act competitively rather than by controlling biosynthesis (Ziegler et al., 1966). Even roots (of grape) become shorter and thicker

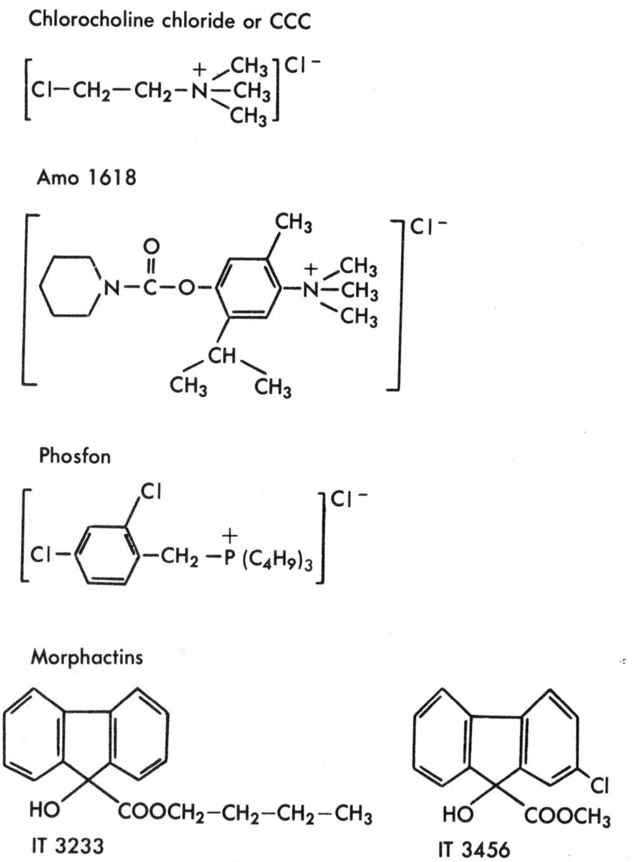

Chlorocholine chloride or CCC

$$\left[Cl-CH_2-CH_2-\overset{+}{N}\begin{array}{c} CH_3 \\ CH_3 \\ CH_3 \end{array} \right] Cl^-$$

Amo 1618

Phosfon

Morphactins

IT 3233

IT 3456

Fig. 19-10. Synthetic growth retardants. (From Jung, J. 1967. Naturwissenschaften **54**:356-360.)

in the presence of chlorocholine chloride, and this is counteracted by gibberellic acid (Skene and Mullins, 1967). However, the retarding effect of chlorocholine chloride may also be overcome by kinetin (Knypl, 1967) so that classifying it as an antigibberellin may be an oversimplification. By retarding vegetative growth these substances often promote flowering and fruiting. Furthermore, even though their general effect is to retard growth, they may result in *increased* yields. In the case of potato plants, B995 (another growth retardant) decreased stem weight but increased tuber weight and, therefore, yield (Bodlaender and Algera, 1966).

Maleic hydrazide is another growth in-

hibitor whose effect may be antagonized by gibberellic acid (Saha, 1966). Another group of synthetic growth inhibitors is called *morphactins* (Fig. 19-10). They induce internode shortening and loss of dormancy of lateral buds in citrus seedlings (Mann et al., 1966); but they neither replace nor antagonize gibberellins. Seedlings grown in their presence lose their capacity to respond to gravity or to unilateral light (Khan, 1967c). Indoleacetic and gibberellic acid did not modify their effect on geotropism. There is, however, an interaction with auxin when another inhibitor, *malformin*, stimulates the abscission of primary leaves of *Phaseolus aureus* because this effect is inhibited by indoleacetic acid (Curtis, 1968). The growth disturbances produced by malformin are apparently caused by the stimulation of ethylene production, at least in some cases. Another synthetic growth regulator (Amchem 66-329) also appears to produce its effect as a result of the stimulation of ethylene production (Warner and Leopold, 1967). Studies of air pollutants have revealed the presence of oxidants (e.g., peroxyacetyl nitrate) that inhibit growth. Auxin can induce recovery from such inhibitors (Ordin et al., 1966).

The different groups of growth regulators have specific effects on growth. For instance, the elongation of seedlings of the grass family is controlled almost exclusively by auxin, but the leaves and leaf sheaths elongate under the exclusive control of gibberellic acid (Thimann, 1965). On the other hand, the frequent similarity in effects of the three groups of growth regulators on plant growth is surprising in view of the pronounced differences in their chemical structures. Auxins and gibberellins induce cell division, although their more characteristic effect is on cell elongation (Digby and Wareing, 1966). Conversely, cytokinins may affect elongation, although they commonly control division. All three may induce parthenocarpy (Crane and van Overbeek, 1965) and retard senescence (Fletcher and Osborne, 1965). However, it must be realized that such effects are judged by artificial applications to plant parts which have unknown quantities of all three regu-

lators. Thus, the observed effect is more likely to depend on the altered *balance* between the three than on the single addition of the one.

The very fact that all four groups of regulators interact with one another (Table 19-2) may indicate that they all affect the same metabolic process. The chemical structure of only one of these groups of regulators—the cytokinins—gives a clue as to which metabolic process may be involved. The cytokinins are chemically related to the purine component of nucleic acids and, therefore, can logically be expected to affect nucleic acid metabolism. For instance, it has been suggested that the growth regulators may help to maintain DNA as a functional template for RNA synthesis (Fletcher and Osborne, 1965). Kinetin-treated tissues do, in fact, show a net synthesis of both RNA and protein, which is totally suppressed by the RNA inhibitor thiouracil (Wallgiehn and Parthier, 1964). All three groups of growth regulators have indeed been found to stimulate RNA synthesis in coconut milk nuclei. This stimulation is inhibited by actinomycin D, and the inhibition is removed by higher concentrations of growth regulators (Roychaudhury and Sen, 1965). However, van Overbeek (1966) believes that this relation between growth regulators and nucleic acids may not be a direct one since it is possible to detect the regulator's action long before the nucleic acid response. The growth process is, of course, dependent

Table 19-2. Interactions between growth regulators*

Regulators	Phenomenon	Effect
1. Auxin as opposed to cytokinin	Apical dominance	Auxin promotes; cytokinin overcomes
2. Auxins plus gibberellins	Activation of cambium	Act together to promote it
3. Auxins, gibberellins, cytokinins	Formation of wood	Interaction of all three
4. Auxin and gibberellin	Internode elongation of dicots	Interdependent
5. Auxin and abscisin	Abscission	Interaction to promote or inhibit
6. Gibberellin, auxin, or cytokinin as opposed to dormin	Dormancy	Dormin promotes; gibberellin, auxin, or cytokinin overcomes its promotion of dormancy

*Adapted from Thimann, K. V. (1965). Recent Progr. Hormone Res. **21**:579-596.

on nucleic acid metabolism; therefore, any growth response to the regulators is sure to be accompanied by a stimulation of nucleic acid metabolism, even if this is not the immediate cause of the growth response.

Other growth regulators have been found in the plant that do not fit into the preceding four classes. Traumatin, for instance, is a wound hormone that accumulates at the cut surface of potatoes. Ethylene has long been known to affect growth, perhaps by inhibiting the transport of auxin (Morgan and Gausman, 1966). Unfortunately, these substances have not been studied sufficiently. Sometimes an unknown mixture may greatly stimulate growth. A good example is coconut milk (van Overbeek, 1942). Its growth-regulating activity is now believed to be caused by auxin, cytokinin, hexitols (e.g., myoinositol), and reduced N-compounds (Kefford, 1963), all of which act synergistically. Myoinositol, for instance, acts synergistically with cytokinin to promote cell division in carrot but not with IAA (Letham, 1966). There must also be some natural growth regulators that have not yet been discovered. In favor of this conclusion is the fact that the tissues of some plants cannot be grown indefinitely in tissue culture. Several other substances, for example, glucobrassicin and helminthosporol (Schantz, 1966), may also stimulate growth. As mentioned earlier, some growth regulators of the auxin type may be used as herbicides. Other substances used to kill plants are not truly growth regulators. The quaternary salts of bipyridyl react with the normal photosynthetic process to produce hydrogen peroxide in concentrations too high for catalase to deal with and therefore kill the plant (Boon, 1967).

In all the preceding discussions, growth regulators have been considered only from the point of view of their effects on the plant that produces them. There are, however, many substances produced by one plant that may regulate the growth of other plants. Normally, this is an inhibition and is then called *allelopathy*. Volatile substances, such as terpenes, produced by one species of plant have in this way inhibited the growth of other species (Muller and Del Moral, 1966). This may sometimes lead to a pure stand of the former (e.g., *Centaurea*, Fletcher and Renney, 1963).

QUESTIONS

1. What is a growth regulator?
2. What is a hormone?
3. What is an auxin?
4. What is the chemical substance in the plant responsible for auxin activity?
5. Have any other auxins been found?
6. What effect does auxin have on growth?
7. What effects do different concentrations of auxin have?
8. Are all plant parts equally sensitive to auxins?
9. What phase(s) of cell development is/are affected?
10. How is cell enlargement induced?
11. What effect does IAA have on metabolism?
12. How is phototropism explained in terms of auxins?
13. How is geotropism explained in terms of auxins?
14. What is an an antiauxin?
15. Does an antiauxim increase or decrease growth?
16. What other effects do auxins have?
17. Name two other kinds of naturally occurring growth regulators.
18. Is their chemistry known?
19. How do they differ from auxins?
20. What organisms are affected by auxins?
21. What amino acid is IAA probably formed from?
22. What practical use are auxins put to?
23. What chemical substances are gibberellins related to?
24. What effects do gibberellins have on shoot growth?
25. What effects do gibberellins have on root growth?

26. What effects do gibberellins have on seed germination?
27. What effects do gibberellins have on enzyme activity?
28. What effects do gibberellin have on leaf growth?
29. What effects do gibberellins have on oat coleoptiles?
30. What phase of cell development are cytokinins believed to affect?
31. Which of the three groups of growth regulators interact with other regulators?
32. Which of the three groups affect senescence?
33. Which of the three groups are hormones?
34. What is the major group of natural growth inhibitors?
35. Which growth regulators do they react with?
36. What is believed to be the mechanism of action of growth regulators?
37. Are there any other natural substances capable of controlling growth beside the four main groups of growth regulators?
38. Are there any synthetic substances?

SPECIFIC REFERENCES

Anderson, J. W., and K. S. Rowan. 1966. Activity of aminoacyl-transfer-ribonucleic acid synthetases in tobacco-leaf tissue in relation to senescence and to the action of 6-furfuryl-aminopurine. Biochem. J. 101:15-18.

Aspinall, D., L. G. Paleg, and F. D. Addicott. 1967. Abscisin II and some hormone-regulated plant responses. Aust. J. Biol. Sci. 20:869-882.

Banerji, D., and M. M. Laloraya. 1967. Correlative studies on plant growth and metabolism. III. Metabolic changes accompanying inhibition of the longitudinal growth of stem and root by kinetin. Plant Physiol. 42:623-627.

Barralis, G. 1965. La germination des folles avoines. Ann. Épiphyt. 6:295-314.

Bernier, G., R. Bronchart, and A. Jacqmard. 1964. Action of gibberellic acid on the mitotic activity of the different zones of the shoot apex of *Rudbeckia bicolor* and *Perilla nankinensis*. Planta 61:236-244.

Bodlaender, K. B. A., and S. Algera. 1966. Influ-

ence of the growth retardant B995 on growth and yield of potatoes. Eur. Potato J. 9:242-258.

Boon, W. R. 1967. The quaternary salts of bipyridyl—a new agricultural tool. Endeavour 26: 27-32.

Brandes, H. 1967. Die Wirkungsmechanismus des Kinetins bei der Induktion von Knospen am Protonema der Laubmoose. Planta 74:55-71.

Brauner, L., and N. Arslan. 1951. Experiments on the auxin reactions of the pulvinus of *Phaseolus multiflorus*. Rev. Fac. Sci. Univ. Istanbul 16B 3:257-300.

Broughton, W. J., and A. J. McComb. 1967. The relation between cell-wall protein synthesis in dwarf pea plants treated with gibberellic acid. Ann. Bot. 31:359-366.

Burdett, A. N., and P. F. Wareing. 1966. The effect of kinetin on the incorporation of labelled orotate into various fractions of ribonucleic acid of excised radish leaf discs. Planta 71:20-26.

Burg, S. P., and E. A. Burg. 1966. The interaction between auxin and ethylene and its role in plant growth. Proc. Nat. Acad. Sci. U. S. A. 55:262-269.

Burg, S. P., and E. A. Burg. 1967. Lateral auxin transport in stems and roots. Plant Physiol. 42: 891-893.

Chrispeels, M. J., and J. E. Varner. 1967. Hormonal control of enzyme synthesis; on the mode of action of gibberellic acid and abscisin in aleurone layers of barley. Plant Physiol. 42:1008-1016.

Cleland, R. 1963. Independence of the effects of auxin on cell wall methylation and elongation. Plant Physiol. 38:12-18.

Cleland, R. 1965. Auxin-induced cell wall loosening in the presence of actinomycin D. Plant Physiol. 40:595-600.

Cleland, R. 1967. A dual role of turgor pressure in auxin-induced cell elongation in Avena coleoptiles. Planta 77:182-191.

Coartney, J. S., D. J. Morré, and J. L. Key. 1967. Inhibition of RNA synthesis and auxin-induced cell wall extensibility and growth by actinomycin D. Plant Physiol. 42:434-439.

Cornforth, J. W., B. V. Milborrow, and G. Ryback. 1966. Identification and estimation of (+) abscisin II ("dormin") in plant extracts by spectropolarimetry. Nature (London) 210:627-628.

Crane, J., and J. van Overbeek. 1965. Kinin-induced parthenocarpy in the fig, *Ficus carica* L. Science 147:1468-1469.

Curtis, R. W. 1968. Mediation of a plant response to malformin by ethylene. Plant Physiol. 43:76-80.

Dennis, F. G., Jr. 1967. Apple fruit-set; evidence for a specific role of seeds. Science 156:71-73.

Dezsi, L., and G. L. Farkas. 1964. Effect of

kinetin on enzymes of glycolic acid metabolism in cereal leaves. Acta Biol. Hung. **14**:325-332.

Digby, J., and P. F. Wareing. 1966. The effect of applied growth hormones on cambial division and the differentiation of the cambial derivatives Ann. Bot. **30**:539-548.

Digby, J., and P. W. Wareing. 1966. The effect of growth hormones on cell division and expansion in liquid suspension cultures of *Acer pseudoplatanus*. J. Exp. Bot. **17**:718-725.

Engelbrecht, L. 1964. Über Kinetinwirkungen bei intakten Blättern von *Nicotiana rustica*. Flora **154**:57-69.

Fan, D. F., and G. A. MacLachlan. 1966. Control of cellulase activity by indoleacetic acid. Canad. J. Bot. **44**:1025-1034.

Fletcher, R. A., and A. J. Renney. 1963. A growth inhibitor found in *Centaurea* spp. Canad. J. Plant Sci. **43**:475-481.

Fletcher, R. A., and D. J. Osborne. 1965. Regulation of protein and nucleic acid synthesis by gibberellin during leaf senescence. Nature (London) **207**:1176-1177.

Flood, A. E., P. P. Rutherford, and E. W. Weston. 1967. Effects of 2,4-dichlorophenoxyacetic acid on enzyme systems in Jerusalem artichoke tubers and chicory roots. Nature (London) **214**:1049-1050.

Fox, J. E. 1966. Incorporation of a kinin, N,6-benzyladenine into soluble RNA. Plant Physiol. **41**:75-82.

Gaspar, T., and A. Xhauflaire. 1967. Effect of kinetin on growth, auxin catabolism, peroxidase and catalase activities. Planta **72**:252-257.

Goodwin, P. B. 1966. An improved medium for the rapid growth of isolated potato buds. J. Exp. Bot. **17**:590-595.

Graebe, J. E. 1967. Isoprenoid biosynthesis in a cell-free system from pea shoots. Science **157**:73-75.

Hemberg, T. 1952. The significance of the acid growth-inhibiting substances for the rest period of the potato tuber. Physiol. Plant. **5**:115-129.

Jackson, G. A. D., and J. B. Blundell. 1966. Effect of dormin on fruit set in *Rosa*. Nature (London) **212**:1470-1471.

Jacobsen, J. V., and J. E. Varner. 1967. Gibberellic acid-induced synthesis of protease by isolated aleurone layers of barley. Plant Physiol. **42**:1596-1600.

Jennings, R. C., and A. J. McComb. 1967. Gibberellins in the red alga, *Hypnea muaciformis* (Wulf.) Lamour. Nature (London) **215**:872-873.

Jones, H. C., H. S. Black, and A. M. Altschul. 1967. Comparison of the effects of gibberellic acid and aflatoxin in germinating seeds. Nature (London) **214**:171-172.

Jung, J. 1967. Synthetische Wachstums regulatoren, insbesondere Chlorcholinchlorid. Naturwissenschaften **54**:356-360.

Kamisaka, S., Y. Masuda, and N. Yanagishima. 1967. Gibberellin-induced yeast sporulation in relation to RNA and protein metabolism. Physiol. Plant. **20**:98-105.

Kefford, N. P. 1963. Natural growth regulators; report on 25th Int. Conf. on Plant Growth Regulation. Science **142**:1495-1505.

Kefford, N. P., and O. H. Caso. 1966. A potent auxin with unique chemical structure—4-amino-3,5,6-trichloropicolinic acid. Bot. Gaz. **127**:159-163.

Key, J. L. 1964. RNA and protein synthesis as essential processes for cell elongation. Plant Physiol. **39**:365-370.

Key, J. L., and J. C. Shannon. 1964. Enhancement by auxin of RNA synthesis in excised soybean hypocotyl tissue. Plant Physiol. **39**:360-364.

Khan, A. A. 1967*a*. Antagonism between cytokinins and germination inhibitors. Nature (London) **216**:166-167.

Khan, A. A. 1967*b*. Dependence of gibberellic acid–induced dark germination of lettuce seed on RNA synthesis. Planta **72**:284-288.

Khan, A. A. 1967*c*. Physiology of morphactins: effect on gravi- and photo-response. Physiol. Plant. **20**:306-313.

Khan, A. A., and M. A. Faust. 1967. Effect of growth retardants on α-amylase production in germinating barley seed. Physiol. Plant. **20**:673-681.

Kitagawa, H., A. Sugiura, and M. Sugiyama. 1966. Effects of gibberellin spray on storage quality of kaki. Hort. Sci. **1**:59-60.

Knypl, J. S. 1966. Specific inhibitors of RNA and protein synthesis as suppressors of the IAA- and coumarin-induced growth responses. Acta Soc. Bot. Pol. **35**:357-373.

Knypl, J. S. 1967. Synergistic inhibition of kale seed germination by coumarin and (2-chloroethyl)-trimethylammonium chloride and its reversal by kinetin and gibberellic acid. Planta **72**:292-296.

Koblitz, H. 1967. Beiträge zur Zellwandforschung in der pflanzlichen Gewebekultur. II. Biochemische Aspekte der Zellwandontogenese in vitro kultivierter Pflanzengewebe. Qualitas Plant. Mater. Veg. **14**:37-69.

Köhler, D. 1966. Veränderungen des Gibberellingehaltes von Salatsamen nach Belichtung. Planta **70**:42-45.

Lam, S. L., and A. C. Leopold. 1966. Role of leaves in phototropism. Plant Physiol. **41**:847-851.

Lagerstedt, H. B., and R. G. Langston. 1967.

Translocation of radioactive kinetin. Plant Physiol. **42**:611-622.

Leike, H., and R. Lau. 1967. Wirkung von Gibberellinsäure und Kinetin auf ruhende Terminalknospen von *Syringa vulgaris* L. Flora **157**:467-470.

Letham, D. S. 1963. Zeatin; a factor inducing cell division isolated from *Zea mays*. Life Sci. **8**:569-573.

Letham, D. S. 1966. Regulators of cell division in plant tissues. II. A cytokinin in plant extracts; isolation and interaction with other growth regulators. Phytochemistry **5**:269-286.

Letham, D. S., and R. K. Ralph. 1967. A cytokinin in soluble RNA from a higher plant. Life Sci. **6**:387-394.

Lewis, L. N., R. A. Khalifah, and C. W. Coggins, Jr. 1965. The existence of the non-indolic citrus auxin in several plant families. Phytochemistry. **4**:203-205.

Libbert, E. 1964. Kontrolliert Auxin die Apikaldominanz? Physiol. Plant. **17**:371-378.

Linser, H. 1966. The hormonal systems of plants. Angew. Chem. Int. Ed. English **5**:776-784.

MacMillan, J., and N. Takahashi. 1968. Proposed procedure for the allocation of trivial names to the gibberellins. Nature (London). **217**:170-171.

Mann, J. D., H. Hield, K. H. Yung, and D. Johnson. 1966. Independence of morphactin and gibberellin effects upon higher plants. Plant Physiol. **41**:1751-1752.

Masuda, Y., E. Tanimoto, and S. Wada. 1967. Auxin-stimulated RNA synthesis in oat coleoptile cells. Physiol. Plant. **20**:713-719.

Moore, T. C. 1967. Gibberellin relationship in the "Alaska" pea *(Pisum sativum)* Amer. J. Bot. **54**:262-269.

Morgan, P. W., and H. W. Gausman. 1966. Effects of ethylene on auxin transport. Plant Physiol. **41**:45-52.

Muller, C. H., and R. Del Moral. 1966. Soil toxicity induced by terpenes from *Salvia leucophylla*. Bull. Torrey Bot. Club **93**:130-137, 332-351.

Newman, I. A. 1963. Electric potentials and auxin translocation in Avena. Aust. J. Biol. Sci. **16**:629-646.

Ninemann, H., J. A. D. Zeevaart, H. Kende, and A. Lang. 1964. The plant growth retardant CCC as inhibitor of gibberellin biosynthesis in *Fusarium moniliforme*. Planta **61**:229-235.

Nitsan, J., and A. Lang. 1966. DNA synthesis in the elongating, non-dividing cells of the lentil epicotyl and its promotion by gibberellin. Plant Physiol. **41**:965-970.

Nooden, L. D., and K. V. Thimann. 1965. Inhibition of protein synthesis and of auxin-induced growth by chloramphenicol. Plant Physiol. **40**:193-201.

Ordin, L., M. J. Garber, B. P. Skee, and G. Rolle. 1966. Role of auxin in growth of inhibitor-treated oat coleoptile tissue. Physiol. Plant. **19**:937-945.

Paleg, L., H. Kende, H. Ninnemann, and A. Lang. 1965. Physiologic effects of gibberellic acid. VIII. Growth retardants on barley endosperm. Plant Physiol. **40**:165-169.

Ray, P. M., and D. B. Baker. 1962. Promotion of cell wall synthesis by indolylacetic acid. Nature (London) **195**:1322.

Roychoudhury, R., and S. P. Sen. 1965. The effect of gibberellic acid on nucleic acid metabolism in coconut milk nuclei. Plant Cell Physiol. **6**:761-765.

Ruddat, F., and R. P. Pharis. 1966. Participation of gibberellin in the control of apical dominance in soybean and redwood. Planta **71**:222-228.

Sachs, T., and K. V. Thimann. 1967. The role of auxins and cytokinins in the release of buds from dominance. Amer. J. Bot. **54**:136-144.

Saha, A. K. 1966. Antagonism between gibberellic acid and maleic hydrazide on the growth of Zinnia (*Zinnia elegans*) plants. Sci. Cult. **32**:548-549.

Schlee, D., H. Reinbothe, and K. Mothes. 1966. Wirkungen von Kinetin auf den Adeninabbau in Chlorophyll defekten Blättern von *Pelargonium zonale*. Z. Pflanzenphysiol. **54**:223-236.

Skene, K. G. M., and M. G. Mullins. 1967. Effect of CCC on the growth of roots of *Vitis vinifera* L. Planta **77**:157-163.

Skoog, F., F. M. Strong, and C. O. Miller, 1965. Cytokinins. Science **148**:532.

Srivastava, B. I. S. 1967. Effect of kinetin on nucleic acid synthesis in barley leaf segments. Biochim. Biophys. Acta. **145**:166-169.

Szalai, I. 1965a. Ueber die quantitative Veraenderung der nativen Gibberellin-Antigen Stoff in mit Rindite stimulierten neuen Kartoffelknollen. Acta Univ. Szeged. **11**:101-106.

Szalai. I. 1965b. Brechung der Keimruhe von Fraxinus-Samen durch Gibberellinsäure. Acta Univ. Szeged. **11**:93-100.

Taylor, H. F., and T. A. Smith. 1967. Production of plant growth inhibitors form xanthophylls; a possible source of dormin. Nature (London) **215**:1513-1514.

Thimann, K. V. 1937. On the nature of inhibitions caused by auxin. Amer. J. Bot. **24**:407-412.

Trewavas, A. J. 1968. Effect of IAA on RNA and protein synthesis. Arch. Biochem. Biophys. **123**:324-335.

van Overbeek, J., J. E. Loeffler, and M. I. R. Mason. 1967. Dormin (abscisin II), inhibitor of plant DNA synthesis? Science **156**:1497-1499.

Vendrig, J. C. 1967. Steroid derivatives as native auxins in *Coleus*. Ann. N. Y. Acad. Sci. **144**:81-93.

Vieitez, E., E. Seoane, D. V. Gesto, C. Mato, A. Vazquez, and A. Carnicer. 1966. Hydroxybenzoic acid, a growth regulator, isolated from woody cuttings of *Ribes rubrum*. Physiol. Plant. **19**: 294-307.

Wallgiehn, R., and B. Parthier. 1964. Der Einfluss des Kinetins auf den RNA und Proteinstoffwechsel in abgeschnittenen, mit Hemmstoffen behandelten Tabakblättern. Phytochemisty **3**:241-248.

Warner, H. L., and A. C. Leopold. 1967. Plant growth regulation by stimulation of ethylene production. BioScience **17**:722-723.

Wightman, F. 1964. Pathways of tryptophan metabolism in tomato plants. Colloq. Int. Cent. Nat. Rech. Sci. **123**:191-212.

Ziegler, H., D. Köhler, and B. Streitz. 1966. Ist 2-chlor-9-fluorenol-9-carbonsäure ein gibberellin antagonist? Z. Pflanzenphysiol. **54**:118-124.

GENERAL REFERENCES

Avery, G. S., E. B. Johnson, R. M. Addoms, and B. F. Thomson. 1947. Hormones and horticulture. McGraw-Hill Book Co., New York.

Boysen-Jensen, P. 1936. Growth hormones in plants. McGraw-Hill Book Co., New York.

Heyn, A. N. J. 1940. The physiology of cell elongation. Bot. Rev. **6**:515-574.

Pilet, P. E. 1961. Les phytohormones de croissance. Masson & Cie, Paris.

Schantz, E. M. 1966. Chemistry of naturally-occurring growth regulating substances. Ann. Rev. Plant Physiol. **17**:409-438.

Skoog, F. (Ed.). 1951. Plant growth substances. University of Wisconsin Press, Madison, Wis.

Thimann, K. V. 1965. Toward an endocrinology of higher plants. Recent Progr. Hormone Res. **21**:579-596.

van Overbeek, J. 1942. Hormonal control of embryo and seedling. Sympos. Quant. Biol. **10**:126-134.

van Overbeek, J. 1966. Plant hormones and regulators. Science **152**:721-731.

Went, F. W., and K. V. Thimann. 1937. Phytohormones. The Macmillan Co., New York.

chapter 20

Development

STAGES OF DEVELOPMENT

Tissues excised (cut) from apical or cambial meristems can be grown in suitable culture media forming so-called callus tissue, which may continue growth indefinitely without forming any specialized, differentiated tissues. It is thus capable of growth without development. The normal plant, however, undergoes both growth and development. The fertilized egg-cell develops into an embryo, the embryo develops into a seedling, and the seedling develops into a flowering and then a fruiting plant. In the case of many annuals (e.g., sunflower), growth ceases with initiation of the reproductive process, and this is followed by senescence and death. Although much longer lived, even perennials eventually senesce and die.

The growing plant thus passes through several stages and phases of development that are associated with definite morpho-

logical as well as physiological changes. They may be classified as follows (adapted from Sax, 1962).

    I. Vegetative stage
       a. Embryonic phase
       b. Juvenile phase
       c. Adolescent phase
    II. Reproductive (adult) stage
       d. Flowering phase (in some preceded by vegetative reproduction)
       e. Fruiting phase
   III. Senescent stage

It is not always easy to recognize all these phases of development. The juvenile phase, for instance, is readily observed only in plants that persist in this phase for some time. The English ivy (*Hedera helix*) and other species of *Hedera* may persist as a vine in the juvenile phase of the vegetative stage or may develop into the arborescent plant, which is the adult, or reproductive stage. It is now possible to induce this adult, arborescent form to produce juvenile shoots by spraying with gibberellin (Robbins, 1960) and to revert from a reproductive to a vegetative condition.

At any one stage the plant as a whole, or parts of it, may be either in the active or the resting (dormant) state. Similarly, a single organ of the plant may pass through several developmental subphases. Even different portions of an organ may be at different stages of development. In the case of the stem, for instance, older internodes are completing their development when younger ones are just beginning theirs. This decreasing rate of elongation in the older internodes of Japanese morning glory *(Pharbitis nil)* has been correlated with an increasing ability of the tissue to destroy indoleacetic acid (IAA) (Yoneda and Stonier, 1966).

The seed goes through a pregermination subphase that involves many changes (Evenari, 1961). This is followed by germination, which requires the synthesis of many enzymes (see Chapter 19). In the case of soybean, for instance, glutamyl-transpeptidase activity was very low during the first 24 hr, but it reached a maximum 40 to 48 hr after the beginning of germination and decreased rapidly after 72 hr (Kasai and Obata, 1967). Accompanying the enzyme changes were a corresponding rise and fall in the respective peptides.

The fruit (and even the leaf) has its own embryonic, juvenile, mature, climacteric, and senescent stages. During the climacteric stage there is no further enlargement, but

(1) ethylene gas is produced, (2) the respiration rate increases, and (3) ripening occurs (see Fig. 14-4). After the climacteric stage the respiration rate begins to drop as senescence progresses. This series of changes has been found to occur in many fruit; but some (e.g., pineapple, Dull et al., 1967), although producing ethylene, do not have a true climacteric stage because the respiration rate does not increase.

TRANSITION FROM VEGETATIVE TO REPRODUCTIVE DEVELOPMENT

The plant physiologist attempts to answer the following two questions about these developmental stages:

1. What are the physiological changes within the plant that give rise to the transitions from one developmental stage to another? Or, what are the internal factors?
2. What are the environmental factors that control the plant's transitions from the vegetative to the reproductive stage? Or, what are the external factors?

The second question has led many of the

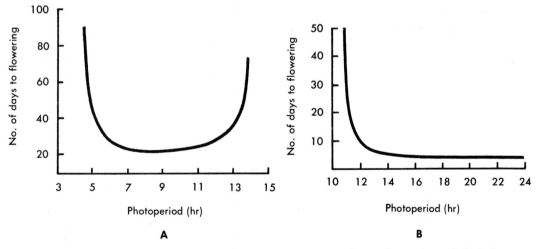

Fig. 20-1. A, Photoperiodic reaction of a short-day plant *(Chrysanthemum)*. A daily light period of 7 to 11 hr is optimal for flowering. No flowering occurs above 14 hr and below 5 hr. **B,** Photoperiodic reaction of a long-day plant *(Hyoscyamus niger)*. The critical day length is 11 hr. Flowering occurs more readily with lengthening of the photoperiod above this value. (From Bünning, E. 1948. *In* A. E. Murneek and R. O. Whyte. Vernalization and photoperiodism. The Ronald Press Co., New York.)

Russian investigators to classify the developmental stages of the plant in terms of the environmental factor which is most important at this stage, that is, the thermostage or phase, the photostage or phase, etc. This kind of classification, however, is not based on specific morphological or physiological characteristics of the plant and is therefore not as fundamental as the preceding morphological and developmental classification. It is also difficult to apply. The physiology of development has been studied intensively for only a few decades. Until recently, nearly all the investigations were confined to the second of the two questions.

Photoperiodism

In 1920, Garner and Allard (1920) showed that the flowering of many plants could be induced or prevented simply by controlling the length of the daylight period (the "photoperiod"). Some plants flowered most rapidly when the day length was about 12 hr or less (short-day plants, Fig. 20-1, *A*), others when the day length was about 12 hr or more (long-day plants, Fig. 20-1, *B*); still others flowered at either day length (day-neutral plants). All types may occur in any one spices, frequently producing so-called early (long-day) and late (short-day) varieties (Table 20-1). If the

day length is unsuited to flowering, the plants may remain vegetative essentially indefinitely, or they may simply take much longer to flower (e.g., Peking and Biloxi soybean, Table 20-1).

The photoperiodic effect has been found in all kinds of plants (herbaceous and woody, annual, biennial, perennial) and even in animals. Beside reproduction, it affects many other phases of the physiology of the plant (Fig. 20-2). Some trees do not enter their rest period in the fall if long days are maintained. As a result, they fail to develop frost resistance and are readily winter killed. Biennials remain in the rosette form if subjetced to short days. Vegetative as well as sexual reproduction is affected. However, the optimum for flowering may be different from the optimum for vegetative reproduction. Maximum tuberization of potatoes occurs during shorter day length than that for maximum flowering.

Table 20-1. Flowering times for soybean varieties grown under different day lengths[*]

Day length (hr)	Time (days) from germination to blossoming		
	Mandarin (day neutral)	Peking (short day)	Biloxi (short day)
5	23	23	27
7	21	21	26
12	21	21	28
Full daylight (12½-15 hr)	26	62	110

[*]Adapted from Garner, W. W., and H. Allard. 1920. J. Agric. Res. **18**:553-606.

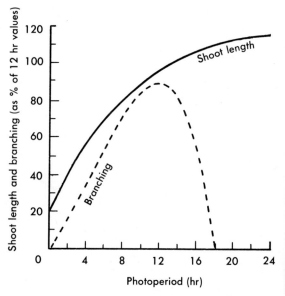

Fig. 20-2. Effect of photoperiod on the vegetative development (shoot growth and branching) in the fiber plant *Crotalaria juncea*. (Redrawn from Singh and Choudri. 1946. *In* R. O. Whyte. Crop production and environment. Faber & Faber, Ltd., London.)

It must be realized that the photoperiodic effect cannot substitute for photosynthesis. Consequently, the optimum day length for flowering will not be the optimum for yield in the case of short-day plants. In order to obtain maximum yield they must first be grown under a long day to enable the ac-

cumulation of reserves and enlargement of the plant. Similarly, the temperature factor cannot be ignored (Fig. 20-3).

In the case of a long-day plant, flowering can be induced in natural short days by lengthening the daylight period with weak light. This subsidiary light may be far too weak to induce a net photosynthesis. Similarly, long-day plants may be forced to flower and short-day plants may be prevented from flowering in short days by interrupting the dark period with a few minutes of light (Fig. 20-4). This and other evidence has proved that it is the dark period that inhibits flowering of long-day plants and stimulates flowering of short-day plants.

For this reason, Lang (1952) has suggested that all plants may be classified as short day or long day, provided that the two groups are defined as follows:

Short-day plants (SDP) are those in which flowering is induced or hastened by a daily dark period. Long-day plants (LDP) are those in which flowering is inhibited or retarded by a daily dark period. This would include day-neutral plants among the long-day plants. However, there are still some

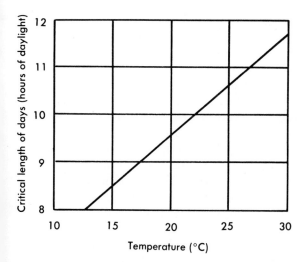

Fig. 20-3. Dependence of critical day length for flowering on temperature in *Hyoscyamus niger.* (Redrawn from Lang, A., and G. Melchers. 1943. Planta 33:653-702.)

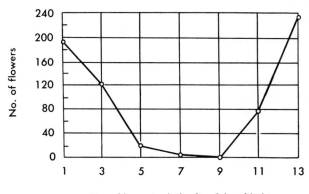

Fig. 20-4. Effect of a 1 min interruption of a long (15 hr) dark period by light on flowering of the short-day plant *Kalanchoe blossfeldiana.* When applied near the beginning or end of the dark period, there is no effect; when applied at the middle, it prevents flowering in spite of the favorable (9 hr) photoperiod. (From Bünning, E. 1948. *In* A. E. Murneek and R. O. Whyte. Vernalization and photoperiodism. The Ronald Press Co., New York.)

that are difficult to classify into one or the other group, for example, some that require both short-day and long-day treatments, the one following the other.

Short-day plants will flower when exposed to long days, provided that they have already been exposed to a sufficient number of short days. Long-day plants will similarly flower during short days if previously exposed to a sufficient number of long days. This carryover is called the *photoperiodic aftereffect,* or *photoperiodic induction.* A plant is said to be photoinduced if it is exposed to a day length favorable for flowering for a sufficient time to result in subsequent flowering in an unfavorable photoperiod.

As in the case of growth itself, there is a hormone control of development (Lang, 1952). The response occurs in the growing points, yet these do not have to be exposed to any particular day length. The region of perception is the leaf; and even a single one, if subjected to the necessary day length, may induce the growing points to flower. The floral stimulus may even be transferred from one plant to another by grafting. It is therefore generated in the leaf and translocated to the growing points.

The floral stimulus or hormone is also able to multiply autocatalytically (Wellensiek, 1966). In this property it resembles certain nucleic acids that may be immortal and rapidly perpetuating. If the floral stimulus is a specific RNA, inhibition of flowering may result from the blocking of a specific chromosomal DNA particle, from which the specific RNA would originate. Induction of flowering may therefore simply be the removal of this inhibition by deblocking (derepressing) the DNA particle. The RNA produced may either be the floral hormone, or it may give rise to its synthesis.

Growth regulators may also affect the flowering of some plants. It was long ago observed that the initiation of floral and fruit development is frequently associated with a decrease in growth. This led to the hypothesis that the same growth regulators that produce growth inhibition may also induce reproductive development. Many investigators have been able to show both an enhancement and a decrease in rate of flowering with increased auxin concentration, depending on the stage of the plant. It has even been possible in some cases (e.g., pineapple) to induce flowering commercially by application of an auxin (indolebutyric acid). But it is now well established that such effects are secondary and cannot explain the primary control. The evidence indicates that auxins, cytokinins, and gibberellins cannot transmit the photoperiodically induced stimulus from the leaves to the apex. A more specific flowering hormone has therefore been postulated, for which the name *florigen* has been proposed. Many attempts have been made to extract and isolate the florigen, and several reports of successful isolation are to be found in the literature (Lincoln et al., 1966).

Nevertheless, there definitely is an effect of photoperiod on growth-regulator content. Short-day conditions that lead to dormancy result in an increase in the level of growth inhibitors (Vegis, 1964, see Chapter 17), although most of these occur in the covering structures (e.g., bud scales). Conversely, long days may lead to the development of growth promoters. Artificial applications of growth regulators can in many cases control flowering. Gibberellin (0.25 μg per plant) inhibits floral initiation in the long-day plant fuchsia, although it increases stem elongation (Sachs et al., 1967). Other growth regulators have no effect on the flowering of fuchsia unless used in high enough concentrations to inhibit vegetative growth. Sachs and associates, therefore, suggest that the action of the long day is to reduce the endogenous gibberellic acid to a level favorable to flowering. In opposition to this concept, the natural gibberellin content of plants increases in the morning, reaching a maximum in daylight, and falls to a minimum at night (Lozhnikova, 1966). This is apparently true of long-day, short-day, and day-neutral plants. It is not surprising, then, that in all cases the amount of gibberellin found in the leaves is greater when the plant is grown on a 16 hr day than on an 8 hr day (Chailakhyan and Lozhnikova, 1966); but other inhibitors of flowering may also be involved. Abscisin inhibited flower induction in the long-day species *Lolium temulentum* and *Spinacea oleracea* when applied to the leaves during exposure to long days (El-Antably et al., 1967). Conversely, it promoted flowering in the short-day plants *Pharbitis nil, Betula nigrum,* and strawberry when applied under long-day conditions. However, it failed to induce flowering in certain other short-day plants. A natural inhibitor of flowering of the long-day plant *Silene armeria* is formed during short days at 20° C (Wellensiek, 1967). Induction by long days (20° C) or by short days (32° or 5° C) is a result of the removal of this inhibitor. Gibberellic acid seemed to have a partially inducing action.

As in any light-controlled process, photoperiodism is possible only if a substance is present that absorbs the light. This substance has now been isolated and is a pale blue protein that has been called *phytochrome*. The chromophore of phytochrome has been separated from the protein and appears to be a bilitriene (Siegelman et al., 1966). Phytochrome differs from all other plant pigments by existing in two light-absorbing forms, each of which can be converted reversibly into the other as a result of light absorption:

$$P_{660} \underset{\text{Far red}}{\overset{\text{Near red}}{\rightleftharpoons}} P_{730}$$

The equation is also written

$$P_r \rightleftharpoons P_{fr} \text{ or } P_R \rightleftharpoons P_{FR}$$

where P_r or P_R = the (near) red absorbing form P_{fr} or P_{FR} = the far red absorbing form.

The P_{660} (near red absorbing) form absorbs light maximally at a wavelength of 660 nm and, as a result, is converted to the P_{730} (far red absorbing) form, which absorbs maximally at 730 nm. The latter absorption process converts it back to the near red absorbing form. The process can apparently be repeated over and over again without limit. The conversion process is very rapid; the half conversion time for P_r to P_{fr} is 2.3 sec (Hendricks and Borthwick, 1967). In contrast to these rapid conversions in the light, the P_{730} form changes back slowly in the dark to the P_{660} form, which is stable in the dark. Thus, in dark-grown maize seedlings, phytochrome is present entirely as P_{660} (Butler et al., 1963). A single brief irradiation with red light converts it into P_{730}. On retransfer to the dark, the P_{730} reverts to P_{660}, and the total amount of photoreversible phytochrome decreases to about 30% of the original amount. Exactly how these reversals are related to the control of the plant's development is not known. The P_{730} is the physiologically active form and is thought to be a highly active enzyme (Mohr, 1962).

The effect of phytochrome has been interpreted in terms of modern concepts of molecular biology. Mohr has explained the phytochrome effect as the result of gene derepression. Scherf and Zenk (1967) suggest that as a consequence of illumination with high-energy blue light, a derepression of formerly repressed genes comes about; but only if there is phytochrome present in its far red absorbing form (P_{730}), can the derepressed gene induce protein synthesis. Some recent results fail to support the gene derepression concept of the phytochrome control system and favor, instead, the older concept of permeability as the control mechanism because an early consequence of phytochrome action is a change in permeability (Hendricks and Borthwick, 1967). In the case of nyctinastic movements in *Albizzia julibrissin,* the closure response is accompanied by an increased rate of electrolyte efflux from the cut pinna base (Jaffe and Galston, 1967). This result is not affected by actinomycin D and is therefore presumably independent of RNA metabolism and gene derepression, yet the nyctinastic movements are dependent on phytochrome. In support of this interpretation the light stimulation of lettuce seed by the action of P_{fr} is almost instantaneous and, therefore, more readily explained by a sudden permeability change than by the much slower gene action (Bewley et al., 1967). The same is true of barley root tips (Tanada, 1968).

Even the growth process itself may be affected by phytochrome. When 100% is in the P_{660} form, this apparently prevents growth of the oat coleoptile; when 100% is in the P_{730} form, maximal growth occurs (Hopkins and Hillman, 1966). This loss in growth capacity must, of course, be related to an observed decrease in diffusible auxin.

It must be realized that the near red⇌far red phytochrome may not be the only one involved in plant development since blue light (which, at least at low intensities, is not preferentially absorbed by phytochrome) may have a very great effect on development in some cases. However, no other phytochromes have been isolated, and the preceding one has been shown to be important in so many processes (seed germination, sleep movements, coloring of fruit, etc) that it is usually called *the* phytochrome of plant development. It occurs in all groups of plants, from algae to monocotyledons (Mohr, 1962). It is synthesized in the maturing leaves (Salisbury, 1965).

The many factors involved in photoperiodic control of flowering may be illustrated by the sequence of events in the flowering of the short-day plant *Pharbitis nil* (Zeevaart, 1962):

1. Exposure to a dark period of sufficient length
2. Change in the phytochrome system in the leaves
3. Translation by an unknown process into the synthesis of florigen (perhaps by the phytochrome acting as an effector)
4. Translocation of the florigen to the bud
5. Triggering of the events associated with the nucleic acid synthesis by florigen, which leads to flowering

In spite of the striking effects of the photoperiod, it is not the only factor that controls flowering; temperature may also be a deciding factor.

Thermoperiodism

Tomatoes are day-neutral plants under ordinary conditions, but their flowering can be controlled by the day and night temperature. A night temperature of 15° C and a day temperature of 25° C are optimum for flowering. If these are changed too greatly from the optimum, flowering is reduced or even prevented. Other plants have shown similar responses. Went has given the name *thermoperiodism* to this control of development by alternation of day and night temperatures. Unlike photoperiodism, there is no evidence that correlation is involved in thermoperiodism. Thus, there is no specific region of perception and no specific hormone known to control the plant's response. One of the major factors is apparently the quantitative relation between the photosynthate accumulated during the day and the respiratory breakdown at night. Other specific temperature effects (e.g., on starch ⇌ sugar equilibrium) may also be involved.

Vernalization

Some winter annuals must be sown in the fall in order to flower the following summer. If spring sown, they either fail to flower the first year or flower much later than when fall sown. If, however, the seeds are moistened (with enough water to increase their weight by about 60%) and kept at 0° to 5° C for about a month, they can then be spring sown and will flower at about the same time as the fall-sown seed. This method of inducing reproductive development by a long exposure to low temperature is known as *vernalization*. After vernalization, plants must still be subjected to a suitable photoperiod (usually a long day) in order to flower.

Some biennials show a similar behavior toward temperature, although at a later stage in their development. If the young seedlings (instead of the swollen seed) are exposed to low temperatures (about 5° to 10° C in the case of several species) for a week or two, they will subsequently flower the first year at a higher temperature, provided that the latter temperature is not too high. The actual temperature requirements vary considerably from species to species.

The mechanism of the temperature effect has not yet been discovered. However, the evidence again indicates a hormonal control, and the name *vernalin* has been coined for the unknown hormone. Lang (1952) suggested that vernalin is either a precursor of florigen or a catalyzer of its formation. According to Chailakhyan (1966), precursors of gibberellic acid are produced and accumulated during vernalization. Under long-day conditions these become converted into gibberellins. These precursors of gibberellins are hormones and are therefore the vernalins, according to his interpretation. But unlike photoperiodism, the region of perception in the case of vernalization always contains dividing (mitotic) cells—the apical region of biennials or the slightly germinated embryo of seeds. Vernalin does not appear to be a true hormone since it seems to move only with dividing cells. Recent results have shown that gibberellin may replace the vernalization treatment in the case of some plants but not in all. Uridylic acid has recently been found to replace vernalization in winter wheat (Suge and Yamada, 1963). Other pyrimidine bases have also been shown to induce flowering (Hirono and Rédei, 1966).

Mechanism of developmental control

As in the case of growth itself, it appears that both the photoperiodic and the temperature control of plant development is dependent on a balance between the growth regulators. Thus, the level of both IAA and gibberellic acid was higher in long-day–grown *Begonia* plants than in short-day–grown plants (Heide, 1967). High temperatures (20° and 25° C), however, increased the level of IAA but not of gibberellic acid. Finally, the regeneration ability of the leaves was apparently dependent on the balance between IAA and cytokinins. This similarity between the dependence of both growth and development on growth regulators led to suggestions of a similar control mechanism, whether caused by light or temperature. According to Heslop-Harrison (1964), flower-forming substances, natural and otherwise, are effectors. They unblock a whole gene system that is quiescent during vegetative growth. This can only happen in active apices possessing an active nucleic acid and protein-synthesizing metabolism. This would explain the flower-inhibiting role of purine and pyrimidine analogs. Direct evidence of a relation to RNA is now available.

The RNA-DNA ratio in the vegetative apex of the long-day plant *Lolium temulentum* is low (Rijven and Evans, 1967b). As a result of photoinduction by a single long day, the RNA content per apex increased during the next 3 days by 46%, whereas residual nitrogen and DNA content per apex increased by only 11% and 16%, respectively. During the next 2 days, RNA content per apex increased 140%, at the same time as morphological differentiation of the spikelets occurred, and residual nitrogen and DNA increased 160% and 95%, respectively. Floral induction in *Lolium temulentum* was followed within 1 day by a detectible increase in RNA at the shoot apex (Rijven and Evans, 1967a). There was a transient increase in the incorporation of P^{32} (applied to the leaves) into the nucleic acids at the apex at about the time when the long-day stimulus was estimated to reach the apex. Most of the P^{32} was incorporated into RNA. In the case of photoperiodically induced *Pharbitis nil,* only messenger RNA (mRNA) was found to change its nucleotide ratio (Yoshida et al., 1967). On the basis of these results it was suggested that production of the floral stimulus may be directed by gene DNA derepressed photoperiodically.

The aforementioned investigations of the environmental factors that control the transition from one developmental stage to another have also thrown some light on the many physiological changes involved within the plant. Wellensiek (1967) has concluded that flower bud formation depends on the relative ratio of promoting and inhibiting substances within the plant, provided that a sufficient amount of food and energy-producing substances is available. Earlier investigators pointed to another kind of balance. It was long suspected that a high carbon-nitrogen ratio within the plant favors flowering, a low ratio favors vegetative growth. Although it has been shown that this ratio by itself is incapable of controlling flowering, it nevertheless may have a great effect on the quantity of flowers produced. It may even have an all-or-none effect. Thus, *Sinapis alba* is a long-day plant; yet, when apices are grown in vitro, they can all be made to flower in short days if supplied with optimal sucrose and low ammonium nitrate (Deltour, 1967).

LONGEVITY, SENESCENCE, AND ABSCISSION
Longevity

The longevity, or total life-span, of a plant or plant part varies between wide extremes. Plant organs such as fruit or leaves may have a life-span from a few weeks to 30 years (Woolhouse, 1967). Whole plants may also live for only a few weeks (ephemerals), for one growing season (annuals), for two growing seasons

(biennials), or for as long as 3,000 years or more (some trees). The longest lived of all plant parts are, of course, the seeds. Recent evidence indicates that seeds may be viable even after 1,700 years (Wesson and Wareing, 1967), although they may need to be stimulated by light in order to germinate. The longevity of seeds may be much greater than this under ideal conditions. Some seeds of *Lupinus arcticus* have been obtained from lemming burrows deeply buried in the permanently frozen silt of an arctic tundra at least 10,000 years old. They readily germinated in the laboratory and grew to normal, healthy plants (Porsild et al., 1967).

In all the foregoing cases, death eventually occurs either because of (1) externally produced stresses (see Chapter 22) leading to "accidental" injury, or (2) internal changes in the normal development of the plant leading to senescence and death. If the stresses can be avoided, it is possible to produce a kind of immortality by vegetative reproduction. This can be achieved by means of stem cuttings, which can be rooted and produce new, vigorous plants, or even by excised roots (i.e., root tips), which can in some cases be maintained in the growing state essentially indefinitely by repeated excision of the tip and subculturing in artificial media. According to Molisch (1938), however, even such "rejuvenated" cuttings cannot be reproduced indefinitely. He believes that senescence continues, although at a slower rate, from one vegetative generation to another. In this way he explains the ultimate weakening and dying out of a variety after being reproduced vegetatively for many generations.

Senescence

Senescence is also called the aging of the plant. It is the stage preceding death or, in the case of organs, preceding abscission, which is normally soon followed by death. In the case of artificially cultivated (excised) roots, aging has been defined as a decrease and ultimate cessation of cell division in the apical meristem (Street, 1967). This does not include effects of starvation or the accumulation of an external inhibitor. It can be measured quantitatively by the decrease in growth rate. More generally, senescence implies the "running down" of an organism or one of its organs, involving changes of a degradative or degenerative nature (Wareing and Seth, 1967). The senescence of a whole perennial plant is usually much more gradual than that of organs (e.g., fruit, leaves) or of whole annual or biennial plants. There seems to be a basic qualitative difference between the two (Woolhouse, 1967). In general, the senescence of perennial plants appears to be dependent on factors external to the living cells.

In the case of trees (e.g., *Pinus sylvestris*), aging is apparently caused by competition for available nutrients between the increasing number of branches and also caused by the increasing distance for translocation between shoot apices and roots (Wareing and Seth, 1967). However, because of the time factor, senescence of perennial plants has not been extensively studied. Even in the more intensively investigated annuals and biennials the evidence seems to point to a competition for nutrients as a cause of senescence. Seed formation occurs at the expense of nutrients drained from the vegetative parts of the plant; and it has long been known that this hastens senescence. Thus, the prevention of seed formation by growing the plant under environmental conditions unsuited to sexual reproduction prevents or postpones senescence. *Perilla frutescens,* a short-day plant, is normally an annual plant and, therefore, dies at the end of a single growing season after flowering and fruiting. When grown in long days, it may survive many years in the vegetative state (Woolhouse, 1967). Similarly, cabbage is normally a biennial plant and therefore dies after flowering and fruiting during

its second year. Yet it has been kept growing vegetatively for several years by maintaining it at warm temperatures.

Simply removing the seeds as they form may delay leaf senescence by as much as 6 weeks. Decapitation of the shoot has the same effect. Detached leaves, if allowed to root at the petiole end, remain alive much longer than if they are left on the plant. These results all show that the senescence of plant organs is often under the control of the whole plant and is not solely caused by the intrinsic characteristics of its own cells (Wareing and Seth, 1967). In this respect plants differ from animals.

Nevertheless, this draining of nutrients from the vegetative plant cannot be the sole cause of senescence. Male annual plants may show normal senescence in the absence of fruiting, and hermaphroditic plants may show it in the absence even of flowering. Thus, in opposition to perennial plants, the senescence of annuals and biennials appears to be programmed by the living cells themselves.

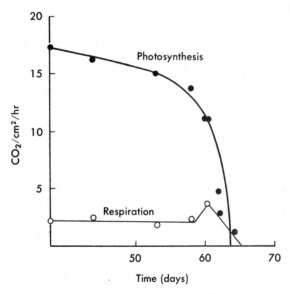

Fig. 20-5. Changes in rates of photosynthesis and respiration of the attached third pair of leaves of *Perilla frutescens* from the time of completion of leaf expansion to abscission. (From Woolhouse, H. W. 1967. Sympos. Soc. Exp. Biol. **21**:179-213.)

More detailed information as to the nature of senescence has been obtained by following the changes in parts of the plant rather than the plant as a whole, for example, in leaves. Measurements made on individual *Perilla* leaves show that the rate of photosynthesis declines steadily over a period of 20 days from the time of completion of expansion (Woolhouse, 1967). The decline then becomes suddenly more rapid and continues during 7 days prior to abscission, when it can no longer photosynthesize (Fig. 20-5). The decrease in rate of photosynthesis parallels the earlier described export of nutrients out of the senescing leaf. Respiration behaves differently. It remains constant until near the end, when it shows a brief, small climacteric rise, followed by a brief senescent drop similar to the changes in fruit (see Chapter 14).

The leaf proteins and pigments also decrease precipitously during senescence (Fig. 20-6). RNA decreases gradually, then more rapidly. DNA remains stable over a long period, followed by a period of rapid loss. The rapid loss in the case of both nucleic acids occurred at the same time as the senescent respiratory drop (Woolhouse, 1967).

The decrease in leaf protein has been shown by many investigators, but it is usually the total leaf protein that is measured. Woolhouse, however, showed that not all the proteins decrease at the same time. Fraction I protein continues to be synthesized up to the time of full leaf expansion, but after this the turnover is very slow. Another protein component of low molecular weight, on the other hand, continues to turn over at a relatively high rate. This may explain the difference between the photosynthesis and respiration curves since the Fraction I protein is from the chloroplast, and the other is not.

The changes in tree leaves are similar to the aging phenomena in leaves of herbaceous plants, but the protein decreases may commence before true senescence. In the case of *Ginkgo* leaves, development can be

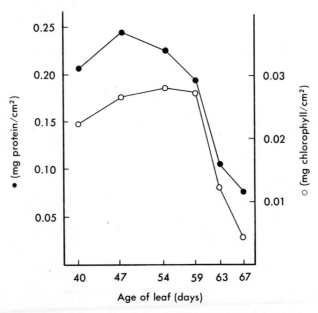

Fig. 20-6. The chlorophyll and protein content per unit area of the third pair of leaves of *Perilla frutescens* from the time of completion of expansion to the time of abscission. (From Woolhouse, H. W. 1967. Sympos. Soc. Exp. Biol. **21**:179-213.)

divided into three different periods (Specht-Jurgensen, 1967): a growing period until the end of June when nitrogen and chlorophyll content increases; a second period until the end of September when nitrogen, protein, and chlorophyll slowly decrease; and a breakdown period from the end of September until leaf abscission, when 66% of the nitrogen is lost and 40% of the remaining nitrogen is soluble (mainly amino acids and amides). Only the last period can be considered true senescence. The decrease in protein observed in tobacco leaves with increasing physiological age is correlated with increase in specific activity of peptidase (Anderson and Rowan, 1965). In the case of pear fruit the ability of mitochondria to incorporate amino acids into their protein declines as the tissues age and reach their climacteric peak (Romani and Fisher, 1966). Senescence may therefore involve both an increase in protein breakdown and a decrease in protein synthesis.

The organelles show a loss of structure during senescence and may be seen to break down. In the case of fully senescent *Phaseolus* mesophyll cells, almost all the cytoplasmic contents are lost (Barton, 1966). Only the plasmalemma and some empty vesicles remain. Even the cell walls may be dissolved (e.g., in senescent fruit). All these changes seem to be caused by the appearance, either by synthesis or activation, of specific enzymes capable of solubilizing these structures. The thylakoids of the chloroplasts, for instance, are decomposed by an enzyme that increases in quantity and moves from the plastids to other organelles at a later stage (Barton, 1966). Similarly, during senescence of barley leaves, polyribosomes and ribosomes are lost (Srivastava and Arglebe, 1967). This is the opposite of what happens during germination, for example, of lily pollen grains. The ribosomes then undergo a change indicating an alteration in the protein translation process. This is an essential step in the developmental process (Linskens, 1967).

It is now obvious that many physiological changes accompany the aging process. However, senescence is not necessarily a single unique process for all plants and plant parts, in which all the aforementioned changes occur in exactly the same sequence. Even in the case of a single organ, senescence may not be a uniform process in all plants. Thus, although these changes in aging leaves are quite general, there are actually two distinct kinds of leaf senescence. *Sequential aging* occurs when the leaves senesce in a definite order, usually the lowest leaf first and the others in an orderly succession. *Synchronous,* or *simultaneous, aging* occurs when they senesce all at once, for example, in the case of annual plants after fruiting or of deciduous plants in the fall. Translocation of substances from the leaf certainly plays a major role in sequential senescence since, as indicated earlier, it is initiated under normal conditions by competition between the mature leaves and the growing regions of the shoot. Translocation may also play a role in simultaneous senescence, for example, in the case of deciduous plants. Even in the case of excised leaves, translocation occurs from the blade to the base of the petiole (Simon, 1967).

The changes accompanying sequential aging of leaves all seem to be readily explained by translocation to the growing regions of the plant. The decrease in protein content would follow logically the export of amino acids since this would shift the equilibrium to hydrolysis of further proteins in the mature leaf. Since the chloroplast proteins serving as photosynthetic enzymes (Fraction I) appear to be the first to suffer such breakdown, this would also explain the early decrease in rate of photosynthesis. The later breakdown of cytoplasmic proteins would similarly lead finally to a decrease in respiration. This would also eventually result from the export of carbohydrate—the substrate for respiration. Even the synchronous aging of leaves of deciduous plants could be explained in the same way, although the translocation would be to storage tissues instead of to growing regions.

Translocation must therefore play an important role in senescence. Yet it was shown earlier that senescence cannot be postponed indefinitely by preventing the translocation of substances from an organ. Furthermore, the export of amino acids and sugars from the mature leaves to the growing (or storage) regions would be expected to continue only until the concentration gradient is eliminated. It seems obvious that the continuation of the changes characteristic of senescence requires something more than a simple decrease in reserves. More recent results have therefore led to a hormonal concept of senescence (Osborne, 1963). According to this point of view, a deficiency of one or more of the growth regulators leads to senescence. Thus, the auxin content declines with the age of the leaf and toward death is too low to measure accurately. This decline in auxin content is accompanied by an export of other substances, perhaps because of a direct effect of auxin on phloem transport (Thrower, 1967). Therefore, older leaves senesce more rapidly in the presence of young auxin-rich leaves above them. Cytokinins also seem to be involved, perhaps because of a diversion away from the mature leaves to developing fruit (Wareing and Seth, 1967). Similarly, a detached leaf that is allowed to root may be kept alive for years. This is believed to result from the synthesis of cytokinins in the roots and their translocation to the leaf blade. Senescence can actually be delayed by treating the leaves with growth regulators. Kinetin delays sequential aging, and auxin delays synchronous aging. Leaf senescence can also be retarded by gibberellic acid applications in the case of *Rumex obtusifolius* (Whyte and Luckwill, 1966). Growth retardants (e.g., CCC) may also be

used successfully (Kessler et al., 1967), even when cytokinins fail (Halevy et al., 1966). In other cases the reverse is true. Consequently, no one growth regulator can be considered a universal senescence inhibitor.

Hormonal control of aging also seems to operate in the case of roots. When excised roots are grown in artificial culture, an externally applied auxin precursor (tryptophan) is important for the maintenance of meristematic activity (Street, 1967). Some other unidentified factor is also involved that critically limits the growth of most strains until eventually it comes to a stop. Street suggests that the shoot is the primary site of synthesis of such limiting factors (e.g., thiamine) and that this may explain the aging of excised roots. This explanation is supported by the antiaging effect of the illuminated shoot on the root of the intact plant.

Many environmental factors such as water, mineral nutrients, light, etc. may alter the rate of senescence. In the case of excised roots of groundsel *(Senecio vulgaris)* or roots of plants grown in the dark, each meristem functions for only a limited time and then dies (Wren and Hannay, 1963). Within one root the lateral meristems continue to grow after the main apex has ceased growth, but they in turn die while newer, secondary laterals continue to extend. When the plants are grown normally in the light, no such rapid aging and senescence occur.

Since growth regulators (or hormones) are now believed to control plant growth by controlling the repression and derepression of genes (see Chapter 19), their control of senescence has been explained in the same way. A disruption of the normal balance between the growth regulators that exists in a growing plant or plant part would lead to a disruption in the normal equilibrium between repression and derepression of the genes. This would alter the nucleic acid metabolism and, therefore, the protein metabolism of the plant or plant part. The result would be senescence and death.

Abscission

One result of senescence of organs (leaves, fruit, flowers) is *abscission*—the natural separation of the organ from the plant. It is under the control of auxin (Thimann, 1965, see Chapter 19). While still active, the leaf secretes auxin, which is transported polarly down the petiole and inhibits formation of the abscission layer. After maturity, auxin production ceases; and when it reaches a low enough level, an abscission layer forms. This layer is also promoted by the protein breakdown in the leaf and the export of amino acids out of it. However, the process of abscission is more complex than this. It can be induced by the growth inhibitor abscisin, or dormin; and auxins may either promote or inhibit abscission (Carns, 1966), although kinetin cannot. At first sight it appears difficult to associate senescence with the abscission zone itself since the protoplasm is denser here (and therefore presumably more active) than in adjacent tissues; yet the zone is really a region of arrested development. A role is indicated by the development of high cellulase activity in the abscission zone before abscission occurs (Horton and Osborne, 1967). Presumably, this enzyme aids in cell separation by dissolving the cellulose portion of the cell walls. The promotion of abscission by inhibitors of protein synthesis (Valdovinos and Ernest, 1967) also supports this concept since it indicates that the enzymes necessary for the degradative process are already present in the abscission zone. In contrast, a continuous synthesis of proteins is required for leaf retention. Leaf abscission is therefore achieved by cellular senescence in the abscission zone (Leopold, 1967). Even this localized senescence is controlled by interacting systems of growth regulators.

Recent evidence has led to the hypothesis that senescence is genetical rather than a physiological wearing out (Curtis, 1963). According to this concept, aging or senescence is caused by somatic mutation. Thus, organs having cells that frequently undergo meiotic cell division show little or no aging, whereas those in which cells seldom, if ever, divide meiotically have no opportunity to throw off harmful mutations. Radiation accelerates the aging process, presumably because of the increase in mutations. This theory of aging is supported in the case of seeds since chromosome damage has been found to accumulate as they get older (Roberts et al., 1967). However, radiation damage seems to be of little importance in the senescence of growing plants (Woolhouse, 1967). Although it may conceivably explain the senescence of long-lived organisms, it certainly cannot explain that of short-lived organs since this can often be reversed. The senescence of old leaves may play a definite role in the physiology of the plant and may have been selected during evolution as a survival factor. In many cases (e.g., cucumber, Hopkinson, 1966) these older leaves may be useful not as photosynthesizing organs but as reserve organs, able to redistribute both their mineral elements and organic substances to younger, growing organs. Shading hastens the senescence of such leaves.

QUESTIONS

1. What are the main stages of development of the plant?
2. Are all parts of the plant simultaneously at the same stage?
3. How are these stages related to growth regulators?
4. What specific regulators have been found to control senescence?
5. What major chemical change occurs during senescence?
6. Is there any observable change in the cells?
7. What changes lead to abscission?
8. How is abscission related to senescence?

9. What is believed to be the origin of senescence?
10. What is meant by photoperiodism?
11. What is a short-day plant?
12. What is a long-day plant?
13. What is a day-neutral plant?
14. Would an alternation of 14 hr light and 14 hr dark act as a short day, a long day or neither?
15. Would 8 hr of light, ½ hr of dark, 7½ hr of light, and 8 hr of dark act as a short day, a long day, or neither?
16. What is a good definition of a short-day plant? Of a long-day plant?
17. Does photoperiodism involve anything beside flowering?
18. Give examples of the effects of photoperiodism.
19. What is meant by photoinduction?
20. What is the region of perception?
21. What evidence is there of hormonal control?
22. What is the relation between florigen and growth regulators?
23. What is phytochrome?
24. In what forms can phytochrome exist?
25. Which is the active form of phytochrome?
26. How is phytochrome believed to control plant responses to light?
27. What is thermoperiodism?
28. What is vernalization?
29. What kinds of plants respond to vernalization?
30. What relation is there between vernalization and photoperiodism?
31. What chemical substances have been found that can replace vernalization?
32. What evidence is there of a relation between nucleic acids and control of plant development?

SPECIFIC REFERENCES

Anderson, J. W., and K. S. Rowan. 1965. Activity of peptidase in tobacco-leaf tissue in relation to senescence. Biochem. J. 97:741-746.

Barton, R. 1966. Fine structure of mesophyll cells in senescing leaves of Phaseolus. Planta 71:314-325.

Bewley, J. D., M. Black, and M. Negbi. 1967. Immediate action, of phytochrome in light-stimulated lettuce seeds. Nature (London) 215:648-649.

Butler, W. L., H. C. Lane, and H. W. Siegelman. 1963. Nonphotochemical transformations of phytochrome in vivo. Plant Physiol. 38:514-519.

Chailakhyan, M. K. 1966. Role of gibberellins in the processes of photoperiodism and vernalization of plants. [Transl. title.] Biol. Zh. Armenii 19:3-14.

Chailakhyan, M. K., and V. N. Lozhnikova. 1966. Effect of interruption of darkness by light and plant gibberellins. Fiziol. Rast. 13:833-841 [734-741, Transl. ed.].

Curtis, H. J. 1963. Biological mechanisms underlying the ageing process. Science 141:686-694.

Deltour, R. 1967. Action de l'azote minéral sur la croissance et la mise à fleurs de plantes issues d'apex de Sinapis alba L. cultivés in vitro. Compt. Rend. Acad. Sci. (Paris) 265:1932-1935.

Dull, G. G., R. E. Young, and J. B. Biale. 1967. Respiratory patterns in fruit of pineapple, Ananas comosus, detached at different stages of development. Physiol. Plant. 20:1059-1065.

El-Antably, H. M. M., P. F. Wareing, and J. Hillman. 1967. Some physiological responses to D,L abscisin (dormin). Planta 73:74-90.

Evenari, M. 1961. A survey of the work done in seed physiology by the department of botany, Hebrew University, Jerusalem (Israel). Proc. Int. Seed Test. Asso. 26:597-658.

Garner, W. W., and H. Allard. 1920. Effect of the relative length of day and night and other factors of the environment on growth and reproduction in plants. J. Agric. Res. 18:553-606.

Halevy, A. H., D. R. Dilley, and S. H. Wittwer. 1966. Senescence inhibition and respiration induced by growth retardants and 6N-benzyladenine. Plant Physiol. 41:1085-1089.

Heide, O. M. 1967. The auxin level of Begonia leaves in relation to their regeneration ability. Physiol. Plant. 20:886-902.

Hendricks, S. B., and H. A. Borthwick. 1967. The function of phytochrome in regulation of plant growth. Proc. Nat. Acad. Sci. U. S. A. 58:2125-2130.

Heslop-Harrison, J. 1964. The control of flower differentiation and sex expression. Colloq. Int. Cent. Nat. Rech. Sci. 123:649-664.

Hirono, Y., and G. P. Rédei. 1966. Early flowering in Arabidopsis induced by DNA base analogs. Planta 71:107-112.

Hopkins, W. G., and W. S. Hillman. 1966. Relationship between phytochrome state and photosensitive growth of Avena coleoptile segments. Plant Physiol. 41:593-598.

Hopkinson, J. M. 1966. Studies on the expansion of the leaf surface. VI. Senescence and the usefulness of old leaves. J. Exp. Bot. **17**:762-770.

Horton, R. F., and D. J. Osborne. 1967. Senescence, abscission and cellulase activity in *Phaseolus vulgaris*. Nature (London) **214**:1086-1088.

Humphries, E. C., and A. W. Wheeler. 1963. The physiology of leaf growth Ann. Rev. Plant Physiol. **14**:385-410.

Jaffe, M. J., and A. W. Galston. 1967. Phytochrome control of rapid nyctinastic movements and membrane permeability in *Albizzia julibrissin*. Planta **77**:135-141.

Kasai, T., and Y. Obata. 1967. Changes in amino acid composition during germination of soybean. III. Changes in γ-glutamyltranspeptidase activity. Agric. Biol. Chem. **31**:127-129.

Kessler, B., S. Spiegel, and Z. Zolotov. 1967. Control of leaf senescence by growth retardants. Nature (London) **213**:311-312.

Lang, A., and G. Melchers. 1943. Die photoperiodische Reaktion von *Hyoscyamus niger*. Planta **33**:653-702.

Leopold, A. C. 1967. The mechanism of foliar abscission. Sympos. Soc. Exp. Biol. **21**:507-516.

Lincoln, R. G., A. Cunningham, B. G. Carpenter, J. Alexander, and D. L. Mayfield. 1966. Florigenic acid from fungal culture. Plant Physiol. **41**:1079-1080.

Linskens, H. F. 1967. Isolation of ribosomes from pollen. Planta **73**:194-200.

Lozhnikova, V. N. 1966. Dynamics of natural gibberellins under various photoperiodic cycle conditions. Dokl. Akad. Nauk S. S. S. R. **168**:223-226.

Mohr, H. 1962. Primary effects of light on growth. Ann. Rev. Plant Physiol. **13**:465-488.

Osborne, D. J. 1963. Hormonal control of plant death. Discovery (London) **24**:31-35.

Porsild, A. E., C. R. Harington, and G. A. Mulligan. 1967. *Lupinus arcticus* Wats. grown from seeds of Pleistocene age. Science **158**:113-114.

Rijven, A. H. G. C., and L. T. Evans. 1967a. Inflorescence initiation in *Lolium temulentum* L. IX. Some chemical changes in the shoot apex at induction. Aust. J. Biol. Sci. **20**:1-12.

Rijven, A. H. G. C., and L. T. Evans. 1967b. Inflorescence initiation in *Lolium temulentum* L. X. Changes in ^{32}P incorporation into nucleic acids of the shoot apex at induction. Aust. J. Biol. Sci. **20**:13-24.

Robbins, W. J. 1960. Further observations on juvenile and adult Hedera. Amer. J. Bot. **47**:485-491.

Roberts, E. H., F. H. Abdalla, and R. J. Owen. 1967. Nuclear damage and the ageing of seeds. Sympos. Soc. Exp. Biol. **21**:65-99.

Romani, R. J., and L. K. Fisher. 1966. Decreased synthesis of mitochondrial protein in the senescent cells of pear fruit. Life Sci. **5**:1187-1190.

Sachs, R. M., D. M. Kofranek, and S. Y. Shyr. 1967. Gibberellin-induced inhibition of floral initiation in Fuchsia. Amer. J. Bot. **54**:921-929.

Salisbury, F. B. 1965. The initiation of flowering. Endeavour **24**:74-80.

Scherf, H., and M. H. Zenk. 1967. Der Einfluss des Lichtes auf die Flavonoidsynthese und die Enzyminduktion bei *Fagopyrum esculentum* Moench. Z. Pflanzenphysiol. **57**:401-418.

Siegelman, H. W., E. C. Turner, and S. B. Hendrix. 1966. The chromophore of phytochrome. Plant Physiol. **41**:1289-1292.

Simon, E. W. 1967. Types of leaf senescence. Sympos. Soc. Exp. Biol. **21**:215-230.

Specht-Jurgensen, I. 1967. Studies on nitrogen compounds and chlorophyll during the senescence of the leaves of *Ginkgo biloba*. I. Leaves ageing on the tree. [Transl. title.] Flora **157**:426-453.

Srivastava, B. I. S., and C. Arglebe. 1967. Studies on ribosomes from barley leaves; changes during senescence. Plant Physiol. **42**:1497-1503.

Street, H. E. 1967. The ageing of root meristems. Sympos. Soc. Exp. Biol. **21**:517-542.

Suge, H., and N. Yamada. 1963. Chemical control of plant growth and development. 4. Promotion of flowering induced by uracil, uridylic acid, and several growth regulators in winter wheat. Proc. Crop Sci. Soc. Japan **32**:77-80.

Tanada, T. 1968. A rapid photoreversible response of barley root tips in the presence of 3-indoleacetic acid. Proc. Nat. Acad. Sci. U. S. A. **59**:376-380.

Thrower, S. L. 1967. The pattern of translocation during leaf ageing. Sympos. Soc. Exp. Biol. **21**:483-506.

Valdovinos, J. G., and L. C. Ernest. 1967. Effect of protein synthesis inhibitors, auxin, and gibberellic acid on abscission. Physiol. Plant. **20**:1027-1038.

Wareing, P. F., and A. K. Seth. 1967. Ageing and senescence in the whole plant. Sympos. Soc. Exp. Biol. **21**:543-558.

Wellensiek, S. J. 1966. The mechanism of flower formation in *Silene armeria* L. Naturwissenschaften **53**:411.

Wellensiek, S. J. 1967. The relations between the flower inducing factors in *Silene armeria* L. Z. Pflanzenphysiol. **56**:33-39.

Wesson, G., and P. F. Wareing. 1967. Light requirement of buried seed. Nature (London) **213**:600-601.

Whyte, P., and L. C. Luckwill. 1966. A sensitive bioassay for gibberellins based on retardation of leaf senescence in *Rumex obtusifolius* (L.) Nature (London) **210**:1360.

Woolhouse, H. W. 1967. The nature of senescence in plants. Sympos. Soc. Exp. Biol. **21**:179-213.

Wren, M. J., and J. W. Hannay. 1963. Ageing in roots of groundsel (*Senecio vulgaris* L.). New Phytol. **62**:249-256.

Yoneda, Y., and T. Stonier. 1966. Elongation of stem internodes in the Japanese morning glory *Pharbitis nil* in relation to auxin destruction. Physiol. Plant. **19**:977-981.

Yoshida, K., K. Umemura, K. Yoshinaga, and Y. Oota. 1967. Specific RNA from photoperiodically induced cotyledons of *Pharbitis nil*. Plant Cell Physiol. **8**:97-108.

Zeevaart, J. A. D. 1962. Physiology of flowering. Science **137**:723-731.

GENERAL REFERENCES

Barton, L. V. 1961. Seed preservation and longevity. Leonard Hill Books, Ltd., London.

Carns, H. R. 1966. Abscission and its control. Ann. Rev. Plant Physiol. **17**:295-314.

Hillman, W. S. 1962. The physiology of flowering. Holt, Rinehart & Winston, Inc., New York.

Hillman, W. S. 1967. The physiology of phytochrome. Ann. Rev. Plant Physiol. **18**:301-324.

Laetsch, W. M., and R. E. Cleland. 1967. Plant growth and development. Little, Brown & Co., Boston.

Lang, A. 1952. Physiology of flowering. Ann. Rev. Plant Physiol. **3**:265-306.

Molisch, H. 1938. The longevity of plants. Science Press, Lancaster, Pa.

Murneek, A. E., and R. O. Whyte. 1948. Vernalization and photoperiodism. The Ronald Press Co., New York.

Salisbury, F. B. 1963. The flowering process. Pergamon Press, Inc., New York.

Sax, K. 1962. Aspects of aging in plants. Ann. Rev. Plant Physiol. **13**:489-506.

Torrey, J. G. 1967. Development in flowering plants. The Macmillan Co., New York.

Whyte, R. O. 1946. Crop production and environment. Faber & Faber, Ltd., London.

chapter 21

Plant rhythms

A. Relation to vibratory, or periodic, motion
B. Annual rhythms
    1. Growth rhythm
    2. Dormancy or rest period rhythm
    3. Hardiness rhythm
    4. Water content rhythm
    5. Reproductive rhythms
C. Lunar rhythms
D. Circadian rhythms
E. Tidal rhythms
F. Exogenous and endogenous rhythms
G. Biological clocks

RELATION TO VIBRATORY, OR PERIODIC, MOTION

The rates and directions of many plant processes show regular patterns of change. The most obvious are the nyctinastic (sleep) movements of many plants—the closing of leaves or flowers at night and their opening at daybreak. Similar, although less obvious, patterns are to be found in many of both the physical and chemical processes in plants. In succulents, for instance, organic acids regularly accumulate at night and just as regularly disappear during daylight (see Chapter 3). The changes may occur daily, weekly, monthly, annually, etc., and they are therefore called rhythms. Since they are repeated at regular intervals of time known as periods, they are also sometimes called periodisms (or periodicities). The latter terminology has the advantage of indicating the similarity between plant rhythms and the periodic motions of physical systems. Because of the similarity, the terminology for plant rhythms has been borrowed from the accepted terminology for physical systems. It is, therefore, first necessary to be familiar with the fundamentals of periodic motion before discussing plant rhythms.

A periodic or vibratory motion may occur in any body that is able to oscillate. When, for instance, a weight is suspended from a spring and a downward force is exerted on the weight, the weight will oscillate in a vertical direction. This oscillation is called vibratory, or periodic, motion, and the quantitative relations are shown in Fig. 21-1.

There are two distinct kinds of oscillation.

1. The capacity to oscillate may be passive—the rhythm being driven by an external forcing agent (the synchronizer). Thus, if an external force repeatedly pushes the weight up and down, a *forced vibration* occurs, resulting in a *forced period*. The periodic motion is then said to be entrained by the external force.

2. On the other hand, the capacity to oscillate may be active, resulting in spontaneously oscillating systems with natural periods. If such a system is allowed to oscillate freely, a *natural vibration* occurs, and the time for one complete cycle will be the *natural period*.

When the impressed force is in unison

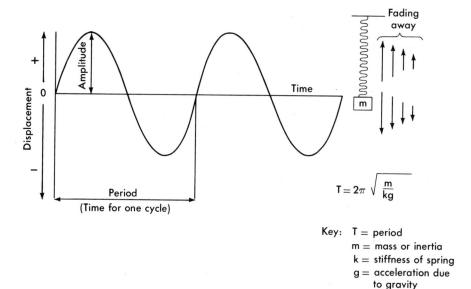

Key: T = period
      m = mass or inertia
      k = stiffness of spring
      g = acceleration due
          to gravity

Fig. 21-1. Periodic or vibratory motion. When a weight is displaced on the end of a spring (top right), a periodic motion is initiated, but it fades away due to frictional resistance.

with the natural vibration, this is called *resonance. Sympathetic vibrations* are produced when there is resonance. The *phase* is the position of the vibrating body at any instant plus the direction in which it is moving. It is measured as an angle with reference to a circle. Two objects are in phase when they are in the same position relative to the circle and moving in the same direction.

In order for a natural vibration to occur, there must be:

1. An impressed force, or an initial displacement
2. A return force
3. Inertia (otherwise the body will come to a stop when the force is zero)
4. Not too large a frictional resistance to movement

Since, however, friction does always exist, the amplitude of the motion in the absence of the impressed force is decreased by the friction in each successive period until the motion comes to a stop. This is called *fading away* (Fig. 21-1). All the preceding terms and concepts are used in describing plant rhythms.

Some plant rhythms (such as the sleep movements of leaves, the opening and closing of flowers) are, indeed, literally periodic motions. Other rhythmic phenomena, such as the luminescence of dinoflagellates, metabolic rhythms, etc. are not motions; nevertheless, the terminology of periodic motion may be applied to them. The plant's environment is also characterized by rhythms, and the plant's rhythms are normally found to follow the environmental rhythms. They are therefore classified according to the environmental rhythm that they parallel.

ANNUAL RHYTHMS

There are many annual plant rhythms, some of which are obviously controlled by the annual environmental temperature rhythm. Others, however, are not so obviously related.

Growth rhythm

In temperate climates, at least, growth must obviously come to a stop during winter when the plant freezes. However, growth actually ceases well before the environ-

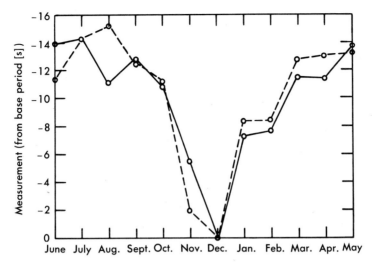

Fig. 21-2. Seasonal pattern in stomatal opening in pinto beans. Ordinate = decrease (sec) in time for the air to flow through the leaf, compared to a minimum opening in December; —, plants grown in a growth chamber and a photoperiod of 16 hr; - - -, plants grown in a greenhouse and the normal photoperiod for the time of the year. (From Seidman, G., and W. B. Riggan. 1968. Nature [London] **217**:684-685.)

mental conditions become unfavorable. This is caused by the following rhythm (dormancy).

Dormancy or rest period rhythm

Many plants enter a rest period every fall and emerge from it in late winter or early spring. This rhythm is not so obviously related to the environment since the plant enters its rest period (and therefore stops growth) while the environment is still capable of supporting growth. Similarly, the rest period is broken before the environmental temperatures are capable of supporting growth. Even the ability of the plant to open its stomata under favorable environmental conditions shows this annual rhythm (Fig. 21-2).

Hardiness rhythm (see Chapter 22)

Again, the hardiness rhythm is not a simple, forced rhythm since a well-adapted plant shows an increase in frost resistance long before any frosts occur. This is accompanied by other annual rhythms, for example, a winter maximum in sugar content,

and therefore in osmotic pressure, and a minimum in starch content. In xylem tissues of some trees there may, however, be a starch maximum in winter. Other substances may also show a maximum in winter, for example, soluble proteins. There are similar annual rhythms in resistance to drought injury, heat injury, and according to recent results (Biebl and Hofer, 1966), even radiation injury.

Water content rhythm

In temperate climates some tree trunks show maximum water contents in winter and minimum water contents in summer (Gibbs, 1939).

Reproductive rhythms

Many plants reproduce only at one season in the year, for example, in spring.

Annual rhythms are only beginning to be investigated under artificial conditions (Fig. 21-2) because of the difficulties in growing plants under controlled conditions for a year and because of the time required to complete an experiment. It is therefore not known to what extent they are forced rhythms and to what extent natural rhythms.

LUNAR RHYTHMS

Lunar rhythms have not been studied extensively and apparently are not as common as annual rhythms. In brown algae (e.g., *Dictyota*) there is a lunar periodicity in the release of sexual cells. This periodicity can be repeated under artificial conditions (Muller, 1962). If the daily light-dark periods are interrupted every 28 days by giving artificial moonlight during one dark period, the production of eggs reaches a maximum on the tenth and twenty-sixth days after the stimulus. A "moonlight" intensity of 3 lux is fully effective. After this periodicity is induced, the plants will show their next maximum release of sexual cells at 16- to 17-day intervals without any further exposure to the moonlight stimulus. This is evidence of the existence of a natural rhythm.

CIRCADIAN RHYTHMS

Circadian rhythms are the commonest and have been studied the most intensively. They have also been called diurnal rhythms. The term *circadian* is preferred because, as will be seen later, the periods are only approximately 24 hr in length (circa means approximately, dies means day).

Leaf (sleep) movements were the earliest studied circadian rhythms (Fig. 21-3). That they are responses to the daily light rhythm can be easily demonstrated by darkening the plant during daylight. This induces the sleep movement that otherwise occurs only at night. On the other hand, the rhythmic movement will continue even in continuous, weak light (Fig. 21-3). When this occurs, the period is seen to be a little longer than 24 hr. The opening of flowers is also a typical circadian rhythm in many plants. Other well-known circadian rhythms are the luminescence of dinoflagellates *(Gonyaulax* sp.) and even metabolic processes such as carbon dioxide evolution.

There are some differences between the rhythmic leaf movements of plants from different geographical regions (Mayer, 1966). In general, under constant conditions the oscillations in arctic plants persist for a shorter time than in plants of other regions. The length of the natural ("free running") period, however, is the same in arctic plants (24 hr) as in plants from Central Europe. On the other hand, the natural periods of some tropical plants are much longer than 24 hr. In several of these tropical species the period decreases greatly with increasing temperature. Consequently, if the tropical plants are tested at 27° C and the Central European and arctic plants at 17° C, the periods always approach 24 hr.

In recent years an artificially induced circadian rhythm has become common. When organisms or tisssues are grown in artificial culture, it is frequently desired to have all, or nearly all, the cells pass through the same stage (e.g., cell division) simultaneously. These are called *synchronous cultures,* and they can be produced by a rhythmic daily alternation of some environmental factor (e.g., light, temperature, etc.).

TIDAL RHYTHMS

Tidal rhythms have a period of 12½ hr and have been investigated in marine plants (Sweeney, 1963). There are many rhythms shorter than the tidal rhythms, such as (1) circular rhythms in twining shoots, tendrils, and some pulvini, with periods of 1 to 2 hr; (2) the reversal of the direction of cyto-

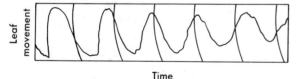

Fig. 21-3. Sleep movements of *Phaseolus* leaves in continuous, weak light—a circadian movement. Guide lines are 24 hr apart, and the length of periods is 27 hr. (From Bünning, E. 1964. The physiological clock. Springer-Verlag, Berlin.)

plasmic streaming, which may have a period of 60 sec. These rhythms have been called ultradian (Bailloud, see Aschoff, 1965).

EXOGENOUS AND ENDOGENOUS RHYTHMS

The first logical question to ask is whether plant rhythms are forced (passive) or natural (active). In most cases, as shown by the preceding classification, the period of plant rhythms seems to be that of an external oscillator arising because of the motions of the earth (annual and circadian rhythms) or of the moon (lunar and tidal rhythms). These external oscillators have their own external rhythms (Sollberger, 1965) and are classified by the strength of their effects on plant rhythms as strong or dominant (light, temperature) and weak synchronizers (sound, vibration, atmospheric humidity, atmospheric pressure).

This would seem to indicate that the plant's rhythms are forced (passive); but it does not reveal whether or not there is an internal oscillator whose period may be entrained by the external oscillator. In order to find out, it is therefore necessary to remove the plant from the action of the external oscillator. Thus, in the case of an external oscillator caused by the daily alternation of light and dark, the plant must be placed in continuous light or continuous dark. A forced (passive) rhythm either stops immediately or fades away after a few oscillations, with the same period as that of the external oscillator. A natural (active) rhythm continues to oscillate, but with a period different from that of the external oscillator. This shows that the external oscillator entrains the plant rhythm, giving it an artificial period; and as soon as the environmental rhythm is removed, the plant reverts to its own natural period. Plant rhythms that are forced are called *exogenous,* and those that are natural are called *endogenous.* An endogenous rhythm can therefore be defined as one capable of self-sustained oscillations after the external entraining rhythm is removed and of having its own natural period.

Some plant rhythms seem to belong clearly to one or other of these two groups. An example of an exogenous rhythm is the daily alternation of carbon dioxide absorption and evolution. If the alternation of light and darkness ceases and the plant is kept in continuous darkness, photosynthetic carbon dioxide absorption also ceases. There are many other examples of such exogenous rhythms. On the other hand, the circadian sleep movements of leaves are as pronounced in continuous weak light as when exposed to normal daily alternations of light and darkness (Fig. 21-3) and have periods close to, although not exactly equal to, 24 hr. They are therefore endogenous. However, even endogenous rhythms must be initiated by an external stimulus. Seedlings raised in complete darkness fail to show their normal rhythms (e.g., sleep movement of leaves). Yet one brief light signal may be sufficient to initiate the rhythms. Even some of the apparently exogenous rhythms may have an endogenous component. The daily rhythm of photosynthetic carbon dioxide assimilation is obviously exogenous since it immediately and completely comes to a stop in continuous darkness and cannot recommence in the absence of light; nevertheless, the *capacity* to photosynthesize follows an endogenous rhythm. The plant will therefore photosynthesize at a greater rate at one time of the day than at another, even under constant environmental conditions. Similarly, the opening and closing of stomata may sometimes be strictly controlled by alternations of light and dark, but the "opening ability" in response to light may be altered in a diurnal periodicity, depending on the length of the preceding dark period. Even in continuous darkness, a diurnal rhythmic opening and closing of the stomata occurs (Stålfelt, 1963). A sharp line between exogenous and endogenous rhythms may, therefore, not always be possible.

There are two possible kinds of endogenous rhythms:

1. The rhythms may be similar to sympathetic vibrations. This would involve a kind of resonance, and the plant would have its own natural rhythm which happened to have the same period as that of the environmental rhythm or synchronizer. Removal of the environmental rhythms would immediately reduce the amplitude of the plant rhythm but would not alter its period.

2. The observed period might be entrained by the environment, and as soon as the environmental rhythm is removed, the plant would revert to its own natural period.

In both cases the amplitude of the rhythm would decrease steadily with time; that is, the rhythm would fade away. Circadian rhythms are commonly of the second kind, and as their name implies, their natural period is only approximately 24 hr in length (Fig. 21-3). The periods of the different

kinds of rhythms may vary from .001 sec to a day, a month, a year, or more (Sollberger, 1965); but they most commonly are close to the period of some environmental factor. This common similarity between the natural periods and those of the entraining external oscillator is not likely to result from mere coincidence. The obvious explanation, of course, is that it arose by natural selection.

Further evidence of the endogenous nature of the plant rhythms can sometimes be obtained after the rhythm has come to a stop because of removal of the environmental rhythm. A single brief exposure to the environmental factor (e.g., a flash of light) may supply the "displacement" to initiate the plant rhythm (e.g., in the case of sleep movements). The unalterability of the plant's natural period can be demonstrated by entraining it to another period (e.g., a 14 hr day). As soon as the environmental rhythm is removed, the plant returns to its own natural period, even after an entrainment for as long as a year.

According to Bünning (1948, see Chapter

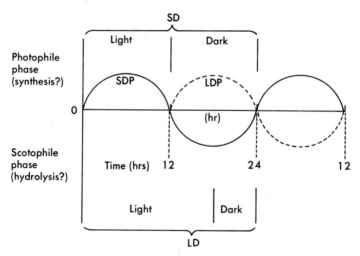

Fig. 21-4. Bünning's explanation of photoperiodism on the basis of an endogenous rhythm of alternating photophile (light-loving) and scotophile (dark-loving) phases. In short-day plants (SDP) the photophile phase begins immediately on illumination; in long-day plants (LDP) it begins only about 12 hr after the beginning of illumination. **SD,** Short day; **LD,** long day. (From Murneek, A. E., and Whyte, R. O. 1948. Vernalization and photoperiodism. The Ronald Press Co., New York.)

20), photoperiodism involves an endogenous rhythm explained in Fig. 21-4. When in the photophile phase, the plant is highly synthetic and when in the scotophile phase, highly hydrolytic. The difference between short-day and long-day plants is that in the former the endogenous rhythm is changed to the photophile phase soon after the beginning of illumination. In the long-day plant the rhythm is not changed until 12 hr after the beginning of illumination. Thus, in the long-day plant a long light period is unnecessary, provided that the plant is illuminated after the photophile phase is initiated—preferably in the middle of the photophile period. Some evidence seems to agree with his concept. Tomatoes, for instance, are indeterminate, yet they will not survive continuous illumination (Highkin and Hanson, 1954). The leaves that develop are small, stiff, and yellow with dark necrotic spots. The same effect is obtained when 24 hr light alternate with 24 hr dark. This is explained by assuming that the endogenous rhythm of alternating photophile and scotophile phases has a period of 24 hr. Only when the photophile phase coincides with the light period and the scotophile phase with the dark period can normal growth and development occur. Other experiments oppose this concept since 4 hr of light during the postulated scotophile phase hastens flowering (Hussey, 1954).

BIOLOGICAL CLOCKS

Since so many rhythms appear to be endogenous, the plant is said to possess a "biological clock," which controls the period of the rhythms. According to most investigators, this clock has an intrinsic mechanism (i.e., within the plant) with a precise and dependable method of time measurement, which does not require continuous information from the environment (Sweeney, 1963). From this point of view the environment supplies only the secondary "impressed force," which may change the period of amplitude of the natural rhythm.

This would lead to the conclusion that the rhythm of plant development is controlled primarily by the internal clock and only secondarily by external rhythms. The latter interpretation is in agreement with the foregoing evidence of natural rhythms in the plant.

Many investigators are now devoting their efforts to an elucidation of the nature of the biological clock. Some say it is physical, others that it is chemical. Hastings (Aschoff, 1965) believes that the rhythmicity is in some way dependent on the cell's normal ability to synthesize a specific RNA. Evidence for this concept is the effect of inhibitors of nucleic acid or protein synthesis on the rhythm. In *Gonyaulax* (Sweeney in Aschoff, 1965) the rhythmic activity of luciferase was abolished by actinomycin D and puromycin—inhibitors of RNA and protein synthesis. A similar result has been obtained with the unicellular green alga *Acetabularia*. The rhythms in photosynthetic capacity and in chloroplast shape were dramatically inhibited by actinomycin D (Driessche, 1966); but this was true only of intact algae. Anucleate algae were able to maintain their rhythms, and in this case actinomycin D did not affect them. These results have been corroborated by Sweeney and associates (1967). In the case of anucleate *Acetabularia*, inhibitors of protein synthesis (actinomycin D, chloramphenicol, puromycin) had no effect on either the period or the phase of the photosynthetic rhythm, in spite of their large effect on incorporation of C^{14}-leucine. According to Driessche's hypothesis, the mechanism is dependent on nuclear DNA, but the oscillations are possible only in the presence of a light-dependent and nucleus-dependent substance that triggers the biological clock. The nucleus controls the synthesis of messenger RNA's (mRNA) carrying the genetic information for the oscillator and for the light-dependent substance. The nucleus is not immediately required for maintenance of the rhythm because the nucleus-dependent mRNA is particularly stable and therefore has a long life.

The endogenous rhythm in the rate of carbon dioxide output by *Bryophyllum* has a different explanation (Wilkins, 1967). It is caused by a rhythm in the dark fixation of carbon dioxide, which continues in prolonged darkness, fading away in 5 days. This is explained by a periodic change in activity of the enzyme phosphoenol-pyruvate carboxylase (which fixes carbon dioxide in the dark). In this case the change in enzyme activity appears to be controlled by an inhibitor that may actually be a product of carbon dioxide fixation.

During synchronous (i.e., uniformly rhythmic) growth of *Chlorella*, there is a periodism in the rate of increase in the cellular level of DNA (Johnson and Schmidt, 1966). A similar periodism in the level of the enzyme deoxythymidine monophosphate kinase precedes the DNA rise by 1 hr, suggesting that this enzyme may be limiting the rate of DNA synthesis. Similar results with another enzyme indicate that perhaps the synthesis of all the enzymes on the deoxythymidine triphosphate (dTTP) pathway may be coordinated and in control of the synchronous growth.

In spite of all the foregoing evidence, we still do not know how the biological clock is timed. We do not even know for certain whether the primary timing oscillators are in the organism or in its environment (Brown, 1960). As indicated earlier, most investigators assume that the organism is an independently oscillating system, with one of its own natural periods closely approximating a day. An alternative hypothesis (Brown in Aschoff, 1965) is that the organism under natural conditions possesses no intrinsic daily rhythmicity but is merely responding to extrinsic (i.e., environmental) rhythms. The organism is exposed not only to the gross rhythms of light and temperature but also to subtler geophysical rhythms of cosmic rays, magnetism, barometric pressure, etc. Unlike the diurnal rhythms, the latter are always present and uncontrolled

by the experimenter. Brown, therefore, suggests that the plant is like a sundial or electric clock: it draws its timing capacity from outside itself. He calls this an extrinsic living clock mechanism. He admits, however, that both kinds of timing mechanism may conceivably occur.

Perhaps the best evidence for the possibility of endogenous rhythms in living organisms is their existence in nonliving systems. An example of the latter are the periodic precipitation rings known as Liesegang rings, which may be seen in ordinary gels. They occur when one electrolyte diffuses into the solution of another with which it interacts to form a precipitate(for instance silver nitrate diffusing into potassium bichromate). The deposition of this precipitate is rhythmical, appearing as bands, which gradually widen and separate outward as the gradient of diffusion falls (Scarth and Lloyd, 1930). Since the rhythmic precipitation is independent of any environmental rhythms outside the gel, it is a true endogenous rhythm. Circadian, Liesegang-like rhythms of zonation in growth have been found in cultures of some fungi (Jerebzoff in Aschoff, 1965). In this case, all the factors that promote the rhythms of zonation are amino acids connected with protein synthesis. It is, therefore, perhaps conceivable that all the rhythms are caused by a rhythmic change in availability of substances required to produce the particular phenomenon. In the case of Liesegang rings, diffusion of one of the reacting substances is the limiting factor; in the case of the plant, perhaps synthesis of RNA, amino acids, etc., or in some cases a physical factor such as diffusion is the limiting factor. When the limiting material is used up, the process stops or is reversed until the deficiency is overcome by diffusion or regeneration of the limiting factor.

Regardless of the origin of plant rhythms, they do exist. Because of their existence, optimal plant development *requires* an alternation in environmental conditions, for example, light, temperatures, etc. They are therefore of practical as well as theoretical importance.

QUESTIONS

1. What are rhythms or periodisms?
2. What is the difference between a forced period and a natural period, a passive rhythm and an active rhythm?
3. When is a rhythm entrained?
4. What is fading away?
5. What annual rhythms occur in plants?
6. Are any lunar rhythms known in plants?
7. What is a circadian rhythm?
8. Give an example.
9. Is the period for a specific circadian rhythm constant or variable? Explain.
10. Are there any rhythms in plants with periods shorter than one day? Explain.
11. What is the difference between exogenous and endogenous rhythms? Give examples of each.
12. What is meant by weak and strong synchronizers of plant rhythms?
13. How can we determine whether the plant rhythm is forced or natural?
14. How can a plant rhythm be initiated?
15. When does it fade away?
16. Can the plant's natural rhythm be changed? Explain.
17. What is meant by a biological clock?
18. Is it physical or chemical in nature?
19. What are the current concepts of its mechanism?
20. How can it be prevented from functioning?
21. Is the timing mechanism internal or external? Explain.
22. For optimum growth and development, should plants be grown under constant or alternating environmental conditions? Why?

SPECIFIC REFERENCES

Biebl, R., and K. Hofer. 1966. Tages-und Jahresperiodik der Strahlenresistenz pflanzlicher Zellen. Radiat. Bot. 6:225-250.

Brown, F. A. 1960. Response to pervasive geophysical factors and the biological clock problem. Sympos. Quant. Biol. 25:57-71.

Driessche, T. Van den. 1966. The role of the nucleus in the circadian rhythms of *Acetabularia mediterranea*. Biochim. Biophys. Acta. 126:456-470.

Gibbs, R. D. 1939. Studies in tree physiology. I. General introduction; water contents of certain Canadian trees. Canad. J. Res. [C] 17:460-482.

Highkin, H. R., and J. B. Hanson. 1954. Possible interaction between light-dark cycles and endogenous daily rhythms. Plant Physiol. 29:301-302.

Hussey, G. 1954. Experiments with two long day plants designed to test Bünning's theory of photoperiodism. Physiol. Plant. 7:253-260.

Johnson, R. A., and R. R. Schmidt. 1966. Enzymatic control of nucleic acid synthesis during synchronous growth of *Chlorella pyrenoidosa*. I. Deoxythymidine monophosphate kinase. Biochim. Biophys. Acta 129:140-144.

Mayer, W. 1966. Besonderheiten der circadianen Rhythmik bei Pflanzen verschiedener geographischer Breiten. Planta 70:237-256.

Moorby, J., and P. F. Wareing. 1963. Ageing in woody plants. Ann. Bot. 27:291-308.

Muller, D. 1962. One year lunar periodicity phenomena in several brown algae. Bot. Marina 4:140-155.

Scarth, G. W., and F. E. Lloyd. 1930. Elementary course in general physiology. John Wiley & Sons, Inc., New York.

Seidman, G., and W. B. Riggan. 1968. Stomatal movements; a yearly rhythm. Nature (London) 217:684-685.

Stålfelt, M. G. 1963. Diurnal dark reactions in the stomatal movements. Physiol. Plant. 16:756-766.

Sweeney, B. M. 1963. Biological clocks in plants. Ann. Rev. Plant Physiol. 14:411-440.

Sweeney, B. M., C. F. Tuffli, Jr., and R. H. Rubin. 1967. The circadian rhythm in photosynthesis in *Acetabularia* in the presence of actinomycin D, puromycin, and chloramphenicol. J. Gen. Physiol. 50:647-659.

Wilkins, M. B. 1967. An endogenous rhythm in the rate of carbon dioxide output of Bryophyllum. V. The dependence of rhythmicity upon aerobic metabolism. Planta 72:66-77.

GENERAL REFERENCES

Aschoff, J. 1965. Circadian clocks. North-Holland Publishing Co., Amsterdam.

Bünning, E. 1964. The physiological clock. Springer-Verlag, Berlin.

Sollberger, A. 1965. Biological rhythm research. Elsevier Publishing Co., Amsterdam.

chapter 22

Stress resistance

ENVIRONMENTAL STRESSES

It follows from the previous chapters that although the growth and development of the plant are internal processes, they are under the control of the environment. Temperature, moisture, radiation, nutrients, and gases can all either enhance or retard growth and development. However, their effects may transcend the quantitative—they may also act as stresses, leading to injury and death of the plant.

In the physical sense a force accompanied by its counterforce constitutes a stress. Quantitatively, the stress is expressed as the force per unit area and therefore has the dimensions of pressure. As a result of the stress, a body undergoes a strain, for example, an increase in length, a change in volume or shape, etc. The strain may be elastic and reversible, or it may be an irreversible set. In the physiological sense a stress is a potentially injurious force or pressure acting on the plant that may lead to a reversible strain or to an irreversible strain (injury or death). The plant is constantly being subjected to the stresses of its environment, and it survives only because of its adaptation to them. There are two main kinds of adaptation possible:

1. *Capacity adaptation* enables the plant to grow and develop under environmental conditions that do not permit normal growth and development in the case of unadapted plants. This may mean simply the development of lower temperature minima or higher temperature maxima for its physiological processes.
2. *Resistance adaptation* enables the plant to survive environmental stresses that kill unadapted plants.

In the second case the stress acts more directly and quickly. In the first case death may also occur, but only after an extended exposure to the environmental stress, sufficient to produce a disturbance in the metabolism of the plant. Capacity adaptation has been studied more extensively in animals than in plants, and resistance adaptation, more in plants than in animals. We will therefore discuss the latter only. Strains must occur as a result of the environmental stress, but in the resistant plants these are always reversible. There are many kinds of environmental stresses, and each may produce its own kind of irreversible strain or stress injury.

Temperature stress

1. Most plants in the growing state are killed by freezing; and even dormant overwintering plants may be injured. This is called *freezing, or frost, injury.*
2. Some plants (mainly from tropical regions) are also injured or killed by exposure to low temperatures above the freezing point. This is called *chilling injury.*
3. Temperatures of 60° C or higher will kill most living organisms, and even more moderate high temperatures may injure some plants. This is called *heat injury.*

Moisture stress

Moisture may injure or kill plants if present in excess, but this is actually due to lack of oxygen and is called a *flooding injury.* A more common injury, however, is caused by a water deficit, and this is called *drought injury.*

Radiation stress

Visible radiations (light) are seldom directly responsible for death, although they are sometimes capable of causing *radiation injury.* Radiation below the wavelength of light may, however, be highly injurious or fatal. This does not commonly occur in the natural habitat of the plant since only the ultraviolet radiations are of shorter wavelength than light in the radiation received on earth from the sun, and they are usually of too low an intensity to injure most plants. However, because of the modern interest in nuclear energy and in outer space, injury by radiations of lower wavelength is now receiving greater attention. Injury caused by infrared radiation is actually heat injury.

Salt stress

Salt stress can be produced by an excess of any one of a large number of salts. It is important in nature since there are many saline (sodium salt) soils. Plants showing salt (sodium) resistance are called *halophytes.*

Gas stress

Because of modern industrialization, pollution injury is daily becoming more important. This is commonly caused by toxic gases and vapors.

KINDS OF STRESS RESISTANCE

In order to survive, the plant must therefore develop resistance toward such potentially injurious environmental stresses. Although there are many kinds of stresses, there are only two basic resistance mechanisms.

1. The plant may exclude the stress from its tissues and therefore possess *stress avoidance.*
2. The plant may survive the penetration of its tissues by the stress and therefore possess *stress tolerance* (or "hardiness").

Thus a cold-avoiding organism is one that remains warm at low temperatures, and a cold-tolerant one becomes as cold as its environment but because of its tolerance or hardiness is not injured by the cold. The same terminology may be used for the other factors (Table 22-1).

The two kinds of resistance may not always be developed to the same degree. Tolerance seems to be the more primitive of the two adaptations because it is the only kind of resistance developed by lower plants in the case of the three factors most intensively studied (Table 22-2). Similarly, drought resistance seems to be the most advanced of the three resistances since higher plants show the highest development of avoidance in the case of the drought stress.

Freezing (or frost) injury and resistance

Plant protoplasm can survive the lowest temperatures attainable (0°K approximately), if no ice forms in the tissues. Even if frozen, it may survive the lowest temperatures, provided that the freezing process is of a specific kind, for example, if the ice crystals are so small as to require x-ray

Table 22-1. Stress resistance

Environmental stress	Condition of resistant plant possessing	
	Avoidance	Tolerance
1. Chilling temperatures (cold but above freezing)	Warm	Cold
2. Freezing temperatures	Unfrozen	Frozen
3. High temperatures	Cool	Hot
4. Drought	High vapor pressure	Low vapor pressure
5. Radiation	Low absorption	High absorption
6. Salt	Low salt concentration	High salt concentration
7. Flooding (oxygen deficiency)	High oxygen content	Low oxygen content

Table 22-2. Relative importance of tolerance and avoidance in lower and higher plants in the vegetative state

Stress	Resistance in lower plants	Resistance in higher plants
Frost	Solely tolerance	Nearly always tolerance
Heat	Solely tolerance	Mainly tolerance
Drought	Solely tolerance	Mainly avoidance

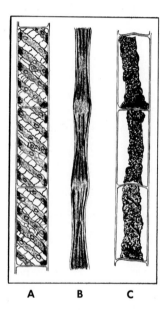

Fig. 22-1. Contraction of *Spirogyra* cell caused by extracellular freezing (and therefore dehydration). Thawed cell shows coagulated (killed) protoplasm. **A,** Normal; **B,** frozen; **C,** thawed. (From Molisch, H. 1897. Untersuchungen über das Erfrieren der Pflanzen. Gustav Fischer Verlag, Jena.)

analysis for detection. Under normal conditions, however, only air-dry plant parts (e.g., seeds) can show the first kind of survival, and the second kind occurs only under artificial conditions (see later). Frost resistance of all except dehydrated plant parts (with a few exceptions) is therefore tolerance since the vast majority of plants cannot avoid freezing on exposure to extreme subfreezing temperatures. But this tolerance exists only toward extracellular (i.e., intercellular) ice formation. No plant can survive the formation of microscopically visible crystals within their living cells (intracellular freezing), at least not if the ice is formed within the protoplasm.

Since the extracellular freezing removes water from the cell, the cell becomes more and more dehydrated as the temperature falls further and further below the freezing point (Fig. 22-1). In many plants it is not until the temperature drops to −20° to −30° C that 95% of the cell's water may be

removed in this way. If the cell cannot survive such a profound dehydration, it will obviously be killed. It is therefore logical to conclude that when plants are frozen extracellularly, any injury that occurs is likely to be a dehydration injury. If dehydration injury and resistance can be explained, this can therefore also be expected

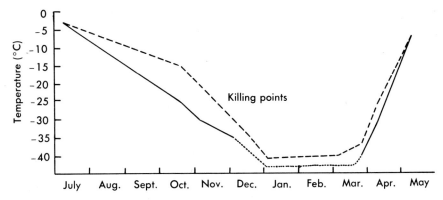

Fig. 22-2. Annual change in freezing resistance of the twigs of two apple varieties. (From Hildreth, see Levitt, J. 1966. *In* H. T. Meryman. Cryobiology. Academic Press, Inc., New York.)

to explain frost injury and resistance. Once the cell has nearly all of its water frozen (at about −30° C) without injury, it can also survive subsequent immersion in liquid air or liquid nitrogen and may logically be expected to survive absolute zero.

Frost resistance varies seasonally, so that even the most resistant plants of midwinter, which survive temperatures of −50° C (or even −190° C under artificial conditions), are killed by about −5° C in early spring (Fig. 22-2). During the fall the plant "hardens"; that is, it slowly develops a greater and greater tolerance or hardiness, until the maximum is reached in midwinter, and then it dehardens slowly until the minimum is reached in spring. Hardening can be induced artificially by exposure to low temperatures (e.g., 0° to 5° C), dehardening by exposure to high temperatures (e.g., above 10° C).

The problem of freezing effects has become so important that in recent years it has led to the development of a whole new science—*cryobiology* (literally, the biology of freezing temperatures). Many biologists from different biological disciplines are cryobiologists: the surgeon who freezes tissues in order to remove them, the microbiologist who freezes cultures in order to keep them alive for long periods without subculturing them, the biophysicist who is interested in the nature of the freezing process, and the biochemist who investigates the effect of freezing on the rates and kinds of chemical reactions. There are, therefore, branches of cryobiology called cryosurgery, cryopreservation, cryobiophysics, cryobiochemistry, etc.

Drought injury and resistance

Unlike frost resistance, drought resistance may be caused by either avoidance or tolerance. Plants such as succulents are "water savers" and may survive long periods of drought by cutting down water loss to a minimum. This is accomplished by a very thick and highly impermeable cuticle and the closure of their stomata during a large part of the day. This forces them to develop a special kind of metabolism (see Chapter 15) in order to survive. Other plants are "water spenders" and maintain their water contents at a high level by means of a very large and active water-absorbing system relative to the water-losing portion (Fig. 22-3). As is true in other societies, the spenders get along better than the savers when liquid water is to be found somewhere (e.g., at considerable depths), but only the savers can survive if there is no water to be absorbed for long periods of time.

Drought avoidance seems to be the major

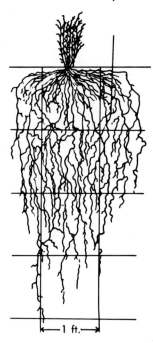

Fig. 22-3. A water spender. Root system is considerably larger than the shoot system. (From Weaver, J. E. 1919. Carnegie Institute of Washington, Pub. No. 286. Washington, D. C.)

mechanism of survival in vegetative higher plants. However, in vegetative lower plants and even in dormant parts of higher plants (e.g., seeds, pollen grains, etc.), the main method of survival is drought tolerance. Even the vegetative parts of higher plants must possess some tolerance, slight though it may be. Thus there are (1) intolerant avoiders, (2) tolerant avoiders, and (3) tolerant nonavoiders. As mentioned earlier, drought tolerance and frost tolerance may depend on similar mechanisms, and, in fact, the two have often proved to be correlated. A plant in midwinter, for instance, has its highest degree of frost tolerance and in some cases, at least, also of drought tolerance.

Heat injury and resistance

The range of high temperatures survived by plants is small compared to the range of low temperatures. The highest temperature that air-dry cells have been able to

survive is about 140° C, and normally moist vegetative cells are usually killed by 40° to 50° C. As in the case of frost resistance, heat resistance is nearly always tolerance. Exceptions have been found among plants of the Sahara desert. Some of these normally survive air temperatures of 50° C by absorbing and transpiring such tremendous quantities of water that their leaves are cooled as much as 10° C below that of the air (Lange, 1959, see Chapter 11). Although heat injury is not a dehydration process, heat tolerance is frequently (but not always) correlated with frost and drought tolerance.

MECHANISM OF STRESS INJURY AND RESISTANCE

Although it is relatively simple to find out whether a plant possesses stress resistance, and if it does, whether or not this results from avoidance or tolerance, it is more difficult to discover the mechanisms by which these two kinds of resistance operate. Even the mechanism of injury is not usually known. The correlations between frost, drought, and heat tolerance encourage attempts to explain the injury and tolerance in all three cases by a single theory. Both freezing and drought injury in plants are the result of the dehydration process. But it is not known how the dehydration produces the injury. Many suggestions have been made. The removal of water may conceivably concentrate the solutes to the point of toxicity to protoplasmic constituents. The collapse of the cell may produce a physical tension on the protoplasm that would injure some sensitive component. Many other possibilities have been proposed. A recent concept suggests that the proteins are denatured reversibly at the low temperature and that the dehydration brings the proteins close enough together to form intermolecular bonds. This results in irreversible aggregation of the proteins and therefore death. Even heat injury may be thought of as a kind of dehydration since the increased kinetic energy of the water molecules may convert previously adsorbed molecules to the free form and may permit the "naked" protein molecules to approach each other close enough for aggregation to occur. All three injuries would then be caused by bringing the protein molecules so close to each other by removing the protective water coats that new chemical bonds would form between them.

Whatever the cause of the dehydration injury, resistance could be caused by either avoidance of dehydration or tolerance of it. Partial avoidance would result if any property of the plant decreased the degree of dehydration at any one freezing temperature. It is not surprising, then, that the cell sap concentration of plants commonly increases during hardening (see Table 6-5). In winter it may be 2 to 4 times as high as in spring and therefore only 50% or even 25% as much ice would form in the plant when frozen in winter as when frozen at the same temperature in spring. This increase in cell sap concentration is primarily caused by a conversion of starch to sugars at low temperatures during the hardening period in the fall.

However, this partial avoidance of dehydration is only one component of the stress resistance. There are plants with high cell sap concentrations that possess no stress resistance (e.g., sugarcane) and plants with low sap concentrations that possess high resistance (some evergreens and some succulents). Tolerance of dehydration is therefore the primary component of such stress resistance. In other words, the protoplasm of the resistant cells possesses a specific ability to become highly dehydrated without suffering injury—the dehydration strain is fully reversible. The exact mechanism of this dehydration tolerance is still unknown.

It has long been known, however, that the physical properties of the protoplasm change during hardening, presumably because of changes in the proteins since these account for the largest fraction of the proto-plasm's dry matter. This interpretation has been supported by biochemical investigations of the proteins. Soluble and even total proteins have been found to increase during the hardening period. Changes in other protoplasmic components (e.g., RNA, lipids) also occur. It appears likely, therefore, that the dehydration tolerance of the hardy plant depends on the properties of its proteins. If, for instance, the injury is caused by denaturation, followed by irreversible aggregation of the proteins, then tolerance that has developed during the hardening process would be caused by changes in the nature of the proteins, which would prevent this irreversible aggregation.

It must be admitted that despite the great deal of information accumulated by investigators of freezing, drought, and heat stresses, neither the mechanism of the stress injury nor that of the stress resistance is fully understood. In the case of other stresses even less is known.

QUESTIONS

1. What is the physiological concept of a stress?
2. What is the physiological concept of a strain?
3. Compare capacity adaptation with resistance adaptation.
4. To what kinds of temperature stress may a plant be exposed?
5. To what other kinds of stress may it be exposed?
6. What is meant by stress resistance?
7. What two kinds are there?
8. Explain the difference between them.
9. Which is the more primitive?
10. What is the maximum possible freezing resistance of plants?
11. How is this related to the moisture content of the plant?
12. Compare the plant's tolerance of extracellular and intracellular freezing?
13. What effect does the normal freezing of plants have on the water content of its cells?

14. What is meant by hardening?
15. How can it be induced artificially?
16. What is cryobiology?
17. Is freezing resistance of plants due to tolerance or avoidance? Explain.
18. Is drought resistance of plants due to tolerance or avoidance? Explain.
19. How high a temperature can plant cells survive?
20. Is heat resistance due to tolerance or avoidance? Explain.
21. Are any kinds of stress resistance correlated with each other? Explain.
22. What is the probable mechanism of injury in the case of freezing stresses?
23. What is probably involved in the mechanism of resistance?

SPECIFIC REFERENCE

Weaver, J. E. 1919. The ecological relation of roots. Carnegie Institute of Washington, Pub. No. 286. Washington, D. C.

GENERAL REFERENCES

Asahina, E. (Ed.). 1967. Cellular injury and resistance in freezing organisms. Institute of Low Temperature Science, Hokkaido University, Sapporo, Japan.

Biebl, R. 1962. Protoplasmatische Ökologie der Pflanzen; Wasser und Temperatur. Protoplasmatologia 12(1). Springer-Verlag, Vienna.

Levitt, J. 1956. The hardiness of plants. Academic Press, Inc., New York.

Levitt, J. 1958. Frost, drought, and heat resistance. Protoplasma-monographien 8(6). Springer-Verlag, Vienna.

Meryman, H. T. (Ed.). 1966. Cryobiology. Academic Press, Inc., New York.

Prosser, C. L. (Ed.). 1967. Molecular mechanism of temperature adaptation. Amer. Assoc. Adv. Sci. Pub. No. 84, Washington, D. C.

Troshin, A. S. (Ed.). 1967. The cell and environmental temperature. Pergamon Press, Inc., New York.

chapter 23

Laws of plant physiology

The aim of plant physiology is to discover and describe the natural laws controlling all life processes in the plant. We are now in a position to ask: What laws have so far emerged? The logical answer is the laws of physics and chemistry since the results have so far upheld the basic assumption (see Chapter 1) that all physiological phenomena are explainable by these laws. But does it follow that there are no additional, specifically physiological laws? If we are seeking laws that can describe life processes quantitatively, in the form of a mathematical equation, we must admit, unfortunately, that none has yet been established. Nevertheless, it must be obvious by now that the explanations of physiological phenomena are not always directly deducible from the laws of physics and chemistry. Physicochemical theory must be supported by experimental evidence. In many cases such evidence has revealed which physicochemical laws must be selected and how they must be combined in order to explain a specific physiological phenomenon. As a result, a number of fundamental relations have emerged that are now beyond the stage of theory and may tentatively be stated in the form of natural laws.

LAWS OF TRANSFER OF MATERIALS

Law of cell permeability. The living protoplasm of a cell is surrounded by a dif-

A. Laws of transfer of materials
  1. Law of cell permeability
  2. Law of water transfer
  3. Law of solute transfer
B. Laws of nutrition and metabolism
  4. Law of autotrophism
  5. Law of nutrient essentiality
  6. Law of enzyme control
  7. Law of electron transfer
  8. Law of energy transfer
  9. Law of cardinal points
  10. Law of template synthesis
C. Laws of growth and development
  11. Law of enlargement
  12. Law of plant movement
  13. Law of regulation of growth and development
  14. Law of plant development
  15. Law of endogenous rhythm
  16. Law of stress resistance

ferentially permeable lipid membrane that controls the diffusion of substances into and out of the cell. It permits some substances to diffuse rapidly into or out of the cell and slows down or prevents the diffusion of others. This law also applies to each individual organelle within the protoplasm.

Law of water transfer. Water movement in the plant is a passive or purely physical process. The water moves into and out of the plant by diffusion and within the plant by diffusion and bulk flow, commonly aided by cohesive and adhesive forces. Active processes (directly dependent on metabolic energy) can affect the movement only indirectly by altering the factors that affect diffusion or bulk flow.

Law of solute transfer. Solute movement in the plant is both active and passive. Apolar solutes move into the plant purely passively; polar substances move into the living part of the plant actively, although aided by passive processes. Movement within the plant is passive in the xylem, active in the phloem.

LAWS OF NUTRITION AND METABOLISM

Law of autotrophism. The normal green plant is capable of synthesizing all the organic substances needed for completion of its life cycle. The abnormal green plant, or a part of it, may be deficient in ability to synthesize one or more of these substances.

Law of nutrient essentiality. All the elements that participate in the metabolism of a plant must be supplied by its environment. In order for the plant to complete its normal life cycle, the quantity must be sufficient to supply all the metabolic needs.

Law of enzyme control. All the metabolic processes in the plant are controlled by enzymes, which are organic catalysts, consisting of a protein and commonly a nonprotein cofactor.

Law of electron transfer. In all living protoplasm, chemical energy is constantly being released by electron transfer between substances undergoing oxidation-reduction reactions.

Law of energy transfer. All the energy requirements of the plant are supplied by the conversion of light energy into chemical energy in the form of high energy phosphates. The light energy may be stored in more stable forms of chemical energy and later again converted into high energy phosphates in the process of respiration.

Law of cardinal points. All factors that affect a plant process may be present in quantities related to the three cardinal points. The minimum quantity of the factor is the smallest amount capable of support-ing a measurable rate of the process. The optimum quantity of the factor is the amount capable of supporting the most rapid rate of the process. The maximum quantity of the factor is the amount above which the process cannot occur at a measurable rate.

Corollary (law of the minimum). The rate of a plant process is controlled by the physical or chemical factor that is present in minimum quantity relative to its optimum quantity.

Law of template synthesis. All the proteins and nucleic acids for each kind of plant are synthesized on a template nucleic acid molecule that can reproduce the same pattern of molecule repeatedly. The patterns of the templates differ for each protein or nucleic acid produced.

LAWS OF GROWTH AND DEVELOPMENT

Law of enlargement. All plant growth is a result of cell enlargement supported by the force of turgor pressure.

Law of plant movement. The movements of plant organs are controlled by growth and plant turgor, and in some cases, by permeability.

Law of regulation of growth and development. The growth and development of a plant is regulated by the presence in very low concentrations of hormones known as growth regulators. These substances may enhance (or inhibit) growth at low concentrations and inhibit it at high concentrations. They may affect cell division, enlargement, or differentiation, depending on the balance between the kinds of substances.

Law of plant development. The plant passes successively through stages or phases of development that differ both morphologically and physiologically. The passage from one stage to another is subject to external control by the environment (e.g., the light and temperature regimen) to which it is exposed and to internal control by the balance of both growth and flowering hormones.

Law of endogenous rhythm. Plant processes exhibit periodic changes in rate, controlled by either an intrinsic or extrinsic clock capable of measuring time. Most of these have approximately diurnal periods, but some have tidal, lunar, or annual periods.

Law of stress resistance. Plants may survive stresses by two kinds of resistance: avoidance or tolerance or a combination of the two. Avoiders prevent the stress from penetrating their tissues; tolerant plants develop an internal resistance that enables them to survive the penetration of their tissues by the stress.

. . .

Evidence is accumulating for other physiological laws, but it is too soon to attempt formulating them. It is obvious that even the suggested laws are not as firmly established as the scientist would desire. Further research is necessary, and a more quantitative, mathematical presentation of these laws is to be hoped for.

Index